T0192843

CROSS-OVER EXPERIMENTS

STATISTICS: Textbooks and Monographs

A Series Edited by

D. B. Owen, Founding Editor, 1972–1991

W. R. Schucany, Coordinating Editor
Department of Statistics
Southern Methodist University
Dallas, Texas

R. G. Cornell, Associate Editor
for Biostatistics
University of Michigan

W. J. Kennedy, Associate Editor
for Statistical Computing
Iowa State University

A. M. Kshirsagar, Associate Editor
for Multivariate Analysis and
Experimental Design
University of Michigan

E. G. Schilling, Associate Editor
for Statistical Quality Control
Rochester Institute of Technology

Additional Volumes in Preparation

CROSS-OVER EXPERIMENTS

Design, Analysis, and Application

DAVID A. RATKOWSKY
MARC A. EVANS
J. RICHARD ALLDREDGE
Washington State University
Pullman, Washington

CRC Press
Taylor & Francis Group
Boca Raton London New York

CRC Press is an imprint of the
Taylor & Francis Group, an **informa** business

First published 1993 by Marcel Dekker, Inc.

Published 2019 by CRC Press
Taylor & Francis Group
6000 Broken Sound Parkway NW, Suite 300
Boca Raton, FL 33487-2742

First issued in paperback 2020

© 1993 by Taylor & Francis Group, LLC
CRC Press is an imprint of Taylor & Francis Group, an Informa business

No claim to original U.S. Government works

ISBN 13: 978-0-367-57991-3 (pbk)
ISBN 13: 978-0-8247-8892-6 (hbk)

This book contains information obtained from authentic and highly regarded sources. Reasonable efforts have been made to publish reliable data and information, but the author and publisher cannot assume responsibility for the validity of all materials or the consequences of their use. The authors and publishers have attempted to trace the copyright holders of all material reproduced in this publication and apologize to copyright holders if permission to publish in this form has not been obtained. If any copyright material has not been acknowledged please write and let us know so we may rectify in any future reprint.

Except as permitted under U.S. Copyright Law, no part of this book may be reprinted, reproduced, transmitted, or utilized in any form by any electronic, mechanical, or other means, now known or hereafter invented, including photocopying, microfilming, and recording, or in any information storage or retrieval system, without written permission from the publishers.

For permission to photocopy or use material electronically from this work, please access www.copyright.com (http://www.copyright.com/) or contact the Copyright Clearance Center, Inc. (CCC), 222 Rosewood Drive, Danvers, MA 01923, 978-750-8400. CCC is a not-for-profit organization that provides licenses and registration for a variety of users. For organizations that have been granted a photocopy license by the CCC, a separate system of payment has been arranged.

Trademark Notice: Product or corporate names may be trademarks or registered trademarks, and are used only for identification and explanation without intent to infringe.

Visit the Taylor & Francis Web site at
http://www.taylorandfrancis.com

and the CRC Press Web site at
http://www.crcpress.com

Library of Congress Cataloging-in-Publication Data

Cross-over experiments : design, analysis, and application / edited by
  David A. Ratkowsky, Marc A. Evans, J. Richard Alldredge.
        p.   cm. -- (Statistics, textbooks and monographs ; v. 135)
    Includes bibliographical references and index.
    ISBN 0-8247-8892-3 (alk. paper)
    1. Crossover trials.   I. Ratkowsky, David A.
  II. Evans, Marc A.   III. Alldredge, J. Richard.
  IV. Series.
  R853.C76C76   1993
  610'.724--dc20   92-35540
                                                            CIP

Preface

A cross–over trial is a special kind of "repeated measurements" experiment. In the traditional repeated measures trial, the experimental units, which are often human or animal subjects or field plots, are measured on more than one occasion, resulting in the likelihood of correlations between the measurements. Generally, an experimental unit is given the same treatment throughout the whole of the trial, although different subjects may, and usually do, have different treatments. A cross–over trial differs from the traditional repeated measures trial in that a subject does not generally have the same treatment from one measuring period to another. Although some subjects may be given the same treatment in two successive periods or more than once over the course of the experiment, the main feature that distinguishes a cross–over trial from the traditional repeated measures trial is that a sequence of two or more treatments is applied to each subject. Multiple use of the same subjects may be justified when an increase in precision results from less variability within subjects than between subjects.

An important feature of cross–over designs is the presence of, and the ability to measure, carryover effects. Carryover effects are commonly viewed as a manifestation of treatment at a future time, and may result from a "late response" to a treatment in a clinical trial, or may represent a "learning effect" or a "fatigue effect," as may happen with human subjects in a psychological experiment. Sometimes, steps are taken by the experimenter to prevent or mitigate the occurrence of carryover effects by use of a waiting period, commonly called a "washout" period, between applications of the drugs or treatments. However, in other experiments, such as psychological tests as well as in certain clinical trials, the ability to estimate carryover may be the main focus of interest in the experiment.

Although cross–over designs are not a particularly new subject in experimental design, with examples in agricultural research dating back to the 1950s, there has been great interest in these designs in the past two decades as a result of increased use of clinical trials in medical research. Thus, among the users of this book will be found researchers in clinical studies, pharmacy and medicine, biology and agriculture, as well as psychology and social science.

The non–mathematicians who wish to apply the methodology presented in this book to the analysis of their data sets should find the method of coding presented in Section 1.1.1 (for data with a quantitative response variate) or the approach presented in Section 7.4 (for data with a categorical response variate) to be directly applicable without requiring specialist knowledge of linear model theory or of matrix operations. On the other hand, students, teachers and researchers in statistics, especially those with an interest in linear models, may find the approach given via matrix operations in Section 1.1.2 and in other sections of this book to be of interest. Similarly, researchers interested in the analysis of correlated categorical response data will find that Chapter 7 provides an entirely new approach to the subject, most of which has not hitherto been published. Hence, this book attempts to target both users of cross–over designs, such as might be employed by pharmaceutical firms (for example, in bioequivalence studies), agricultural experiment stations and some university social science departments, as well as statisticians, biostatisticians, and research workers in psychology and related fields who are working towards the development of suitable statistical methodology for the analysis of cross–over trials.

A major aim of this book is to present a unified approach towards the design and analysis of cross–over trials, thereby removing *ad hoc* decision-making from consideration. We have endeavored to develop a model–based approach that enables any cross–over trial, of any degree of imbalance, to be analyzed for both "direct" treatment effects and for "residual" treatment effects ("carryover") using consistent procedures employing commercially–available statistical computing software. Throughout the book, we emphasize the importance of choosing designs that are highly efficient from the point of view of separating treatment and carryover effects. We emphasize the use of efficient designs despite the fact that badly unbalanced designs may still be analyzed using the method of coding that we have developed, because designs that do not allow a good separation of treatment and carryover effects can lead to misinterpretations in the inference stage.

We would like to offer our appreciation to Dr. John W. Cotton, Departments of Psychology and Education, University of California at Santa Barbara, for having introduced us to the subject of cross–over designs, and for many stimulating discussions since. The joint paper of Ratkowsky, Alldredge and Cotton (1990) can accurately be said to be the basis upon which this book has been founded.

David A. Ratkowsky
Marc A. Evans
J. Richard Alldredge

Contents

CROSS-OVER
EXPERIMENTS

1
Introduction to the Analysis of Cross-over Designs: Basic Principles and Some Useful Tools

The goal of this chapter is to establish some basic tools that will enable the user to analyze any cross–over design for which the effects are estimable. The two major tools are: (1) a method of coding, by which any cross–over design can be prepared for parameter estimation and subsequent statistical analysis; (2) a measure of "separability" of treatment and carryover effects in a model. In addition, this chapter deals with a variety of other issues and principles which are necessary in order to analyze cross–over trials in a systematic and correct way, making appropriate use of statistical testing procedures. An important feature of the approach is the fact that it is model–based, as the viewpoint is taken that appropriate statistical tests can only be valid when the estimation procedure is an attempt to estimate parameters previously defined by a model, prior to the data having been collected. In this way, there ceases to be any ambiguity or confusion about which statistical tests are the appropriate ones to use. They are determined in advance and are not influenced by the actual numerical results.

1.1. A General Method for Coding Treatment and Carryover Effects

We adopt the view in this book that carryover effects are a manifestation of treatment in subsequent periods of time. Although this is not the only viewpoint for carryover, it is one that is consistent with the notion that a treatment may persist into a later time period and influence or modify the effect of subsequently applied treatments. Carryover effects can have a variety of forms. They can be learning effects or fatigue effects, having, respectively, a positive or negative effect on the response, or they can be of a psychological, rather than of a physical, form. The latter may occur with human subjects, when the outcome of the treatment that was first applied, if it had no effect or a negative effect, could condition the subject into producing a poor result (that is, a negative effect or no effect) in the second treatment period. Conversely, a positive response in the first treatment period may cause the subject to experience an improvement in the second treatment period, that is, a type of placebo effect, even if the treatment applied in the second period lacks efficacy.

It is easier to illustrate the method of coding using an example than to present it in abstract form. We emphasize that the coding method is applicable to unbalanced as well as balanced data; data sets with missing values are discussed in Chapter 9. Consider the pair of 3 x 3 Latin squares given in Table 1.1, taken from Cochran and Cox (1957), p. 135. The response variable is milk yield. This pair of squares is an example of a "digram–balanced" pair, in which each treatment is preceded and followed equally often by each other treatment. As is widely known (see Chapter 2 for a detailed discussion of these designs), for an even number n of treatments, n also being the number of rows and columns of the Latin square, digram balance can be achieved with a single square, but for odd n, a pair of squares is necessary to achieve this condition (but see Exercise 2.3 of Chapter 2 for some exceptions to this rule).

Table 1.1. Pair of Balanced Latin Squares for Milk Yield Example, Cochran and Cox (1957), reproduced here with the permission of John Wiley & Sons

	Square 1				Square 2		
Period→	1	2	3	Period→	1	2	3
Cow↓				Cow↓			
1	A 38	B 25	C 15	4	A 86	C 76	B 46
2	B 109	C 86	A 39	5	B 75	A 35	C 34
3	C 124	A 72	B 27	6	C 101	B 63	A 1

We follow Cochran and Cox in considering that the periods, numbered 1, 2 and 3, do not necessarily represent the same time frame in each of the two squares, just as the cows numbered 1, 2,..., 6, represent physically different animals in the two squares. The model for the response variable may be written, modifying a notation used by Kershner and Federer (1981), and others, as

$$Y_{ij} = \mu + \gamma_i + \pi_j + \tau_t + \delta_j \lambda_r + \epsilon_{ij} \tag{1.1}$$

where

Y_{ij} = observed milk yield in period j, $j=1$, 2 or 3, for the cow in sequence i, $i=1$, 2 or 3 in Square 1 and $i=4$, 5 or 6 in Square 2, given treatment t, $t=$A, B or C in period j and experiencing carryover from treatment r, $r=$A, B or C, in period $j-1$;

μ = effect of an overall mean;

γ_i = effect due to sequence i, $i=1,2,...,6$;

π_j = effect due to period j, $j=1$, 2 or 3, which is nested within Squares 1 and 2;

τ_t = direct effect of treatment t, $t=$A, B or C;

λ_r = first-order "residual" or "carryover" effect due to treatment r, $r=$A, B or C, having been applied in the previous period;

δ_j = an indicator variable whose value depends upon the period j such that $\delta_j = 1$, if $j > 1$, and $\delta_j = 0$, if $j = 1$;

ϵ_{ij} = random experimental error effect corresponding to sequence i and period j.

Note that cow and sequence are synonymous for the design in Table 1.1, as each of the six cows has a unique sequence of treatments. In other applications, there may be several randomly assigned sampling units per sequence, resulting in two error terms in the model. One of these error terms will be found in the "Between–subjects stratum" and the other in the "Within–subjects stratum". The error term in the Within–subjects stratum is used to test for treatment, carryover and period effects, whereas the error term in the Between–subjects stratum may be used to test for the presence of sequence effects, although an exact test is seldom possible when carryover parameters are in the model.

The indicator variable δ_j in (1.1) makes it possible to distinguish between the first treatment period, for which there is no carryover, and subsequent treatment periods, for which there will be carryover from the treatment applied in the period preceding the period in question. In this example, only first–order carryover is considered, that is, carryover which extends only into the subsequent treatment period. Second–order carryover, which extends two periods beyond the period in which the treatment is applied, can be handled in a similar fashion, and will be illustrated later, although the general view adopted here is that the existence of second–order carryover is much less likely than that of first–order carryover. Therefore, throughout most of this book, carryover will be viewed as affecting only the period immediately following the application of a treatment.

We choose here to illustrate the method using the statistical computing package SAS©, which makes both the coding and the subsequent statistical analysis a straightforward matter. The use of this package simplifies the task of presenting a unified approach to analyzing cross–over designs. However, it is important to realize that the coding method presented here is not specific to SAS, and may be implemented by any system which translates a "nominal" (non–numeric or categorical) variable into a set of dummy variables. This includes a wide variety of statistical computing packages, provided that the package offers a "language" capability rather than just merely being a set of executable modules to perform specific statistical calculations. To show that the coding method is not specific to SAS, the same calculations will also be shown using matrix operations in Section 1.1.2 and for some other statistical packages in Appendix 1.1.

1.1.1. Coding method illustrated using the data step of SAS©

The response, milk yield, is labelled Y, and square, cow, and period, labelled SQUARE, COW and PERIOD, respectively, are coded using integers in the ranges 1–2, 1–6, and 1–3, respectively, corresponding to the appropriate square, cow number, and time period at which the reading was taken. Treatment, represented by the character variable TREAT, is conveniently, but not of necessity, defined with values "A", "B" or "C", depending upon which treatment was applied to the experimental animal during that time period. Carryover is conveniently defined by a character variable CARRY, which is coded as "0" (although any other unique character may be used) if the reading is in the first period, where there is no preceding treatment to cause a carryover effect, as "A" if the reading is preceded by treatment A, as "B" if the reading is preceded by treatment B, and as "C" if the reading is preceded by treatment C. Hence, the DATA step in a SAS program for this example might look as follows:

```
DATA MILKYLD;
INPUT Y SQUARE COW PERIOD TREAT $ CARRY $ @@;
CARDS;
    38   1   1   1   A   0      25   1   1   2   B   A      15   1   1   3   C   B
   109   1   2   1   B   0      86   1   2   2   C   B      39   1   2   3   A   C
   124   1   3   1   C   0      72   1   3   2   A   C      27   1   3   3   B   A
    86   2   4   1   A   0      76   2   4   2   C   A      46   2   4   3   B   C
    75   2   5   1   B   0      35   2   5   2   A   B      34   2   5   3   C   A
   101   2   6   1   C   0      63   2   6   2   B   C       1   2   6   3   A   B
RUN;
```

The dollar signs ($) after TREAT and CARRY in the INPUT statement are present to declare those variables to be character variables, and the double "at" symbol @@ enables more than one observation, in this case three of them, to be read in a single line of input data. It will be seen later that for identifiability it will be necessary to add an additional line to the DATA step, but for the moment, the above commands are all that are necessary to represent the data for subsequent processing by the SAS procedure PROC GLM, which will be used to carry out the statistical analysis.

An important feature of the manner in which PROC GLM is used is the order in which the terms of the model are entered into the subsequent regression analysis. This is controlled by the order in which the variables are specified in the MODEL statement. Since the design under consideration consists of two Latin squares, it is sensible to enter first those factors which specify or define the "structural" aspects of the design, that is, the "row" and "column" factors, which are the cow effect and periods within squares effect, represented by COW and PERIOD(SQUARE), respectively. Then follows treatment, coded by TREAT, and carryover, coded by CARRY. The full SAS code for this example is

```
PROC GLM;
CLASS COW PERIOD SQUARE TREAT CARRY;
MODEL Y=COW PERIOD(SQUARE) TREAT CARRY/SOLUTION
    SS1 SS2 E1 E2;
RANDOM COW;
LSMEANS TREAT CARRY/PDIFF;
RUN;
```

The option SOLUTION causes the parameter estimates to be printed, the options SS1 and SS2 result in analysis of variance tables of the sequential sum of squares (the so–called Type I) and the adjusted sum of squares (the so–called Type II), respectively, to be printed, and the options E1 and E2 cause the Type I and Type II estimable functions to be printed as well as the tables of Type I and Type II expected mean squares.   It is the adjusted sum of squares that is relevant for testing whether there are significant differences between treatment levels and between carryover levels. For the above example, the SAS output will look as follows (in part):

SOURCE	DF	SUM OF SQUARES	MEAN SQUARE	F VALUE
MODEL	13	20163.19444	1551.01496	31.14
ERROR	4	199.25000	49.81250	
TOTAL	17	20362.44444		

(Note that the "Model" sum of squares is equal to the sum of all the Type I sum of squares in the following table, and the "Model" degrees of freedom is equal to the sum of the corresponding degrees of freedom. The "Total" sum of squares is corrected for the overall mean.)

Source	df	Type I SS	Mean Square	F value	Pr > F
COW	5	5781.1	1156.2	23.21	0.0047
PERIOD (SQUARE)	4	11489.1	2872.3	57.66	0.0009
TREAT	2	2276.8	1138.4	22.85	0.0065
CARRY	2	616.2	308.1	6.19	0.0597

Source	df	Type II SS	Mean Square	F value	Pr > F
COW	4	3817.9	954.5	19.16	0.0071
PERIOD (SQUARE)	3	3177.7	1059.3	21.26	0.0064
TREAT	2	2854.5	1427.3	28.65	0.0043
CARRY	2	616.2	308.1	6.19	0.0597

All of the sums of squares in the above table may be found in the analysis of variance table of Cochran and Cox (1957), p. 135, with the exception of the COW and PERIOD(SQUARE) Type II sums of squares, since those adjusted sums of squares are not appropriate for statistical significance testing. Testing for whether or not period effects are significant, when a carryover effect term is in the model, is not possible, because carryover into Period 1, indicated here by the code "0", is completely confounded with Period 1 effects. In fact, the Type 2 sum of squares of 3177.7 for period is really the unadjusted (sequential) sum of squares when Period 1 is ignored and only the last two periods are considered. In Section 1.4 and in Chapter 2, we will discuss how the user may attempt to answer questions about whether or not there are period effects.

That the adjusted sums of squares are the appropriate ones for testing the significance of treatment and carryover effects follows from the expected mean squares produced by SAS, which contain the following information on the effects of interest:

SOURCE TYPE I EXPECTED MEAN SQUARE

TREAT VAR(ERROR) + Q(TREAT,CARRY)
CARRY VAR(ERROR) + Q(CARRY)

SOURCE TYPE II EXPECTED MEAN SQUARE

TREAT VAR(ERROR) + Q(TREAT)
CARRY VAR(ERROR) + Q(CARRY)

The Type I treatment sum of squares cannot be used for a test of treatment effects because its expected mean square has a quadratic form [the Q(TREAT, CARRY) term] in which treatment and carryover are entangled. The Type II sum of squares, on the other hand, has a quadratic form [the Q(TREAT) term] containing only the effect of treatment, providing a valid test of differential treatment effects against the error mean square, VAR(ERROR). For this example, VAR(ERROR), usually denoted as σ^2, has the estimate 49.8125 (see SAS output). Thus, the F–value of 28.65 reported in the table is the ratio of the treatment mean square 1427 ($= 2854.5/2$) to 49.8125. The Type I and Type II expected mean squares for carryover are the same, because CARRY was the last term entered in the model, and hence its Type I SS is adjusted for the presence of all other terms in the model. The probability value of 0.0597 provides only marginal evidence of a carryover effect, although for this design the carryover effect may be underestimated due to relatively poor separability of treatment and carryover effects (see Sections 1.3 and 2.1).

The statement

LSMEANS TREAT CARRY/PDIFF;

in the PROC GLM code produces the response that the least squares means for treatment and carryover are not estimable. This is due to the fact that there is no information on carryover effect into the first treatment period, from the hypothetical or non-existent period "zero". This is easily verified from the output generated by use of the E2 option:

TYPE II ESTIMABLE FUNCTIONS FOR

EFFECT		COW COEFFS.	PERIOD (SQUARE) COEFFS.	TREAT COEFFS.	CARRY COEFFS.
INTERCEPT		0	0	0	0
COW	1	L2	0	0	0
	2	L3	0	0	0
	3	–L2–L3	0	0	0
	4	L5	0	0	0
	5	L6	0	0	0
	6	–L5–L6	0	0	0
PERIOD					
(SQUARE)	1 1	0	L8	0	0
	2 1	0	L9	0	0
	3 1	0	–L8–L9	0	0
	1 2	0	–L8	0	0
	2 2	0	L12	0	0
	3 2	0	L8–L12	0	0
TREAT					
	A	0	0	L14	0
	B	0	0	L15	0
	C	0	0	–L14–L15	0
CARRY					
	A	0	0	0	L17
	B	0	0	0	L18
	C	0	0	0	–L17–L18
	0	0	0	0	0

The interpretation of estimable functions involves some experience; they are described in the SAS User's Guide: Statistics (1990) [Chapter 9: The Four Types of Estimable Functions, pp. 77–92]. Briefly, the key to deciding whether a function, such as the difference between means of two levels of a factor, is estimable or not involves setting one of the "L" coefficients equal to unity, with the remaining "L" coefficients being set equal to zero. If a coefficient of a level of an effect is already zero, then no function containing that level can be estimable. Hence, if we look at the coefficients for CARRY in the Table, we see that the carryover effect labelled "0" is not estimable. That is, there is no information available by which one can determine a differential effect between that level, which occurs only in the first treatment period, and any of the levels representing the residual effects of treatments A, B or C, respectively. By setting L17=1 and L18=0, we see that the contrast A–C is estimable, and then by setting L18=1 and L17=0, the contrast B–C is estimable. Of course, this also implies that the contrast A–B is estimable. Thus, from the four levels of CARRY, there are only two *independent* estimable contrasts, and we would expect the CARRY term to have only two degrees of freedom, as indeed the ANOVA table shows.

With respect to the estimable functions for treatment (TREAT), by letting L14=1 and L15=0, the contrast A–C is seen to be estimable. Similarly, so is B–C, and by implication, so is A–B. There are but two degrees of freedom. The estimable functions for period within squares tell an interesting story. One cannot get a meaningful estimable contrast involving the first period, since letting L8=1 and L9=L12=0 results in periods 1 and 3 in both squares being involved in the contrast, which violates the assumption made earlier in the analysis, following Cochran and Cox (1957), that the two squares have different time frames. However, letting L9=1 and L8=L12=0, and then L12=1 and L8=L9=0, it appears that one should be able to estimate contrasts between the second and third periods of each square separately. This question will be examined in Chapter 2. Some contrasts, accounting for four degrees of freedom, can be made amongst cows, but only within the individual squares. The remaining degree of freedom is accounted for by differences between squares, and indeed it is possible to have written the model statement in PROC GLM as follows:

```
MODEL Y=SQUARE COW(SQUARE) PERIOD(SQUARE) TREAT
        CARRY/SOLUTION SS1 SS2 E1 E2;
```

which would have resulted in the Type I SS for COW of 5781.1 to be partitioned into 18.0 for SQUARE (1 df) and 5763.11 for COW(SQUARE) (4 df). It is important to realize, however, that this manipulation and others involving SQUARE, COW and PERIOD do not affect the adjusted (Type II) sum of squares for TREAT and CARRY. That is, for inferential purposes on treatment and carryover effects, it is of no consequence which of various alternatives one uses to specify the relationship between sequences and subjects within sequences. We will return to this point later.

It is the presence of the non–estimable level "0" of CARRY that prevents the LSMEANS statement from producing estimable contrasts for CARRY and TREAT. This is remedied by adding the following statement to the data step of the above example,

IF CARRY='0' THEN CARRY='C';

prior to the CARDS statement. This modification produces the following least squares means for treatment and carryover:

LEAST SQUARES MEANS

TREAT	Y LSMEAN	PROB > |T| H0: LSMEAN(I)=LSMEAN(J)		
		I/J 1	2	3
A	38.417	1 .	0.0403	0.0017
B	52.042	2 0.0403	.	0.0106
C	72.667	3 0.0017	0.0106	.

CARRY	Y LSMEAN	PROB > |T| H0: LSMEAN(I)=LSMEAN(J)		
		I/J 1	2	3
A	46.333	1 .	0.5605	0.0296
B	50.208	2 0.5605	.	0.0553
C	66.583	3 0.0296	0.0553	.

The above least–square means for treatment differ from those presented in Cochran and Cox (1957) by a fixed constant, reflecting different conventions about the sum of the carryover effects. Nevertheless, the *differences* between the means of the various levels are the same, and would also be the same if the line inserted in the data step had read

IF CARRY='0' THEN CARRY='A';

using level A rather than level C. It makes no difference to any aspect of the analysis which of these statements is used. All parameter estimates, sum of squares, and inferences will be the same, as is also the case if level 'B' is used in place of level 'C' or 'A' in the above IF statements. In Section 1.5, we will examine another technique by which one can test for whether effects are estimable or not, by putting the design matrix into row echelon form.

1.1.2. Coding using matrix operations

We will now code the data for the milk yield example using matrix operations such as can be realized with the package GAUSS© or with SAS IML©, for example. This will demonstrate that the method of coding that is being advocated is not specific to any package, and requires only that the appropriate number of dummy vectors be set up. The number of dummy vectors needed is 14, made up as follows: one for the overall mean μ, five to account for the six cows (alternatively, we could have one vector to contrast the two squares, plus four others for cows within squares, but, as emphasized above, these alternative choices will not affect the estimation of the treatment and carryover effects), two each for periods in the first square and second square, respectively, two for treatments and two for carryover. Note that although there appears to be four levels of carryover, one for each treatment and one for level "0" in the first period, the latter is non-estimable so that there are only two degrees of freedom, requiring only two dummy variables for coding. As there are 18 observations in the two squares, the design matrix X will contain 18 rows and 14 columns. Writing the linear equations as

$$X\beta = Y,$$

their matrix representation is shown at the top of the next page, where a "sum to zero" convention (see Searle, 1987, or Hocking, 1985, for a discussion of these conventions) has been employed. The solution vector $\hat{\beta} = (X'X)^{-1}X'Y$ is

$$\hat{\beta}' = (\hat{\mu},\hat{\gamma}_1,\hat{\gamma}_2,\hat{\gamma}_3,\hat{\gamma}_4,\hat{\gamma}_5,\hat{\pi}_1,\hat{\pi}_2,\hat{\pi}_3,\hat{\pi}_4,\hat{\tau}_1,\hat{\tau}_2,\hat{\lambda}_1,\hat{\lambda}_2)' =$$

$$(58.444,-28.375,16.875,14.500,9.500,-6.375,30.889,1.556,$$

$$29.889,0.556,-15.958, -2.333,-8.042,-4.167)' \,.$$

$$\begin{pmatrix}
1 & 1 & 0 & 0 & 0 & 0 & 1 & 0 & 0 & 0 & 1 & 0 & 0 & 0 \\
1 & 1 & 0 & 0 & 0 & 0 & 0 & 1 & 0 & 0 & 0 & 1 & 1 & 0 \\
1 & 1 & 0 & 0 & 0 & 0 & -1 & -1 & 0 & 0 & -1 & -1 & 0 & 1 \\
1 & 0 & 1 & 0 & 0 & 0 & 1 & 0 & 0 & 0 & 0 & 1 & 0 & 0 \\
1 & 0 & 1 & 0 & 0 & 0 & 0 & 1 & 0 & 0 & -1 & -1 & 0 & 1 \\
1 & 0 & 1 & 0 & 0 & 0 & -1 & -1 & 0 & 0 & 1 & 0 & -1 & -1 \\
1 & 0 & 0 & 1 & 0 & 0 & 1 & 0 & 0 & 0 & -1 & -1 & 0 & 0 \\
1 & 0 & 0 & 1 & 0 & 0 & 0 & 1 & 0 & 0 & 1 & 0 & -1 & -1 \\
1 & 0 & 0 & 1 & 0 & 0 & -1 & -1 & 0 & 0 & 0 & 1 & 1 & 0 \\
1 & 0 & 0 & 0 & 1 & 0 & 0 & 0 & 1 & 0 & 1 & 0 & 0 & 0 \\
1 & 0 & 0 & 0 & 1 & 0 & 0 & 0 & 0 & 1 & -1 & -1 & 1 & 0 \\
1 & 0 & 0 & 0 & 1 & 0 & 0 & 0 & -1 & -1 & 0 & 1 & -1 & -1 \\
1 & 0 & 0 & 0 & 0 & 1 & 0 & 0 & 1 & 0 & 0 & 1 & 0 & 0 \\
1 & 0 & 0 & 0 & 0 & 1 & 0 & 0 & 0 & 1 & 1 & 0 & 0 & 1 \\
1 & 0 & 0 & 0 & 0 & 1 & 0 & 0 & -1 & -1 & -1 & -1 & 1 & 0 \\
1 & -1 & -1 & -1 & -1 & -1 & 0 & 0 & 1 & 0 & -1 & -1 & 0 & 0 \\
1 & -1 & -1 & -1 & -1 & -1 & 0 & 0 & 0 & 1 & 0 & 1 & -1 & -1 \\
1 & -1 & -1 & -1 & -1 & -1 & 0 & 0 & -1 & -1 & 1 & 0 & 0 & 1
\end{pmatrix}
\begin{pmatrix}
\mu \\ \gamma_1 \\ \gamma_2 \\ \gamma_3 \\ \gamma_4 \\ \gamma_5 \\ \pi_1 \\ \pi_2 \\ \pi_3 \\ \pi_4 \\ \tau_1 \\ \tau_2 \\ \lambda_1 \\ \lambda_2
\end{pmatrix}
=
\begin{pmatrix}
38 \\ 25 \\ 15 \\ 109 \\ 86 \\ 39 \\ 124 \\ 72 \\ 27 \\ 86 \\ 76 \\ 46 \\ 75 \\ 35 \\ 34 \\ 101 \\ 63 \\ 1
\end{pmatrix}$$

The above estimates are different from those produced by the SAS procedure PROC GLM, because that procedure uses the "set to zero" convention. To obtain the set–to–zero estimates, one simply replaces all the –1 entries in the design matrix by zeros. Hence, one obtains

$$X =
\begin{pmatrix}
1 & 1 & 0 & 0 & 0 & 0 & 1 & 0 & 0 & 0 & 1 & 0 & 0 & 0 \\
1 & 1 & 0 & 0 & 0 & 0 & 0 & 1 & 0 & 0 & 0 & 1 & 1 & 0 \\
1 & 1 & 0 & 0 & 0 & 0 & 0 & 0 & 0 & 0 & 0 & 0 & 0 & 1 \\
1 & 0 & 1 & 0 & 0 & 0 & 1 & 0 & 0 & 0 & 0 & 1 & 0 & 0 \\
1 & 0 & 1 & 0 & 0 & 0 & 0 & 1 & 0 & 0 & 0 & 0 & 0 & 1 \\
1 & 0 & 1 & 0 & 0 & 0 & 0 & 0 & 0 & 0 & 1 & 0 & 0 & 0 \\
1 & 0 & 0 & 1 & 0 & 0 & 1 & 0 & 0 & 0 & 0 & 0 & 0 & 0 \\
1 & 0 & 0 & 1 & 0 & 0 & 0 & 1 & 0 & 0 & 1 & 0 & 0 & 0 \\
1 & 0 & 0 & 1 & 0 & 0 & 0 & 0 & 0 & 0 & 0 & 1 & 1 & 0 \\
1 & 0 & 0 & 0 & 1 & 0 & 0 & 0 & 1 & 0 & 1 & 0 & 0 & 0 \\
1 & 0 & 0 & 0 & 1 & 0 & 0 & 0 & 0 & 1 & 0 & 0 & 1 & 0 \\
1 & 0 & 0 & 0 & 1 & 0 & 0 & 0 & 0 & 0 & 0 & 1 & 0 & 0 \\
1 & 0 & 0 & 0 & 0 & 1 & 0 & 0 & 1 & 0 & 0 & 1 & 0 & 0 \\
1 & 0 & 0 & 0 & 0 & 1 & 0 & 0 & 0 & 1 & 1 & 0 & 0 & 1 \\
1 & 0 & 0 & 0 & 0 & 1 & 0 & 0 & 0 & 0 & 0 & 0 & 1 & 0 \\
1 & 0 & 0 & 0 & 0 & 0 & 0 & 0 & 1 & 0 & 0 & 0 & 0 & 0 \\
1 & 0 & 0 & 0 & 0 & 0 & 0 & 0 & 0 & 1 & 0 & 1 & 0 & 0 \\
1 & 0 & 0 & 0 & 0 & 0 & 0 & 0 & 0 & 0 & 1 & 0 & 0 & 1
\end{pmatrix} .$$

This design matrix leads to the following estimates of the parameters, which are identical to those produced by PROC GLM of SAS:

$$\hat{\beta}' = (\hat{\mu}, \hat{\gamma}_1, \hat{\gamma}_2, \hat{\gamma}_3, \hat{\gamma}_4, \hat{\gamma}_5, \hat{\pi}_1, \hat{\pi}_2, \hat{\pi}_3, \hat{\pi}_4, \hat{\tau}_1, \hat{\tau}_2, \hat{\lambda}_1, \hat{\lambda}_2)' =$$

$$(52.375, -24.250, 21.000, 18.625, 15.625, -0.250, 51.125, 34.000,$$

$$48.125, 31.000, -34.250, -20.625, -20.250, -16.375)' \; .$$

Despite the fact that the estimates obtained from the two conventions are quite different, they lead to the same "differential" effects for each of the factors. Consider treatment differences, for example. From the "sum–to–zero" results, one obtains the estimates $\hat{\tau}_1 = -15.9583$, $\hat{\tau}_2 = -2.3333$, from which it follows that $\hat{\tau}_3 = 18.2916$ (the three estimates must sum to zero, hence the name of the convention). Therefore, $\hat{\tau}_1 - \hat{\tau}_2 = -13.625$, telling the user that the mean of treatment A is 13.625 units less than the mean of treatment B. Similarly, $\hat{\tau}_1 - \hat{\tau}_3 = -34.250$ and $\hat{\tau}_2 - \hat{\tau}_3 = -20.625$. From the results for the "set–to–zero" convention, the estimates are $\hat{\tau}_1 = -34.25$, $\hat{\tau}_2 = -20.625$ and $\hat{\tau}_3 = 0.0$, the latter estimate being "set" equal to zero, as the name of the convention suggests. The differences $\hat{\tau}_1 - \hat{\tau}_2 = -13.625$, $\hat{\tau}_1 - \hat{\tau}_3 = -34.250$ and $\hat{\tau}_2 - \hat{\tau}_3 = -20.625$ are identical to those obtained using the sum–to–zero convention. Furthermore, these differences are the same as those obtained from the LSMEANS statement reported in the table of least–squares means in Section 1.1.1, which also gave levels of significance associated with each of the pairwise differences.

In a similar fashion, the same conclusions will be drawn, irrespective of which of the two conventions is employed, when making comparisons within each of the other factors. It is easy to confirm that the fitted or predicted values for any unit will be the same irrespective of which set of estimates is used. For example, suppose one wishes to calculate the fitted response for Cow 6 in Period 2 (noting that this cow received treatment B in that period and had carryover from treatment C in the previous period). From the sum–to–zero estimates, one obtains

$$\hat{Y} = 58.444 + (28.375 - 16.875 - 14.5 - 9.5 + 6.375) + 0.556 - 2.333 +$$
$$(8.042 + 4.167) = 62.75,$$

agreeing with the set-to-zero prediction, calculated as

$$\hat{Y} = 52.375 + 0.0 + 31.0 - 20.625 + 0.0 = 62.75 .$$

The residual sum of squares can be calculated from the following matrix equation,

$$\text{RSS} = Y'Y - \hat{\beta}'X'Y = 199.25,$$

which leads to the estimate of the residual variance of

$$\hat{\sigma}^2 = 199.25/4 = 49.8125,$$

as there are four degrees of freedom (being the difference between the 18 observations and the 14 parameters estimated). The variances and covariances of the period, treatment and carryover parameters, that is, those effects estimated in the within–subjects stratum, are contained in the matrix

$$\text{Cov}(\hat{\beta}) = (X'X)^{-1}\hat{\sigma}^2.$$

Although $\hat{\sigma}^2$ does not depend upon which of the sum–to–zero or set–to–zero conventions has been employed, the inverse of $X'X$ *does* depend upon the convention. For the sum–to–zero convention, the portion of the covariance matrix relating to the treatments and carryover parameters is as follows:

$$\text{Cov}(\hat{\tau}_1,\hat{\tau}_2,\hat{\lambda}_1,\hat{\lambda}_2) = 49.8125 \begin{pmatrix} 0.13889 & -0.06944 & 0.08333 & -0.04167 \\ -0.06944 & 0.13889 & -0.04167 & 0.08333 \\ 0.08333 & -0.04167 & 0.25000 & -0.12500 \\ -0.04167 & 0.08333 & -0.12500 & 0.25000 \end{pmatrix}$$

$$= \begin{pmatrix} 6.9184 & -3.4592 & 4.1510 & -2.0755 \\ -3.4592 & 6.9184 & -2.0755 & 4.1510 \\ 4.1510 & -2.0755 & 12.4531 & -6.2266 \\ -2.0755 & 4.1510 & -6.2266 & 12.4531 \end{pmatrix} .$$

Likewise, for the set–to–zero convention, the corresponding portion of the matrix relating to the treatments and carryover parameters is

$$\text{Cov}(\hat{\tau}_1,\hat{\tau}_2,\hat{\lambda}_1,\hat{\lambda}_2) \quad = \quad \begin{pmatrix} 20.7552 & 10.3776 & 12.4531 & 6.2266 \\ 10.3776 & 20.7552 & 6.2266 & 12.4531 \\ 12.4531 & 6.2266 & 37.3594 & 18.6797 \\ 6.2266 & 12.4531 & 18.6797 & 37.3594 \end{pmatrix}.$$

The standard errors of the estimated parameters are given as the square–roots of the diagonal elements of the above matrices. Hence, for the set–to–zero convention, the estimates and their standard errors for the treatment and carryover parameters are

$$\tau_1: -34.25 \pm 4.556, \quad \tau_2: -20.625 \pm 4.556, \quad \lambda_1: -20.25 \pm 6.112,$$
$$\lambda_2: -16.375 \pm 6.112.$$

The above covariance matrices also enable one to calculate correlation coefficients between the estimators of various parameters. For example, using the results from the sum–to–zero convention, the correlation between $\hat{\tau}_1$ and $\hat{\tau}_2$ is $-3.4592/(6.9184 \cdot 6.9184)^{1/2} = -0.5$, whereas the correlation between $\hat{\tau}_1$ and $\hat{\lambda}_1$ (and also between $\hat{\tau}_2$ and $\hat{\lambda}_2$) is $4.1510/(6.9184 \cdot 12.4531)^{1/2} = 0.4472$. For the set–to–zero convention, the corresponding correlation coefficients are $+0.5$ and 0.4472, respectively. If one's purpose is to calculate correlation coefficients without needing to calculate variances and covariances, then one can determine the correlation coefficients from knowledge only of the design matrix $\mathbf{X}$, without reference to a vector of observed responses $\mathbf{Y}$. The advantage of this will become obvious when we discuss the question of the relative efficiency of various designs, where one wishes to choose the most efficient design *before* conducting an experiment, that is, before values of the response variable become known. For example, for the sum–to–zero convention, the correlation between $\hat{\tau}_1$ and $\hat{\lambda}_2$ can be computed from the $(\mathbf{X}'\mathbf{X})^{-1}$ matrix, without the variance estimate $\hat{\sigma}^2$ being needed. Hence, the required correlation coefficient is $0.08333/(0.13889 \cdot 0.25)^{1/2} = 0.4472$, which is, of course, the same as that calculated above.

For this problem, each sequence consisted of a single cow only, and there is no error term against which a test of sequence effects may be attempted. More generally, there is more than one experimental subject per sequence, and it is usual that the ratio of sequence mean square to the subjects within sequence mean square is not distributed as central F, using either the Type I or Type II sums of squares.

1.2. An Omnibus Test of the Separability of Treatment and Carryover Effects

Cross–over designs are examples of designs in which the effects to be estimated are, in general, not "orthogonal". This is in contrast to many experimental designs, including Latin square designs where carryover effects are not in the model, where, as a result of orthogonality, the magnitude of the explained sum of squares due to the presence of a particular effect does not depend upon the order in which the effect enters the model. To illustrate this, let us consider the milk yield data of Cochran and Cox (1957) that was presented earlier in this chapter. The effects COW, PERIOD(SQUARE), and TREAT are orthogonal to each other, so the magnitude of their contributions to the explained sum of squares, viz. 5781.1, 11489.1, and 2276.8, respectively, as given in the ANOVA table, remains the same even if the order of adding these terms to the model is changed to any of its other five permutations. An important consequence of this is that the adjusted (Type II) sum of squares are also identical to the sequential (Type I) sum of squares in this situation. Thus, the effect of treatment (TREAT) is the same whether one puts it into the model first or whether one introduces it after the other terms are in the model. This is a major consequence of orthogonality, resulting in the contribution of an effect being independent of the order in which it is included in the model. Balanced designs generally have the property of orthogonality, whereas unbalanced designs do not.

Introducing carryover effects into a balanced Latin square design destroys the orthogonality property, resulting in a situation where the magnitude of a contribution depends upon the order in which the effects are introduced into the model. In this book, we take a particular viewpoint towards carryover effects which views them as manifestations of treatment at a later period of time, a type of "late response", which can also be a learning effect, or conversely, a "fatigue" effect, as might happen in a psychological experiment. Thus, carryover effects would always appear in the model *after* treatment effects. Indeed, many workers in cross–over designs refer to treatment effects as "direct" treatment effects, and to carryover effects as "residual" treatment effects. Furthermore, it seems sensible to precede the inclusion of the two forms of treatment effect in the model by the "structural" aspects of the design, that is, the "rows" and "columns" effects, when a Latin square design is the object of consideration. In the case of the milk yield data, the rows are made up of the subjects or sequences (cows) and the columns constitute the periods. These effects should appear first in the model statement, then treatments, and finally carryover effects.

If there were "second–order" carryover effects, that is, residual effects extending two periods, rather than one period, beyond the application of the treatment, that would logically appear last, after first–order carryover. One might anticipate that, except for rare circumstances, direct treatment effects should dominate over first–order carryover, and first–order carryover should, in turn, be greater than second–order carryover. Thus, the model statement in SAS [omitting the options after the solidus (/) for brevity] would be written as follows, if both first– and second–order carryover were being considered, in a typical Latin square design situation, as

MODEL Y=SUBJECT PERIOD TREAT CARRY1 CARRY2

where CARRY1 and CARRY2 are first–order and second–order carryover, respectively.

Because carryover destroys orthogonality, it is capable of causing problems of interpretation as it can in multiple regression situations, where a high degree of non–orthogonality, known as multicollinearity [see Neter *et al.* (1990) or Gunst and Mason (1980)], can make it appear that an explanatory variable is unimportant, when in fact its contribution towards explaining the response variable is being masked by its relationship to one or more other explanatory variables. We propose a measure of "separability", which enables one to determine the extent to which treatment and carryover effects are "correlated" with each other. The measure applies when there is balance in the sequence–by–treatment sub–table and minimal imbalance in the sequence–by–carryover sub–table. This will occur in Latin square designs and in some other designs.

The measure is based upon a contingency table between treatment and carryover, and is best illustrated by an example. Consider the following 4 x 4

Latin square design:

		PERIOD			
		1	2	3	4
	1	A	B	C	D
SEQUENCE	2	B	C	D	A
(or SUBJECT)	3	C	D	A	B
	4	D	A	B	C

It can be seen immediately that this design is a rather poor one for separating treatment effects from carryover effects, since Treatment A is always followed by Treatment B (expect when A falls in the last period), Treatment B is always followed by Treatment C, etc. Thus, if there is a significant treatment effect when the above design is applied to a real situation, an apparently higher response from Treatment A (say) may not really be due to A being more efficacious than the other treatments, but may be due to a significant positive carryover effect of Treatment D, which always precedes A in the above design. We construct a contingency table from the above design by recording the number of times that each level of treatment is associated with each level of carryover, including the "0" level. For example, because Treatment D precedes Treatment A three times, the cell representing the combination of Treatment "A" and Carryover "D" contains a "3". The counts for each of the other 19 cells are calculated in a similar fashion, resulting in the following contingency table:

		CARRYOVER					
		0	A	B	C	D	
	A	1	0	0	0	3	4
TREATMENT	B	1	3	0	0	0	4
	C	1	0	3	0	0	4
	D	1	0	0	3	0	4
		4	3	3	3	3	16

It is from this contingency table that the omnibus measure of separability is calculated. First, the usual Pearson chi–square is calculated as

$$\chi^2 = \underset{i\,j}{\Sigma\Sigma}[(O_{ij} - E_{ij})^2/E_{ij}],$$

where O_{ij} are the observed frequencies in the above table and E_{ij} are the expected frequencies under an independence model. Independence means that the probability of an observation falling into the jth column is not a function of the row in which the observation occurs. Thus, under the null hypothesis, the expected frequencies for the above table are

		CARRYOVER					
		0	A	B	C	D	
	A	1	0.75	0.75	0.75	0.75	4
TREATMENT	B	1	0.75	0.75	0.75	0.75	4
	C	1	0.75	0.75	0.75	0.75	4
	D	1	0.75	0.75	0.75	0.75	4
		4	3	3	3	3	16

which leads to a Pearson chi–square of 36.0. In order to compare values of chi–square between designs, it is helpful to be able to convert this value into a standardized form. Such a standardization, which ranges between zero and unity, is provided by Cramèr's V (Cramèr, 1951), defined as

$$V = \left[\frac{\chi^2/N}{\min(r-1,\,c-1)}\right]^{1/2},$$

where N is the total number of incidences in the table (N=16 in the above table, the grand total in the lower right–hand corner of the table, which is also the number of sequences times the number of periods), r is the number of rows and c is the number of columns. For the above table, r=4, c=5, so

$$V = [(36/16)/\min(3,4)]^{1/2} = 0.866 \ .$$

It follows that a measure of "efficiency" of separation of treatment and carryover effects can be given by:

$$S = 100(1 - V),$$

which, for this example, is 13.4%. The extremely low efficiency of separation of direct and residual treatment effects in this design is not surprising, in view of its structure. Consider now the following 4 x 4 Latin square:

| | | PERIOD | | | |
		1	2	3	4
	1	A	B	C	D
SEQUENCE	2	B	D	A	C
(or SUBJECT)	3	C	A	D	B
	4	D	C	B	A

This design is the most efficient possible amongst the 4 x 4 Latin squares, as each treatment is preceded by, and followed by, each of the other treatments, with the exception of itself. The contingency table relating treatment and carryover incidences is, for this design,

| | | CARRYOVER | | | | | |
		0	A	B	C	D	
	A	1	0	1	1	1	4
TREATMENT	B	1	1	0	1	1	4
	C	1	1	1	0	1	4
	D	1	1	1	1	0	4
		4	3	3	3	3	16

The chi–square value associated with this design is 4.0. Converting to Cramèr's V gives $V = 0.289$, so that the measure of separability is $S = 100(1-V) = 71.1\%$, a much greater "efficiency" of separation than that for the 4 x 4 Latin square presented earlier. This design can be made 100% efficient by the addition of a fifth period which is identical to that of the fourth period, resulting in the following design:

		PERIOD				
		1	2	3	4	5
	1	A	B	C	D	D
SEQUENCE	2	B	D	A	C	C
(or SUBJECT)	3	C	A	D	B	B
	4	D	C	B	A	A

This leads to the following contingency table relating treatment and carryover effects:

		CARRYOVER					
		0	A	B	C	D	
	A	1	1	1	1	1	5
TREATMENT	B	1	1	1	1	1	5
	C	1	1	1	1	1	5
	D	1	1	1	1	1	5
		4	4	4	4	4	20

Clearly, the chi-square value corresponding to this contingency table is zero, leading to a separability S of 100% between treatment and carryover effects. Note that this extra-period design is fully balanced in its sequence-by-carryover sub-table, with minimal imbalance in the sequence-by-treatment sub-table.

The measure of separability proposed is this section is valid for all Latin square designs because each sequence contains each treatment the same number of times (once). In addition, the sub-table of sequence-by-carryover has minimal imbalance, that is, its chi-square exceeds zero only because the nature of carryover is such that there cannot be any carryover from the non-existent "zero" period into the first period. Examples of other designs, to be dealt with in detail in Chapter 4, where the proposed measure can be used, are the two–treatment, four–period, two–sequence designs

		PERIOD			
		1	2	3	4
SEQUENCE	1	A	B	A	B
(or SUBJECT)	2	B	A	B	A

and

		PERIOD			
		1	2	3	4
SEQUENCE	1	A	B	B	A
(or SUBJECT)	2	B	A	A	B

,

in which each treatment appears twice in each of the two sequences, and in which the sequence–by–carryover sub–table is unbalanced only because of the inevitable "zero" period effect that is a feature of cross–over designs. The separability measure proposed in this section may also be used for such minimally unbalanced designs.

An example of a design lacking the required balance is

		PERIOD			
		1	2	3	4
SEQUENCE	1	A	B	B	B
(or SUBJECT)	2	B	A	A	A

where each treatment does not appear the same number of times in each sequence. Hence the proposed separability measure cannot be employed.

We shall now describe a more traditional measure of efficiency for separating treatment and carryover effects, and we shall compare it with the proposed omnibus measure. The traditional method is more difficult to calculate, and will be demonstrated using matrix operations. We consider first the 4 x 4 Latin square design that was a very poor one for separating treatment and carryover effects:

		PERIOD			
		1	2	3	4
	1	A	B	C	D
SEQUENCE	2	B	C	D	A
(or SUBJECT)	3	C	D	A	B
	4	D	A	B	C

As we wish to obtain a full–rank matrix, the parameter set is chosen to include an overall mean μ, three sequence parameters γ_1, γ_2 and γ_3, three period parameters π_1, π_2 and π_3, three treatment parameters τ_1, τ_2 and τ_3, and three carryover parameters λ_1, λ_2 and λ_3. Employing the "sum to zero" convention,

$$\gamma_4 = -\gamma_1-\gamma_2-\gamma_3, \quad \pi_4 = -\pi_1-\pi_2-\pi_3, \quad \tau_4 = -\tau_1-\tau_2-\tau_3, \quad \text{and } \lambda_4 = -\lambda_1-\lambda_2-\lambda_3,$$

results in the following set of 16 linear equations:

$$\mu_{11} = \mu + \gamma_1 + \pi_1 + \tau_1$$
$$\mu_{12} = \mu + \gamma_1 + \pi_2 + \tau_2 + \lambda_1$$
$$\mu_{13} = \mu + \gamma_1 + \pi_3 + \tau_3 + \lambda_2$$
$$\mu_{14} = \mu + \gamma_1 - \pi_1 - \pi_2 - \pi_3 - \tau_1 - \tau_2 - \tau_3 + \lambda_3$$
$$\mu_{21} = \mu + \gamma_2 + \pi_1 + \tau_2$$
$$\mu_{22} = \mu + \gamma_2 + \pi_2 + \tau_3 + \lambda_2$$
$$\mu_{23} = \mu + \gamma_2 + \pi_3 - \tau_1 - \tau_2 - \tau_3 + \lambda_3$$
$$\mu_{24} = \mu + \gamma_2 - \pi_1 - \pi_2 - \pi_3 + \tau_1 - \lambda_1 - \lambda_2 - \lambda_3$$
$$\mu_{31} = \mu + \gamma_3 + \pi_1 + \tau_3$$
$$\mu_{32} = \mu + \gamma_3 + \pi_2 - \tau_1 - \tau_2 - \tau_3 + \lambda_3$$
$$\mu_{33} = \mu + \gamma_3 + \pi_3 + \tau_1 - \lambda_1 - \lambda_2 - \lambda_3$$
$$\mu_{34} = \mu + \gamma_3 - \pi_1 - \pi_2 - \pi_3 + \tau_2 + \lambda_1$$
$$\mu_{41} = \mu - \gamma_1 - \gamma_2 - \gamma_3 + \pi_1 - \tau_1 - \tau_2 - \tau_3$$

$$\mu_{42} = \mu - \gamma_1 - \gamma_2 - \gamma_3 + \pi_2 + \tau_1 - \lambda_1 - \lambda_2 - \lambda_3$$
$$\mu_{43} = \mu - \gamma_1 - \gamma_2 - \gamma_3 + \pi_3 + \tau_2 + \lambda_1$$
$$\mu_{44} = \mu - \gamma_1 - \gamma_2 - \gamma_3 - \pi_1 - \pi_2 - \pi_3 + \tau_3 + \lambda_2,$$

where the population cell means, μ_{11}, μ_{12},..., μ_{44}, relate to the cells of the sequence–by–period matrix as follows:

PERIOD

		1	2	3	4
	1	μ_{11}	μ_{12}	μ_{13}	μ_{14}
SEQUENCE	2	μ_{21}	μ_{22}	μ_{23}	μ_{24}
(or SUBJECT)	3	μ_{31}	μ_{32}	μ_{33}	μ_{34}
	4	μ_{41}	μ_{42}	μ_{43}	μ_{44}

The above equations can be written in matrix form as
$$X\beta = \mu,$$
that is, as

$$
\begin{pmatrix}
1 & 1 & 0 & 0 & 1 & 0 & 0 & 1 & 0 & 0 & 0 & 0 & 0 \\
1 & 1 & 0 & 0 & 0 & 1 & 0 & 0 & 1 & 0 & 1 & 0 & 0 \\
1 & 1 & 0 & 0 & 0 & 0 & 1 & 0 & 0 & 1 & 0 & 1 & 0 \\
1 & 1 & 0 & 0 & -1 & -1 & -1 & -1 & -1 & -1 & 0 & 0 & 1 \\
1 & 0 & 1 & 0 & 1 & 0 & 0 & 0 & 1 & 0 & 0 & 0 & 0 \\
1 & 0 & 1 & 0 & 0 & 1 & 0 & 0 & 0 & 1 & 0 & 1 & 0 \\
1 & 0 & 1 & 0 & 0 & 0 & 1 & -1 & -1 & -1 & 0 & 0 & 1 \\
1 & 0 & 1 & 0 & -1 & -1 & -1 & 1 & 0 & 0 & -1 & -1 & -1 \\
1 & 0 & 0 & 1 & 1 & 0 & 0 & 0 & 0 & 1 & 0 & 0 & 0 \\
1 & 0 & 0 & 1 & 0 & 1 & 0 & -1 & -1 & -1 & 0 & 0 & 1 \\
1 & 0 & 0 & 1 & 0 & 0 & 1 & 1 & 0 & 0 & -1 & -1 & -1 \\
1 & 0 & 0 & 1 & -1 & -1 & -1 & 0 & 1 & 0 & 1 & 0 & 0 \\
1 & -1 & -1 & -1 & 1 & 0 & 0 & -1 & -1 & -1 & 0 & 0 & 0 \\
1 & -1 & -1 & -1 & 0 & 1 & 0 & 1 & 0 & 0 & -1 & -1 & -1 \\
1 & -1 & -1 & -1 & 0 & 0 & 1 & 0 & 1 & 0 & 1 & 0 & 0 \\
1 & -1 & -1 & -1 & -1 & -1 & -1 & 0 & 0 & 1 & 0 & 1 & 0
\end{pmatrix}
\begin{pmatrix}
\mu \\
\gamma_1 \\
\gamma_2 \\
\gamma_3 \\
\pi_1 \\
\pi_2 \\
\pi_3 \\
\tau_1 \\
\tau_2 \\
\tau_3 \\
\lambda_1 \\
\lambda_2 \\
\lambda_3
\end{pmatrix}
=
\begin{pmatrix}
\mu_{11} \\
\mu_{12} \\
\mu_{13} \\
\mu_{14} \\
\mu_{21} \\
\mu_{22} \\
\mu_{23} \\
\mu_{24} \\
\mu_{31} \\
\mu_{32} \\
\mu_{33} \\
\mu_{34} \\
\mu_{41} \\
\mu_{42} \\
\mu_{43} \\
\mu_{44}
\end{pmatrix} .
$$

After finding the inverse of $X'X$, it is only the elements that appear jointly in the last six rows and last six columns of $(X'X)^{-1}$ that relate to the question of efficiency of treatment and carryover effects. These elements are

$$
\begin{pmatrix}
1.03125 & -0.34375 & -0.34375 & 0.375 & 0.375 & 0.375 \\
-0.34375 & 1.03125 & -0.34375 & -1.125 & 0.375 & 0.375 \\
-0.34375 & -0.34375 & 1.03125 & 0.375 & -1.125 & 0.375 \\
0.375 & -1.125 & 0.375 & 1.5 & -0.5 & -0.5 \\
0.375 & 0.375 & -1.125 & -0.5 & 1.5 & -0.5 \\
0.375 & 0.375 & 0.375 & -0.5 & -0.5 & 1.5
\end{pmatrix} .
$$

Thus, the variances of the treatment and carryover parameters, except for a constant multiplier, the unknown variance σ^2, are

$$\text{Var}(\tau_1) = \text{Var}(\tau_2) = \text{Var}(\tau_3) = 1.03125$$

and

$$\text{Var}(\lambda_1) = \text{Var}(\lambda_2) = \text{Var}(\lambda_3) = 1.5,$$

with the correlation coefficient between treatment parameter τ_2 and carryover parameter λ_1 being calculated as

$$\text{Corr}(\tau_2,\lambda_1) = -1.125/(1.03125 \cdot 1.5)^{1/2} = -0.9045.$$

This very high correlation shows the consequence of having Treatment B (parameter τ_2) always following Treatment A and thus receiving carryover from that treatment (parameter λ_1). $\text{Corr}(\tau_3,\lambda_2)$ relates to Treatment C following Treatment B, and has the same correlation coefficient as above. Similarly, the correlation between Treatment D and carryover from Treatment C can also be shown by reparameterization or by symmetry arguments to have the same high value. The much lower correlations of $0.375/(1.03125 \cdot 1.5)^{1/2} = 0.3015$ relate to treatments not following each other.

To obtain a measure of efficiency of this design, one can now consider the same design matrix X as above, but omit the carryover parameters. That is, the design matrix now under consideration is as follows,

$$
\begin{pmatrix}
1 & 1 & 0 & 0 & 1 & 0 & 0 & 1 & 0 & 0 \\
1 & 1 & 0 & 0 & 0 & 1 & 0 & 0 & 1 & 0 \\
1 & 1 & 0 & 0 & 0 & 0 & 1 & 0 & 0 & 1 \\
1 & 1 & 0 & 0 & -1 & -1 & -1 & -1 & -1 & -1 \\
1 & 0 & 1 & 0 & 1 & 0 & 0 & 0 & 1 & 0 \\
1 & 0 & 1 & 0 & 0 & 1 & 0 & 0 & 0 & 1 \\
1 & 0 & 1 & 0 & 0 & 0 & 1 & -1 & -1 & -1 \\
1 & 0 & 1 & 0 & -1 & -1 & -1 & 1 & 0 & 0 \\
1 & 0 & 0 & 1 & 1 & 0 & 0 & 0 & 0 & 1 \\
1 & 0 & 0 & 1 & 0 & 1 & 0 & -1 & -1 & -1 \\
1 & 0 & 0 & 1 & 0 & 0 & 1 & 1 & 0 & 0 \\
1 & 0 & 0 & 1 & -1 & -1 & -1 & 0 & 1 & 0 \\
1 & -1 & -1 & -1 & 1 & 0 & 0 & -1 & -1 & -1 \\
1 & -1 & -1 & -1 & 0 & 1 & 0 & 1 & 0 & 0 \\
1 & -1 & -1 & -1 & 0 & 0 & 1 & 0 & 1 & 0 \\
1 & -1 & -1 & -1 & -1 & -1 & -1 & 0 & 0 & 1
\end{pmatrix},
$$

leading to an $(X'X)^{-1}$ matrix which has the following elements for the treatment parameters.

$$
\begin{pmatrix}
0.1875 & -0.0625 & -0.0625 \\
-0.0625 & 0.1875 & -0.0625 \\
-0.0625 & -0.0625 & 0.1875
\end{pmatrix}.
$$

Notice that the terms proportional to the variances of the treatment parameters, 0.1875, are much less than the corresponding terms when carryover parameters were also included in the model, 1.03125. Thus, as a result of a poor design for separating treatment and carryover effects, the variances of the treatment parameters experience an inflation factor of $1.03125/0.1875 = 5.5$. The concept of a "variance inflation factor" is well known in the case of multiple regression analysis using continuous variables [see Neter *et al.* (1990) or Gunst and Mason (1980)], but we have been unable to find examples of its usage with categorical variables in the ANOVA context, as with the models under consideration [see Appelbaum and Cramer (1974) for an approach to the problem of nonorthogonal ANOVA]. It is important to realize that a design in which treatment and carryover effects are completely separable is also a design in which the variance inflation factor for treatments is 1.0. That is, the analysis suffers no reduction in the precision by which treatment effects can be estimated in such designs.

It follows that one can define a measure of efficiency to be the reciprocal of the variance inflation factor. Thus, for the 4 x 4 Latin square design in question, the efficiency, defined as the ratio of the variance of the treatment parameters when carryover is not in the model to the variance of the treatment parameters when it is in the model, is

$$E = 0.1875/1.03125 = 0.1818,$$

which is, of course, the reciprocal of 5.5. Jones and Kenward (1989) denote the above efficiency as E_d, the "direct" treatment effect efficiency, and define a "carryover" effect efficiency, denoted E_c, as the ratio of the variance of the treatment parameters when carryover is absent to the variance of the carryover parameters with both treatment and carryover in the model. These efficiencies are identical to those used by Patterson and Lucas (1962) and denoted E_d and E_r (for direct and residual, respectively). Thus, for the above design,

$$E_c = 0.1875/1.5 = 0.1250.$$

This efficiency will always be lower than the efficiency E_d for treatment effects. When these two efficiencies are compared to the measure of separability S calculated earlier in this section for this design, 13.4%, or 0.134 as a fraction, it is seen that the value of S is intermediate between the values of E_d and E_c.

Now let us turn our attention to the good 4 x 4 Latin square design for separating treatment and carryover effects, namely

| | | PERIOD | | | |
		1	2	3	4
	1	A	B	C	D
SEQUENCE	2	B	D	A	C
(or SUBJECT)	3	C	A	D	B
	4	D	C	B	A

For this design, the design matrix **X** is as follows:

$$
\begin{pmatrix}
1 & 1 & 0 & 0 & 1 & 0 & 0 & 1 & 0 & 0 & 0 & 0 & 0 \\
1 & 1 & 0 & 0 & 0 & 1 & 0 & 0 & 1 & 0 & 1 & 0 & 0 \\
1 & 1 & 0 & 0 & 0 & 0 & 1 & 0 & 0 & 1 & 0 & 1 & 0 \\
1 & 1 & 0 & 0 & -1 & -1 & -1 & -1 & -1 & -1 & 0 & 0 & 1 \\
1 & 0 & 1 & 0 & 1 & 0 & 0 & 0 & 1 & 0 & 0 & 0 & 0 \\
1 & 0 & 1 & 0 & 0 & 1 & 0 & -1 & -1 & -1 & 0 & 1 & 0 \\
1 & 0 & 1 & 0 & 0 & 0 & 1 & 1 & 0 & 0 & -1 & -1 & -1 \\
1 & 0 & 1 & 0 & -1 & -1 & -1 & 0 & 0 & 1 & 1 & 0 & 0 \\
1 & 0 & 0 & 1 & 1 & 0 & 0 & 0 & 0 & 1 & 0 & 0 & 0 \\
1 & 0 & 0 & 1 & 0 & 1 & 0 & 1 & 0 & 0 & 0 & 0 & 1 \\
1 & 0 & 0 & 1 & 0 & 0 & 1 & -1 & -1 & -1 & 1 & 0 & 0 \\
1 & 0 & 0 & 1 & -1 & -1 & -1 & 0 & 1 & 0 & -1 & -1 & -1 \\
1 & -1 & -1 & -1 & 1 & 0 & 0 & -1 & -1 & -1 & 0 & 0 & 0 \\
1 & -1 & -1 & -1 & 0 & 1 & 0 & 0 & 0 & 1 & -1 & -1 & -1 \\
1 & -1 & -1 & -1 & 0 & 0 & 1 & 0 & 1 & 0 & 0 & 0 & 1 \\
1 & -1 & -1 & -1 & -1 & -1 & -1 & 1 & 0 & 0 & 0 & 1 & 0
\end{pmatrix} .
$$

The elements in the last six rows and last six columns, relating to the treatment and carryover parameters, are

$$
\begin{pmatrix}
0.20625 & -0.06875 & -0.06875 & 0.075 & -0.025 & -0.025 \\
-0.06875 & 0.20625 & -0.06875 & -0.025 & 0.075 & -0.025 \\
-0.06875 & -0.06875 & 0.20625 & -0.025 & -0.025 & 0.075 \\
0.075 & -0.025 & -0.025 & 0.3 & -0.1 & -0.1 \\
-0.025 & 0.075 & -0.025 & -0.1 & 0.3 & -0.1 \\
-0.025 & -0.025 & 0.075 & -0.1 & -0.1 & 0.3
\end{pmatrix} .
$$

Comparing this matrix to the corresponding matrix obtained for the poor 4 x 4 Latin square, it is seen that the variances (except for a constant multiplier) of the treatment parameters, 0.20625, and the corresponding values for the carryover parameters, 0.3, are one–fifth of those in the poor design. Furthermore, the correlation coefficient between treatment parameter τ_2 and carryover parameter λ_1, and in general between any treatment parameter τ_i and any carryover parameter λ_j, where $i \neq j$, is

$$\text{Corr}(\tau_2, \lambda_1) = -0.025/(0.20625 \cdot 0.3)^{1/2} = -0.1005,$$

very much lower in absolute magnitude than that for these parameters in the poor design. The highest correlation coefficients between τ_i and λ_j occur when $i = j$, and that coefficient is only 0.3015.

If the carryover parameters are now removed from the model, the variances and covariances of the treatment parameters becomes

$$\begin{pmatrix} 0.1875 & -0.0625 & -0.0625 \\ -0.0625 & 0.1875 & -0.0625 \\ -0.0625 & -0.0625 & 0.1875 \end{pmatrix},$$

which is identical to the covariance matrix obtained from the poor design. Thus, in the absence of carryover effects, both designs are equally efficient. However, when carryover effects are a possibility and carryover parameters need inclusion in the model, the efficiency of estimating treatment effects in the better design is

$$E_d = 0.1875/0.20625 = 0.9091,$$

and that for estimating carryover effects is

$$E_c = 0.1875/0.3 = 0.625.$$

The separability measure S, defined earlier in this section, is 0.711 for this design, if expressed as a fraction. Thus once again its value lies between E_c and E_d. These results seem to provide some justification for the use of the omnibus measure S.

Despite the fact that the better of the two 4 x 4 Latin square designs, the one in which each treatment is preceded and followed by each other treatment, has an efficiency for estimation treatment effects of 0.9091, it is not a fully efficient design. To achieve this, one needs to introduce an extra period containing the same treatments in each sequence as those in the last period. Thus, the design

PERIOD

		1	2	3	4	5
	1	A	B	C	D	D
SEQUENCE	2	B	D	A	C	C
(or SUBJECT)	3	C	A	D	B	B
	4	D	C	B	A	A

which was previously shown to have a separability S of 100%, can now be dealt with using matrix operations. The 20 linear equations can be readily obtained using the same procedures as illustrated for the two 4 x 4 Latin square designs. Details are omitted here. The elements in the last six rows and last six columns of $(X'X)^{-1}$ are as follows:

$$
\begin{pmatrix}
0.15625 & -0.05208 & -0.05208 & 0 & 0 & 0 \\
-0.05208 & 0.15625 & -0.05208 & 0 & 0 & 0 \\
-0.05208 & -0.05208 & 0.15625 & 0 & 0 & 0 \\
0 & 0 & 0 & 0.1875 & -0.0625 & -0.0625 \\
0 & 0 & 0 & -0.0625 & 0.1875 & -0.0625 \\
0 & 0 & 0 & -0.0625 & -0.0625 & 0.1875
\end{pmatrix} .
$$

When the variances of the treatment parameters are compared to those obtained using the same design but without the extra period, their ratio is 0.7576 (=0.15625/0.20625), and when the carryover parameters are compared, a ratio of 0.625 (=0.1875/0.3) is obtained. The extra period design is thus seen to have a much better efficiency. Furthermore, the treatment parameters and carryover parameters are uncorrelated, so that treatment effects and carryover effects are fully separable. If the carryover parameters are now excluded from the design, the resulting covariance matrix for the treatment parameters is

$$
\begin{pmatrix}
0.15625 & -0.05208 & -0.05208 \\
-0.05208 & 0.15625 & -0.05208 \\
-0.05208 & -0.05208 & 0.15625
\end{pmatrix} ,
$$

which is identical to the covariance matrix for the treatment parameters when carryover parameters are included in the model. This is to be expected, since the treatment and carryover parameters are uncorrelated when both effects are included in the design.

This shows that a fully efficient cross-over design is one in which the following properties may be observed:

(1) Treatment and carryover effects are uncorrelated; the presence of one of these effects in the model does not influence the estimates of the other effects. The variance inflation factor is unity.

(2) Variances and covariances of parameters of both effects will be lower than those in designs of lesser efficiency. The ratio of the variance of treatment parameters in fully efficient designs to those in other designs will always be less than unity.

Thus, fully efficient designs combine separability of treatment and carryover effects with high precision (low variance) in the estimation of the parameters of the model. The consequences of poor separability between treatment and carryover parameters is investigated in the next section.

1.3. Consequences of Poor Separability of Treatment and Carryover Effects

There are two major reasons for concern about one's ability to separate direct treatment effects from carryover effects. Both reasons concern the interpretability of the results of the analysis of variance of a cross-over design. The first reason relates to circumstances where the investigator will not know whether an apparently significant treatment effect is truly a "direct" treatment effect or whether it is a "residual" effect of some other treatment. This is of undoubted importance in the interpretation of the efficacy of treatments in all areas of research. Clearly, if treatment C is exhibiting a late response, it is important that that not be confused with treatment A (say) delivering a direct effect.

The second reason for wishing to separate direct and carryover effects unambiguously relates to a phenomenon akin to "multicollinearity" in multiple

regression applications with continuous variables.   There, the presence of two
multicollinear explanatory (regressor) variables in the model may lead to the
erroneous interpretation that there are neither significant direct treatment effects
nor carryover effects.   This is illustrated by the following example, used by
Edwards (1985).   The design consisted of five repetitions of the following 5 x 5
Latin square so that each sequence appears five times.

		PERIOD				
		1	2	3	4	5
	1	A	D	B	E	C
	2	B	E	C	A	D
SEQUENCE	3	C	A	D	B	E
	4	D	B	E	C	A
	5	E	C	A	D	B

It should be obvious that this is an extremely poor design for the examination of
carryover effects. Treatment A, for example, is always followed by Treatment D
(except, of course, when it appears in the last period), and in no circumstance is
any treatment preceded by, or followed by, more than one other treatment. This
results in the worst possible contingency table between treatment and carryover
effects, viz.

		CARRYOVER						
		0	A	B	C	D	E	
	A	1	0	0	4	0	0	5
	B	1	0	0	0	4	0	5
TREATMENT	C	1	0	0	0	0	4	5
	D	1	4	0	0	0	0	5
	E	1	0	4	0	0	0	5
		5	4	4	4	4	4	25 .

The chi–square value for this table is 80.0, leading to V = 0.894 and a separability of S=100(1−V)=10.6%. With actual data for 25 subjects for this trial (Edwards, 1985, p. 373), the analysis of variance table, containing treatment and carryover effects only, is as follows:

Source of variation	Degrees of freedom	Type I (sequential sum of squares)		Type II (adjusted sum of squares)	
Treatments	4	232.05	(P<0.0001)	27.41	(P=0.3632)
Carryover	4	5.61	(P=0.9241)	5.61	(P=0.9241) .

Since the Type II sums of squares are the appropriate ones for testing treatment and carryover effects, one would draw the conclusion, from the P-values obtained (0.3632 and 0.9241, respectively), that there were neither significant treatment nor carryover effects in that trial. This contradicts the conclusion that would be drawn if one looks at the sequential (Type I) sum of squares just after treatment has been added but carryover effects has not yet been included. Up to this point, the 5 x 5 Latin square design that was employed is a perfectly good orthogonal design. Effects such as those for subjects, periods, and treatments can be entered in any order, and the Type I sums of squares for each effect will always be the same. The inclusion of the treatment term results in a reduction in the residual sum of squares by an amount equal to 232.05, which is highly significant (P<0.0001). This result, by itself, would suggest that there are important differences between treatments.

In fact, in this psychological experiment involving the assessment by the subjects of the position of targets when they appeared on screens of five different sizes (the treatments), there seems to be little doubt that the number of correct judgments (the response variable) increases monotonically with aperture size. Adding the carryover term to the model only reduces the residual sum of squares by an insignificant (P=0.9187) amount. Thus, taking the view that a carryover effect is a manifestation of treatment at a later time, and that a significant carryover effect is less likely than a significant direct treatment effect, one would be content to conclude that treatment was important and that carryover was not. The fact that the adjusted sum of squares lead to the conclusion that neither treatment nor carryover effects are important must be seen to be a spurious result attributable to a design that is extremely sensitive to this phenomenon which is akin to multicollinearity. This illustrates that although the adjusted sums of squares are the appropriate ones for testing these effects, it is important to look also at the sequential sum of squares, if a poor design for separating treatment and carryover effects has been used.

More importantly, to avoid the possibility of misleading interpretations about the significance of direct and/or residual effects, one should choose efficient designs that separate these effects unambiguously. The measure of separability proposed in Section 1.2 can aid users in deciding the extent to which designs separate these effects.

1.4. Considering Period Effects as Continuous, Rather than Categorical, Variables

In Section 1.1, it was seen that a problem arose due to a confounding of the effects of the first period and of carryover into the first period. As a result of this confounding in the milk yield data reported by Cochran and Cox (1957), the degrees of freedom corresponding to the adjusted sum of squares for period effects was only three instead of four. Carryover, instead of having three degrees of freedom, one less than its four levels 0, A, B, and C, had only two degrees of freedom, since carryover "0" was not independently estimable of Period 1 effects. A serious consequence of this is that the Type II SS for periods is not an assessment of differences between periods for all periods, but only for the second and third periods. Thus, if the first period is the most different period with the others being similar, one can obtain a non–significant period effect assessment.

The problem of confounding can be solved if it is permissible to consider period to be a continuous variable. Often one has to make an assumption about the implied distances between the period levels. Thus, if period is coded using the numbers 1, 2 and 3, fitting period as a continuous variable with these values implies that the time interval between 1 and 2 is identical to the time interval between 2 and 3. Of course, the user can choose any set of intervals, corresponding to whatever belief there may be about how the response variable changes with real time. Thus, coding periods as 1, 2 and 4, respectively, implies that there is twice the distance between the second and third interval as between the first and second interval.

We now present a reanalysis of the milk yield data, using SAS PROC GLM. The code is presented below.

```
DATA MILKYLD;
INPUT Y SQUARE COW PERIOD TREAT $ CARRY $ @@;
CARDS;
    38   1   1   1   A   0      25   1   1   2   B   A      15   1   1   3   C   B
   109   1   2   1   B   0      86   1   2   2   C   B      39   1   2   3   A   C
   124   1   3   1   C   0      72   1   3   2   A   C      27   1   3   3   B   A
    86   2   4   1   A   0      76   2   4   2   C   A      46   2   4   3   B   C
    75   2   5   1   B   0      35   2   5   2   A   B      34   2   5   3   C   A
   101   2   6   1   C   0      63   2   6   2   B   C       1   2   6   3   A   B
RUN;

PROC GLM;
CLASS COW SQUARE TREAT CARRY;
MODEL Y=COW PERIOD(SQUARE) TREAT CARRY/SOLUTION
    SS1 SS2 E1 E2;
RANDOM COW;
LSMEANS TREAT CARRY/PDIFF;
RUN;
```

The reader can check this code against that presented in Section 1.1.1 for these data, and observe that the only difference between them is the absence of PERIOD in the "CLASS" statement in the above code. Any variable included in the CLASS statement is treated as a qualitative variable, resulting in the creation of a set of dummy vectors. A variable not included in the CLASS statement is automatically treated as a continuous variable. The output from the above code is presented below, after some editing:

SOURCE	DF	SUM OF SQUARES	MEAN SQUARE	F VALUE
MODEL	12	20160.94444	1680.0787	41.69
ERROR	5	201.50000	40.3000	
TOTAL	17	20362.44444		

SOURCE	DF	TYPE I SS	F VALUE	PR > F
COW	5	5781.1	28.69	0.0011
PERIOD (SQUARE)	2	11476.8	142.39	0.0001
TREAT	2	2276.8	28.25	0.0019
CARRY	3	626.2	5.18	0.0541

(continued on next page)

SOURCE	DF	TYPE II SS	F VALUE	PR > F
COW	5	3834.0	19.03	0.0029
PERIOD (SQUARE)	2	3175.5	39.40	0.0009
TREAT	2	2854.5	35.42	0.0011
CARRY	3	626.2	5.18	0.0541 .

The comparison between these results and those with period as a qualitative variable is interesting. The effect of period is almost completely explained by a linear term in each square, accounting for an adjusted sum of squares of 3175.5 compared to 3177.7 with period qualitative. More importantly, the degrees of freedom for period is two, one from each square, and there is no confounding of period effects with carryover effects. The adjusted sum of squares for the latter is 626.2, slightly more than before, but now it has three degrees of freedom, since contrasts involving Carryover "0" are now estimable. The adjusted sum of squares for cow is also slightly higher than before, but now its five degrees of freedom are what one would expect, considering that there are six animals. The only term that remains unchanged is that for treatment, its sum of squares being the same as previously.

The tests of significance for all effects leads to somewhat lower P–values, due to the fact that the residual mean square is lower than it was previously. This is because the error term has an extra degree of freedom, five instead of four, resulting from the ability to explain almost all of the period effect with a linear term. Had one considered period to be a continuous variable but added a quadratic effect as well as a linear effect, the sum of squares for Period would have been identical to that in Section 1.1.1. A quadratic effect may be constructed by squaring the period values (in the milk yield example one would obtain 1, 4 and 9 by squaring 1, 2 and 3, respectively) and incorporating a variable having these values, say PERIOD2, in the model as well as PERIOD. In other examples and exercises in this book, there will be occasions where the use of a quadratic effect, and perhaps even higher order effects, may be warranted.

1.5. <u>Testing Whether Effects Are Estimable in Cross-over Designs</u>

Elswick *et al.* (1991) have presented a method by which a design matrix X can be tested to see if it is of full rank, and hence whether all of its parameters are estimable. This is done by reducing the design matrix to "row echelon form", and checking whether the upper p x p portion of this n x p matrix, where n is the number of data points and p is the number of parameters, is an identity matrix. The computations may be achieved by using the ECHELON function of SAS/IML software.

We illustrate the results by using the milk yield example of Cochran and Cox (1957) discussed in Section 1.1. It does not matter whether the design matrix for that problem conforms to the sum-to-zero convention or to the set-to-zero convention; the result will be the same. Let us consider the matrix derived in Section 1.1.2 using the set-to-zero convention, except that we include three parameters for carryover, corresponding to the levels A, B and C, as was the original intention. The 15 columns of the design matrix correspond to 15 parameters listed in the following order, corresponding to (1) a grand mean, (2) five sequence (cow) parameters, (3) four period parameters (two for each square), (4) two treatment parameters, and (5) three carryover parameters. The matrix is given at the top of the following page.

The first 15 rows of the matrix, after being put into row echelon form, is given below the original matrix on the following page. Because the row echelon matrix of X is not an identity matrix, some parameters are confounded with others. From this matrix, one can conclude that carryover is confounded with the grand mean and with the parameters of Period 1. Only linear combinations of the three carryover parameters are estimable, and these are $\lambda_1-\lambda_3$, $\lambda_2-\lambda_3$, and, by subtracting the second last row from the third last row, $\lambda_1-\lambda_2$. By eliminating the last column of the design matrix and putting it in row echelon form, one obtains an identity matrix, thereby showing that all parameters are estimable. The consequence of this is that the design matrix should have at most two carryover parameters, and that carryover into Period 1 and carryover from Treatment C are both coded identically, having λ_1 and λ_2 both equal to zero in the design matrix. That is why the statement

IF CARRY = '0' THEN CARRY = 'C'

Original matrix:

$$
X =
\begin{pmatrix}
1 & 1 & 0 & 0 & 0 & 0 & 1 & 0 & 0 & 0 & 1 & 0 & 0 & 0 & 0 \\
1 & 1 & 0 & 0 & 0 & 0 & 0 & 1 & 0 & 0 & 0 & 1 & 1 & 0 & 0 \\
1 & 1 & 0 & 0 & 0 & 0 & 0 & 0 & 0 & 0 & 0 & 0 & 0 & 1 & 0 \\
1 & 0 & 1 & 0 & 0 & 0 & 1 & 0 & 0 & 0 & 0 & 1 & 0 & 0 & 0 \\
1 & 0 & 1 & 0 & 0 & 0 & 0 & 1 & 0 & 0 & 0 & 0 & 0 & 1 & 0 \\
1 & 0 & 1 & 0 & 0 & 0 & 0 & 0 & 0 & 0 & 1 & 0 & 0 & 0 & 1 \\
1 & 0 & 0 & 1 & 0 & 0 & 1 & 0 & 0 & 0 & 0 & 0 & 0 & 0 & 0 \\
1 & 0 & 0 & 1 & 0 & 0 & 0 & 1 & 0 & 0 & 1 & 0 & 0 & 0 & 1 \\
1 & 0 & 0 & 1 & 0 & 0 & 0 & 0 & 0 & 0 & 0 & 1 & 1 & 0 & 0 \\
1 & 0 & 0 & 0 & 1 & 0 & 0 & 0 & 1 & 0 & 1 & 0 & 0 & 0 & 0 \\
1 & 0 & 0 & 0 & 1 & 0 & 0 & 0 & 0 & 1 & 0 & 0 & 1 & 0 & 0 \\
1 & 0 & 0 & 0 & 1 & 0 & 0 & 0 & 0 & 0 & 0 & 1 & 0 & 0 & 1 \\
1 & 0 & 0 & 0 & 0 & 1 & 0 & 0 & 1 & 0 & 0 & 1 & 0 & 0 & 0 \\
1 & 0 & 0 & 0 & 0 & 1 & 0 & 0 & 0 & 1 & 1 & 0 & 0 & 1 & 0 \\
1 & 0 & 0 & 0 & 0 & 1 & 0 & 0 & 0 & 0 & 0 & 0 & 1 & 0 & 0 \\
1 & 0 & 0 & 0 & 0 & 0 & 0 & 0 & 1 & 0 & 0 & 0 & 0 & 0 & 0 \\
1 & 0 & 0 & 0 & 0 & 0 & 0 & 0 & 0 & 1 & 0 & 1 & 0 & 0 & 1 \\
1 & 0 & 0 & 0 & 0 & 0 & 0 & 0 & 0 & 0 & 1 & 0 & 0 & 1 & 0
\end{pmatrix} .
$$

Echelon form of matrix:

$$
\begin{pmatrix}
1 & 0 & 0 & 0 & 0 & 0 & 0 & 0 & 0 & 0 & 0 & 0 & 0 & 0 & 1 \\
0 & 1 & 0 & 0 & 0 & 0 & 0 & 0 & 0 & 0 & 0 & 0 & 0 & 0 & 0 \\
0 & 0 & 1 & 0 & 0 & 0 & 0 & 0 & 0 & 0 & 0 & 0 & 0 & 0 & 0 \\
0 & 0 & 0 & 1 & 0 & 0 & 0 & 0 & 0 & 0 & 0 & 0 & 0 & 0 & 0 \\
0 & 0 & 0 & 0 & 1 & 0 & 0 & 0 & 0 & 0 & 0 & 0 & 0 & 0 & 0 \\
0 & 0 & 0 & 0 & 0 & 1 & 0 & 0 & 0 & 0 & 0 & 0 & 0 & 0 & 0 \\
0 & 0 & 0 & 0 & 0 & 0 & 1 & 0 & 0 & 0 & 0 & 0 & 0 & 0 & -1 \\
0 & 0 & 0 & 0 & 0 & 0 & 0 & 1 & 0 & 0 & 0 & 0 & 0 & 0 & 0 \\
0 & 0 & 0 & 0 & 0 & 0 & 0 & 0 & 1 & 0 & 0 & 0 & 0 & 0 & -1 \\
0 & 0 & 0 & 0 & 0 & 0 & 0 & 0 & 0 & 1 & 0 & 0 & 0 & 0 & 0 \\
0 & 0 & 0 & 0 & 0 & 0 & 0 & 0 & 0 & 0 & 1 & 0 & 0 & 0 & 0 \\
0 & 0 & 0 & 0 & 0 & 0 & 0 & 0 & 0 & 0 & 0 & 1 & 0 & 0 & 0 \\
0 & 0 & 0 & 0 & 0 & 0 & 0 & 0 & 0 & 0 & 0 & 0 & 1 & 0 & -1 \\
0 & 0 & 0 & 0 & 0 & 0 & 0 & 0 & 0 & 0 & 0 & 0 & 0 & 1 & -1 \\
0 & 0 & 0 & 0 & 0 & 0 & 0 & 0 & 0 & 0 & 0 & 0 & 0 & 0 & 0
\end{pmatrix} .
$$

had to be used in Section 1.1.1 to enable the LSMEANS statement to produce estimable differences among the means for treatment and carryover effects.

We now examine the case described in Section 1.4, where period is considered to be a continuous, rather than a qualitative, variable. The design matrix will now contain two less parameters than before, as follows, with the time span in Square 1 being considered to be different from the time span in Square 2.

$$
X = \begin{pmatrix}
1 & 1 & 0 & 0 & 0 & 0 & 1 & 0 & 1 & 0 & 0 & 0 & 0 \\
1 & 1 & 0 & 0 & 0 & 0 & 2 & 0 & 0 & 1 & 1 & 0 & 0 \\
1 & 1 & 0 & 0 & 0 & 0 & 3 & 0 & 0 & 0 & 0 & 1 & 0 \\
1 & 0 & 1 & 0 & 0 & 0 & 1 & 0 & 0 & 1 & 0 & 0 & 0 \\
1 & 0 & 1 & 0 & 0 & 0 & 2 & 0 & 0 & 0 & 0 & 1 & 0 \\
1 & 0 & 1 & 0 & 0 & 0 & 3 & 0 & 1 & 0 & 0 & 0 & 1 \\
1 & 0 & 0 & 1 & 0 & 0 & 1 & 0 & 0 & 0 & 0 & 0 & 0 \\
1 & 0 & 0 & 1 & 0 & 0 & 2 & 0 & 1 & 0 & 0 & 0 & 1 \\
1 & 0 & 0 & 1 & 0 & 0 & 3 & 0 & 0 & 1 & 1 & 0 & 0 \\
1 & 0 & 0 & 0 & 1 & 0 & 0 & 1 & 1 & 0 & 0 & 0 & 0 \\
1 & 0 & 0 & 0 & 1 & 0 & 0 & 2 & 0 & 0 & 1 & 0 & 0 \\
1 & 0 & 0 & 0 & 1 & 0 & 0 & 3 & 0 & 1 & 0 & 0 & 1 \\
1 & 0 & 0 & 0 & 0 & 1 & 0 & 1 & 0 & 1 & 0 & 0 & 0 \\
1 & 0 & 0 & 0 & 0 & 1 & 0 & 2 & 1 & 0 & 0 & 1 & 0 \\
1 & 0 & 0 & 0 & 0 & 1 & 0 & 3 & 0 & 0 & 1 & 0 & 0 \\
1 & 0 & 0 & 0 & 0 & 0 & 0 & 1 & 0 & 0 & 0 & 0 & 0 \\
1 & 0 & 0 & 0 & 0 & 0 & 0 & 2 & 0 & 1 & 0 & 0 & 1 \\
1 & 0 & 0 & 0 & 0 & 0 & 0 & 3 & 1 & 0 & 0 & 1 & 0
\end{pmatrix} .
$$

Note that we are using three carryover parameters, as in the original plan. This time none of them will be confounded with other parameters, as the following row echelon form of the above matrix shows:

$$
\begin{pmatrix}
1 & 0 & 0 & 0 & 0 & 0 & 0 & 0 & 0 & 0 & 0 & 0 & 0 \\
0 & 1 & 0 & 0 & 0 & 0 & 0 & 0 & 0 & 0 & 0 & 0 & 0 \\
0 & 0 & 1 & 0 & 0 & 0 & 0 & 0 & 0 & 0 & 0 & 0 & 0 \\
0 & 0 & 0 & 1 & 0 & 0 & 0 & 0 & 0 & 0 & 0 & 0 & 0 \\
0 & 0 & 0 & 0 & 1 & 0 & 0 & 0 & 0 & 0 & 0 & 0 & 0 \\
0 & 0 & 0 & 0 & 0 & 1 & 0 & 0 & 0 & 0 & 0 & 0 & 0 \\
0 & 0 & 0 & 0 & 0 & 0 & 1 & 0 & 0 & 0 & 0 & 0 & 0 \\
0 & 0 & 0 & 0 & 0 & 0 & 0 & 1 & 0 & 0 & 0 & 0 & 0 \\
0 & 0 & 0 & 0 & 0 & 0 & 0 & 0 & 1 & 0 & 0 & 0 & 0 \\
0 & 0 & 0 & 0 & 0 & 0 & 0 & 0 & 0 & 1 & 0 & 0 & 0 \\
0 & 0 & 0 & 0 & 0 & 0 & 0 & 0 & 0 & 0 & 1 & 0 & 0 \\
0 & 0 & 0 & 0 & 0 & 0 & 0 & 0 & 0 & 0 & 0 & 1 & 0 \\
0 & 0 & 0 & 0 & 0 & 0 & 0 & 0 & 0 & 0 & 0 & 0 & 1
\end{pmatrix} .
$$

Since the matrix is an identity matrix, all parameters are estimable. The results of fitting that model to the milk yield data are shown in Section 1.4.

Consider now the case where period is considered to be a continuous variable but where one wishes to fit both linear and quadratic effects. The design matrix is as follows:

$$
X = \begin{pmatrix}
1 & 1 & 0 & 0 & 0 & 0 & 1 & 1 & 0 & 0 & 1 & 0 & 0 & 0 & 0 \\
1 & 1 & 0 & 0 & 0 & 0 & 2 & 4 & 0 & 0 & 0 & 1 & 1 & 0 & 0 \\
1 & 1 & 0 & 0 & 0 & 0 & 3 & 9 & 0 & 0 & 0 & 0 & 0 & 1 & 0 \\
1 & 0 & 1 & 0 & 0 & 0 & 1 & 1 & 0 & 0 & 0 & 1 & 0 & 0 & 0 \\
1 & 0 & 1 & 0 & 0 & 0 & 2 & 4 & 0 & 0 & 0 & 0 & 0 & 1 & 0 \\
1 & 0 & 1 & 0 & 0 & 0 & 3 & 9 & 0 & 0 & 1 & 0 & 0 & 0 & 1 \\
1 & 0 & 0 & 1 & 0 & 0 & 1 & 1 & 0 & 0 & 0 & 0 & 0 & 0 & 0 \\
1 & 0 & 0 & 1 & 0 & 0 & 2 & 4 & 0 & 0 & 1 & 0 & 0 & 0 & 1 \\
1 & 0 & 0 & 1 & 0 & 0 & 3 & 9 & 0 & 0 & 0 & 1 & 1 & 0 & 0 \\
1 & 0 & 0 & 0 & 1 & 0 & 0 & 0 & 1 & 1 & 1 & 0 & 0 & 0 & 0 \\
1 & 0 & 0 & 0 & 1 & 0 & 0 & 0 & 2 & 4 & 0 & 0 & 1 & 0 & 0 \\
1 & 0 & 0 & 0 & 1 & 0 & 0 & 0 & 3 & 9 & 0 & 1 & 0 & 0 & 1 \\
1 & 0 & 0 & 0 & 0 & 1 & 0 & 0 & 1 & 1 & 0 & 1 & 0 & 0 & 0 \\
1 & 0 & 0 & 0 & 0 & 1 & 0 & 0 & 2 & 4 & 1 & 0 & 0 & 1 & 0 \\
1 & 0 & 0 & 0 & 0 & 1 & 0 & 0 & 3 & 9 & 0 & 0 & 1 & 0 & 0 \\
1 & 0 & 0 & 0 & 0 & 0 & 0 & 0 & 1 & 1 & 0 & 0 & 0 & 0 & 0 \\
1 & 0 & 0 & 0 & 0 & 0 & 0 & 0 & 2 & 4 & 0 & 1 & 0 & 0 & 1 \\
1 & 0 & 0 & 0 & 0 & 0 & 0 & 0 & 3 & 9 & 1 & 0 & 0 & 1 & 0 \\
\end{pmatrix} .
$$

The echelon form of the above matrix is as follows:

$$
\begin{pmatrix}
1 & 0 & 0 & 0 & 0 & 0 & 0 & 0 & 0 & 0 & 0 & 0 & 0 & 0 & -2 \\
0 & 1 & 0 & 0 & 0 & 0 & 0 & 0 & 0 & 0 & 0 & 0 & 0 & 0 & 0 \\
0 & 0 & 1 & 0 & 0 & 0 & 0 & 0 & 0 & 0 & 0 & 0 & 0 & 0 & 0 \\
0 & 0 & 0 & 1 & 0 & 0 & 0 & 0 & 0 & 0 & 0 & 0 & 0 & 0 & 0 \\
0 & 0 & 0 & 0 & 1 & 0 & 0 & 0 & 0 & 0 & 0 & 0 & 0 & 0 & 0 \\
0 & 0 & 0 & 0 & 0 & 1 & 0 & 0 & 0 & 0 & 0 & 0 & 0 & 0 & 0 \\
0 & 0 & 0 & 0 & 0 & 0 & 1 & 0 & 0 & 0 & 0 & 0 & 0 & 0 & 3 \\
0 & 0 & 0 & 0 & 0 & 0 & 0 & 1 & 0 & 0 & 0 & 0 & 0 & 0 & -1 \\
0 & 0 & 0 & 0 & 0 & 0 & 0 & 0 & 1 & 0 & 0 & 0 & 0 & 0 & 3 \\
0 & 0 & 0 & 0 & 0 & 0 & 0 & 0 & 0 & 1 & 0 & 0 & 0 & 0 & 0 \\
0 & 0 & 0 & 0 & 0 & 0 & 0 & 0 & 0 & 0 & 1 & 0 & 0 & 0 & 0 \\
0 & 0 & 0 & 0 & 0 & 0 & 0 & 0 & 0 & 0 & 0 & 1 & 0 & 0 & 0 \\
0 & 0 & 0 & 0 & 0 & 0 & 0 & 0 & 0 & 0 & 0 & 0 & 1 & 0 & -1 \\
0 & 0 & 0 & 0 & 0 & 0 & 0 & 0 & 0 & 0 & 0 & 0 & 0 & 1 & -1 \\
0 & 0 & 0 & 0 & 0 & 0 & 0 & 0 & 0 & 0 & 0 & 0 & 0 & 0 & 0 \\
\end{pmatrix} .
$$

The fact that the matrix is not the identity matrix indicates that some parameters are confounded with some other parameters. Carryover is confounded with the grand mean and period, and only linear combinations involving the carryover parameters are estimable. This is exactly the same state of affairs that prevailed when period was considered to be qualitative and three carryover parameters were included in the model. Clearly, considering period to be a continuous variable will only have utility when there are less fitted terms than degrees of freedom. Since period has only two degrees of freedom for the milk yield data, confounding between carryover and period effects can only be eliminated if a linear term is fitted. For data sets that involve four or more periods, it will be possible to include both linear and quadratic effects without causing confounding. For designs with a larger number of periods, one might consider higher order terms, although these might cause problems of interpretation.

1.6. Testing Whether the Residuals Are Normally Distributed

In fitting the linear model described in Section 1.1 to continuous data from cross-over designs, the modeler is implicitly making a number of assumptions about the nature of the data. One of the most important assumptions is that the data obtained from each subject conform to a multivariate normal distribution with a covariance matrix that satisfies the "Type H" structure described by Huynh and Feldt (1970). This structure will be described in detail in Chapter 8. Suffice it to say here that it is permissible for the observations on any one subject over time to be correlated, provided that the covariances satisfy the Type H structure. Special cases of the Type H structure are the uniform covariance or compound symmetry structure, where all diagonal elements (the variances) are equal and all off-diagonal elements (the covariances) are equal, and the independent error case, where all covariances are equal to zero. Although either of these two special cases is sufficient to justify the use of the ordinary least squares (OLS) procedures described in Section 1.1, neither of them is necessary. Much of the data obtained from cross-over designs in practice proves to be consistent with the Type H structure, or at least being not too far from it, justifying OLS. When that condition is not satisfied, specialized methodology that is touched upon in Chapter 8 may have to be used. In this section, we will assume that the covariances satisfy the Type H structure, and turn attention to other considerations.

If the observations on the response variable from a cross-over trial are corrected for all model terms that can be estimated, such as a grand mean, subject, period, treatment and carryover effects, the resulting residuals should estimate the random experimental error effect corresponding to the measurement of any subject during any period. It is useful to check these residuals for normality as an aid in deciding whether or not there are any "outliers", that is, unusual observations, in the data set. Various tests of normality are available.

One of these tests is due to Shapiro and Wilk (1965), and operates upon the residuals obtained as described above. Another procedure, described by Cook and Weisberg (1982), is to plot standardized residuals against their expected normal order statistics. The standardization is carried out by dividing the above residuals by their standard errors. These "studentized" residuals are then treated as though they were standard normal variates. When plotted against the expected normal order statistics, they should produce a straight line if the residuals are normally distributed. It is customary that the assessment of whether the line is straight be made visually, rather than by using a formal test.

We now illustrate the procedure employing the milk yield data set of Section 1.1. The "raw" residuals and "studentized" residuals are readily obtained by use of PROC GLM of SAS, making some straightforward additions to the code that appeared in that section. Consider the following code:

```
PROC GLM;
CLASS COW PERIOD SQUARE TREAT CARRY;
MODEL Y=COW PERIOD(SQUARE) TREAT CARRY/SOLUTION
    SS1 SS2 E1 E2;
RANDOM COW;
OUTPUT OUT=DRES STUDENT=STDRES R=RESID P=FITVAL;
LSMEANS TREAT CARRY/PDIFF;
RUN;  .
```

This is the same code as in Section 1.1 except that now the OUTPUT statement has been added to produce a new data set named DRES which will contain all the variables in the original data set including the studentized residuals STDRES, the ordinary residuals RESID, and the fitted values of the model FITVAL. Any or all of these variables may be used in subsequent data steps and procedures. PROC UNIVARIATE produces information about the moments of the residuals and calculates the Wilk–Shapiro statistic. It also produces a rough normal probability plot. Following is some code to produce the required test and plots:

```
PROC UNIVARIATE DATA=DRES PLOT NORMAL;
 VAR RESID;
RUN;  .
```

For the milk yield data, the Wilk–Shapiro statistic was 0.9421, with an associated P–value of 0.3183. This statistic can be viewed as a correlation coefficient between the standardized residuals and the expected normal order statistics. The closer to unity that value is, the better the data conform to the normal assumption. The procedure also produces, *inter alia*, a "box plot" and lists quantiles and the five most extreme positive and negative values. For this data set, the largest negative standardized residual is –1.794 and the largest positive one is 1.314. These are not very large for a set of 18 observations, and rather than there being any potential "outliers" in the data set, the distribution, if anything, is slightly "short–tailed".

It is always good practice in a regression analysis to plot the residuals against the fitted values to see if there is any indication of a systematic departure of the data from the model (see, for example, Draper and Smith, 1981). Such a plot can be made using the following code, recalling that RESID contains the raw residuals and FITVAL holds the fitted values from the model:

```
PROC PLOT DATA=DRES HPCT=90 VPCT=50;
     PLOT RESID*FITVAL;
RUN;  .
```

If the variance is not homogeneous, the heterogeneity will be visually obvious from such a graph.

To obtain a more accurate normal probability plot than that produced by PROC UNIVARIATE, the following SAS/GRAPH code, with options selected to produce a good quality graph, calculates the expected normal order statistics and then plots the standardized residuals STDRES against those expected values.

```
PROC SUMMARY DATA=DRES;
     VAR STDRES; /* This procedure determines the sample size n. */
     OUTPUT OUT=DN (KEEP=N) N=N;
PROC SORT DATA=DRES;
     BY STDRES;    /* Sorts studentized residuals in increasing order. */
DATA DPLOT (KEEP=STDRES ENOSTAT);
     IF _N_ EQ 1 THEN SET DN;
     SET DRES;  /* Calculates expected normal order statistics. */
     EOS = PROBIT((_N_ - 0.363)/(N - 0.726 + 1));
PROC GPLOT DATA=DPLOT;
```

```
/* The following statement sets the appropriate plotting device.  This will differ
depending upon the plotter to which the user has access. */
```

```
GOPTIONS DEVICE=HP7550 GACCESS='SASGASTD>LPT1';
    TITLE 'NORMAL PROBABILITY PLOT';
    AXIS1 WIDTH=3 label=(height=2 'Expected Normal Order Statistic')
        value=(h=2);
    AXIS2 WIDTH=3 label=(height=2 angle=90 'Residual')
        value=(h=2);
    PLOT STDRES*EOS='DOT'/HAXIS=AXIS1 VAXIS=AXIS2;
RUN;
PROC PRINT;
    VAR COW PERIOD TREAT CARRY RESID STDRES;
RUN;
```

The output from PROG GPLOT will be a good quality graph which should enable the user to decide whether or not the residuals are normally distributed. If the graph is a reasonably good straight line, the user can feel confident that the residuals are close to being normally distributed and that at least that part of the underlying assumptions holds. If the graph is not a good straight line, particularly if there is one or more points with a large residual (say greater than 2.5 or so), then one has to doubt at least part of the statistical assumptions. As an aid to discovering which measurements have the large absolute residuals, the PROC PRINT procedure (see above code) will print the residuals in increasing order, and list the values of the other variables along with them. That should enable the user to identify the experimental units in which the largest deviations occur. For the milk yield data, the graph shown on the following page was obtained.

As noted earlier, there are no apparent outliers in this data set but rather a tendency to have shorter tails than in a normal distribution. As the Wilk–Shapiro statistic is non–significant, suggesting that the residuals are close to having a normal distribution, one should feel satisfied that the normality assumption made when PROC GLM was used is a sufficiently good approximation to the truth.

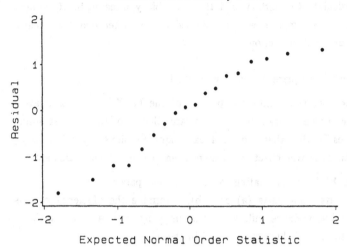

Normal Probability Plot

Exercises

1.1. Consider the following 4 x 4 Latin square, which is not balanced with respect to the separability of treatment and carryover effects.

		PERIOD			
		1	2	3	4
	1	A	B	C	D
SEQUENCE	2	B	A	D	C
(or SUBJECT)	3	C	D	B	A
	4	D	C	A	B

Write down a contingency table representing the incidences of treatment and carryover, and calculate Cramer's V and the separability measure S. Comment on how this value of S compares with S obtained for the other two 4 x 4 Latin square designs considered in Section 1.2.

1.2. Consider the Latin square design of Exercise 1.1.

(a) Using matrix operations, find the portion of the $(X'X)^{-1}$ matrix relating treatment and carryover parameters.   Compare these to the corresponding covariance matrices for the other two 4 x 4 Latin square designs of Section 1.2.

(b) Now drop the carryover effect parameters from the model and calculate the portion of the $(X'X)^{-1}$ matrix relating to the treatment parameters.

(c) Using the results from parts (a) and (b), determine the efficiency of this design.   What is the variance inflation factor in going from a model with no carryover effects to one in which carryover *is* in the model?

1.3.   Consider the following set of data on milk production (kg milk/day) from an experiment where 12 Guernsey cows were randomly allocated to a whey-supplemented ration A, with another 12 cows given control ration B.   The two sequences were ABAB and BABA [data of Gill (1978), Exercise 8.9, reproduced here with the permission of the Iowa State University Press].

Sequence ABAB					Sequence BABA			
	Period					Period		
Cow↓	1	2	3	4	Cow↓	1	2	3	4
1	30.8	31.0	29.6	21.4	13	31.2	31.3	29.2	28.1
2	35.4	33.3	29.2	30.2	14	35.3	36.7	34.4	31.9
3	35.5	35.5	37.0	29.9	15	28.2	26.8	24.2	22.0
4	31.3	28.0	29.0	26.3	16	26.4	26.7	27.0	25.8
5	30.2	27.3	26.2	24.9	17	21.7	21.4	20.3	19.1
6	30.2	32.2	30.4	29.5	18	25.6	27.1	25.8	25.1
7	22.7	22.6	20.4	19.1	19	31.3	30.2	26.8	25.5
8	31.7	30.3	27.3	24.5	20	29.7	27.2	25.6	22.2
9	34.1	29.5	27.4	25.7	21	30.0	28.6	26.8	23.5
10	32.4	29.5	28.9	25.8	22	37.6	36.4	33.5	29.2
11	36.3	33.8	30.0	26.7	23	32.6	27.9	26.5	22.7
12	25.6	23.9	20.1	19.2	24	25.0	25.4	25.7	24.1

(a) Since each sequence contains each treatment the same number of times, one may use the omnibus measure of Section 1.2 to determine the separability of treatment and carryover effects. Calculate S for the above design, and confirm that "switchback" or "reversal" designs like the above, in which the two treatments alternate in successive periods, are poor ones for separating treatment and carryover effects.

(b) Calculate the variance inflation factor VIF, that is, the ratio between the variance of the treatment effects in the presence of the carryover parameters to that when the carryover parameters are removed from the model.

(c) Carry out the analysis of variance of the above data set. In view of your answer to the question in part (a), is there evidence that a serious multicollinearity situation prevails here?

(d) The analysis of variance should have shown that although there is little evidence of treatment or carryover effects, there is considerable evidence of period effects. Using continuous period effects, describe how the milk production changes with time.

Appendix 1.1: Use of statistical packages other than SAS to analyze data from cross–over trials.

[1] BASS, Version 85 (Bass Institute, P.O. Box 349, Chapel Hill, North Carolina, 27514, U.S.A). [Note: Later versions may have different procedures or require different coding.]

The following code creates a data set named "cochcox":

```
run create >cochcox;
input y cow period treat $ carry $ square @;
if carry='0' then carry='C';
cards;
    38 1 1 A 0 1   25 1 2 B A 1   15 1 3 C B 1
   109 2 1 B 0 1   86 2 2 C B 1   39 2 3 A C 1
   124 3 1 C 0 1   72 3 2 A C 1   27 3 3 B A 1
    86 4 1 A 0 2   76 4 2 C A 2   46 4 3 B C 2
    75 5 1 B 0 2   35 5 2 A B 2   34 5 3 C A 2
   101 6 1 C 0 2   63 6 2 B C 2    1 6 3 A B 2
```

The analysis of variance can be performed using the following code:

```
run anova <cochcox;
model y=cow period period*square treat carry;
 class cow period square treat carry;
```

The above code produces the following output (slightly edited):

Dependent Variable:
Y
   Mean: 58.4444444   Std. Deviation: 34.6091151

Levels found for class variables:
   CARRY:C,A,B
      The first 2 value(s) listed will have associated dummy variables.
      The last value listed is the "omitted level"; it occurred in 4 cases.
   TREAT:A,B,C
      The first 2 value(s) listed will have associated dummy variables.
      The last value listed is the "omitted level"; it occurred in 6 cases.
   SQUARE:1,2
      The first 1 value(s) listed will have associated dummy variables.
      The last value listed is the "omitted level"; it occurred in 9 cases.
   PERIOD:1,2,3
      The first 2 value(s) listed will have associated dummy variables.
      The last value listed is the "omitted level"; it occurred in 6 cases.
   COW:1,2,3,4,5,6
      The first 5 value(s) listed will have associated dummy variables.
      The last value listed is the "omitted level"; it occurred in 3 cases.

Model Statistics:
   NOBS: 18   Multiple Correlation: 0.99509539
   Std Error: 7.05779711   R–Squared: 0.99021483
   Dataset Type: DATA   Adjusted R–Squared: 0.95841302

Analysis of Variance:

Source	DF	Sum of Squares	Mean Square	F Value	Prob > F
Model	13	20163.19444	1551.01496	31.13706	0.00225
Error	4	199.25000	49.81250		
Total	17	20362.44444			

Parameter Estimates:

Variable	Coeffic.	Std. Error	Toler.	Std.Coeff.	T	Prob
CONSTANT	54.375	2.0374105		0.00000000	26.69	0.000
_D7_ COW=1	–28.375	3.9013441	0.545	–0.48707579	–7.27	0.002
_D8_ COW=2	16.875	3.9013441	0.545	0.28967062	4.33	0.012
_D9_ COW=3	14.500	3.9013441	0.545	0.24890216	3.72	0.021
_D10_ COW=4	9.500	3.9013441	0.545	0.16307383	2.44	0.072
_D11_ COW=5	–6.375	3.9013441	0.545	–0.10943112	–1.63	0.178
_D5_ PERIOD=1	22.250	3.3270775	0.375	0.54013918	6.69	0.003
_D6_ PERIOD=2	5.125	2.6302857	0.600	0.12441408	1.95	0.123
_I1_ _D12_ * _D5_ == SQUARE=1*PERIOD=1	0.500	2.3525990	0.750	0.01213796	0.21	0.842
_I2_ _D12_ * _D6_ == SQUARE=1*PERIOD=2	0.500	2.3525990	0.750	0.01213796	0.21	0.842

_D3_	−15.958	2.6302857	0.600	−0.38740320	−6.07	0.004
TREAT=A						
_D4_	−2.333	2.6302857	0.600	−0.05664381	−0.89	0.425
TREAT=B						
_D1_	12.208	3.5288986	0.333	0.29636850	3.46	0.026
CARRY=C						
_D2_	−8.042	3.5288986	0.500	−0.15939552	−2.28	0.085
CARRY=A						

Analysis of Variance for Classification Effects:

Source	DF	Sum of Squares	Mean Square	F Value	Prob > F
COW	5	3835.95000	767.19000	15.40156	0.01015
PERIOD	2	5396.53125	2698.26563	54.16844	0.00127
PERIOD* SQUARE					
	2	9.00000	4.50000	0.09034	0.91543
TREAT	2	2854.55000	1427.27500	28.65295	0.00426
CARRY	2	616.19444	308.09722	6.18514	0.05970
Error	4	199.25000	49.81250		

[2] MINITAB, Version 7 (Minitab, Inc., 3081 Enterprise Drive, State College, PA 16801, U.S.A.)

The following file called 'cochcox.dat' has been previously prepared and stored:

```
38 1 1 1 3 1
25 1 2 2 1 1
15 1 3 3 2 1
109 2 1 2 3 1
86 2 2 3 2 1
39 2 3 1 3 1
124 3 1 3 3 1
72 3 2 1 3 1
27 3 3 2 1 1
86 1 1 1 3 2
76 1 2 3 1 2
46 1 3 2 3 2
75 2 1 2 3 2
35 2 2 1 2 2
34 2 3 3 1 2
101 3 1 3 3 2
63 3 2 2 3 2
1 3 3 1 2 2
```

Now follows the results of an interactive session using Minitab:

```
MTB > read 'cochcox.dat' c1–c6
       18 ROWS READ
```

```
ROW    Y    COW   PERIOD  TREAT  CARRY  SQUARE
 1     38    1      1       1       3      1
 2     25    1      2       2       1      1
 3     15    1      3       3       2      1
 4    109    2      1       2       3      1
       .     .
```

MTB > print c1–c6

```
ROW    Y    COW   PERIOD  TREAT  CARRY  SQUARE
 1     38    1      1       1       3      1
 2     25    1      2       2       1      1
 3     15    1      3       3       2      1
 4    109    2      1       2       3      1
 5     86    2      2       3       2      1
 6     39    2      3       1       3      1
 7    124    3      1       3       3      1
 8     72    3      2       1       3      1
 9     27    3      3       2       1      1
10     86    1      1       1       3      2
11     76    1      2       3       1      2
12     46    1      3       2       3      2
13     75    2      1       2       3      2
14     35    2      2       1       2      2
15     34    2      3       3       1      2
16    101    3      1       3       3      2
17     63    3      2       2       3      2
18      1    3      3       1       2      2
```

MTB > glm y=square cow(square) period(square) treat carry;
SUBC> means treat carry.

Factor	Levels	Values		
SQUARE	2	1	2	
COW(SQUARE)	3	1	2	3
PERIOD(SQUARE)	3	1	2	3
TREAT	3	1	2	3
CARRY	3	1	2	3

Analysis of Variance for Y

Source	DF	Seq SS	Adj SS	Adj MS	F	P
SQUARE	1	18.0	18.0	18.0	0.36	0.580
COW(SQUARE)	4	5763.1	3817.9	954.5	19.16	0.007
PERIOD(SQUARE)	4	11489.1	5405.5	1351.4	27.13	0.004
TREAT	2	2276.8	2854.6	1427.3	28.65	0.004
CARRY	2	616.2	616.2	308.1	6.19	0.060
Error	4	199.3	199.3	49.8		
Total	17	20362.4				

Means for Y

TREAT	Mean	Stdev
1	38.42	3.529
2	52.04	3.529
3	72.67	2.881
CARRY		
1	46.33	4.556
2	50.21	4.556
3	66.58	2.881

MTB > stop

[3] GENSTAT, Version 5 (Numerical Algorithms Group Ltd., Mayfield House, 256 Banbury Road, Oxford, U.K. OX2 7DE)

Code for producing the correct analysis of the milk yield data is presented below. Since Genstat only produces sequential sums of squares (rather than the adjusted sums of squares needed for testing treatment and carryover effects), the regression model is fitted twice, once with the carryover term last (the usual order), and once with the treatment term last (to give the adjusted sum of squares for treatment). Note also that since there is no option for nesting of terms in a regression model, Cows are labeled 1, 2,..., 6 (rather than 1, 2, 3 in Square 1 and 1, 2, 3 in Square 2) and Periods are labeled 1, 2,..., 6 to achieve a pooled estimate of Periods within Squares, rather than use the more natural labeling 1, 2, 3 in Period 1 and 1, 2, 3 in Period 2.

```
units [nvalues=18]
variate y
factor [level=6] cow
factor [level=6] period
factor [level=3] treat,carry
factor [level=2] square
read [print=data,errors] y,cow,period,treat,carry,square
 38 1 1 1 3 1
 25 1 2 2 1 1
 15 1 3 3 2 1
109 2 1 2 3 1
 86 2 2 3 2 1
 39 2 3 1 3 1
124 3 1 3 3 1
 72 3 2 1 3 1
 27 3 3 2 1 1
 86 4 4 1 3 2
 76 4 5 3 1 2
 46 4 6 2 3 2
 75 5 4 2 3 2
 35 5 5 1 2 2
```

```
34 5 6 3 1 2
101 6 4 3 3 2
 63 6 5 2 3 2
  1 6 6 1 2 2 :
model y
fit square,cow,period,treat,carry
rdisplay [print=a]
fit square,cow,period,carry,treat
rdisplay [print=a]
```

The output from the above code is as follows (somewhat edited). Note that Genstat complains that there is aliasing of some of the effects. Nevertheless, the correct sequential sums of squares result.

Genstat 5   Release 2.2   Thu Oct 10 09:44:33 1991
Copyright 1990, Lawes Agricultural Trust (Rothamsted Experimental Station)

```
READ>   8    38 1 1 1 3 1
READ>   9    25 1 2 2 1 1
READ>  10    15 1 3 3 2 1
READ>  11   109 2 1 2 3 1
READ>  12    86 2 2 3 2 1
READ>  13    39 2 3 1 3 1
READ>  14   124 3 1 3 3 1
READ>  15    72 3 2 1 3 1
READ>  16    27 3 3 2 1 1
READ>  17    86 4 4 1 3 2
READ>  18    76 4 5 3 1 2
READ>  19    46 4 6 2 3 2
READ>  20    75 5 4 2 3 2
READ>  21    35 5 5 1 2 2
READ>  22    34 5 6 3 1 2
READ>  23   101 6 4 3 3 2
READ>  24    63 6 5 2 3 2
READ>  25     1 6 6 1 2 2 :
```

```
>    * MESSAGE: Term cow cannot be fully included in the model
        because 1 parameter is aliased with terms already in the model
     * MESSAGE: Term period cannot be fully included in the model
        because 1 parameter is aliased with terms already in the model
```

***** Regression Analysis
***** Response variate: y
    Fitted terms: Constant, square, cow, period, treat, carry

*** Summary of analysis ***

	d.f.	s.s.	m.s.	v.r.
Regression	13	20163.2	1551.02	31.14
Residual	4	199.3	49.81	
Total	17	20362.4	1197.79	

Percentage variance accounted for 95.8

```
*** Estimates of regression coefficients ***
                estimate        s.e.           t
Constant        24.75          8.48          2.92
square 2       -26.87          7.71         -3.48
cow 2           45.25          6.11          7.40
cow 3           42.88          6.11          7.01
cow 4           15.63          6.11          2.56
cow 5          -0.25           6.11         -0.04
cow 6           0              *             *
period 2       -17.12          6.76         -2.53
period 3       -51.13          6.76         -7.57
period 4        48.13          6.76          7.12
period 5        31.00          5.76          5.38
period 6        0              *             *
treat 2         13.63          4.56          2.99
treat 3         34.25          4.56          7.52
carry 2         3.87           6.11          0.63
carry 3         20.25          6.11          3.31
```

***** Regression Analysis
******** Accumulated analysis of variance ***

Change	d.f.	s.s.	m.s.	v.r.
+ square	1	18.00	18.00	0.36
+ cow	4	5763.11	1440.78	28.92
+ period	4	11489.11	2872.28	57.66
+ treat	2	2276.78	1138.39	22.85
+ carry	2	616.19	308.10	6.19
Residual	4	199.25	49.81	
Total	17	20362.45	1197.79	

(After fitting "carry" before "treat", the following accumulated analysis of variance table is produced:)

***** Regression Analysis
******** Accumulated analysis of variance***

Change	d.f.	s.s.	m.s.	v.r.
+ square	1	18.00	18.00	0.36
+ cow	4	5763.11	1440.78	28.92
+ period	4	11489.11	2872.28	57.66
+ carry	2	38.42	19.21	0.39
+ treat	2	2854.55	1427.28	28.65
Residual	4	199.25	49.81	
Total	17	20362.45	1197.79	

2

Latin Square Designs

2.1. Optimum Designs for Separating Treatment and Carryover Effects

There are both good and bad Latin square designs from the point of view of the estimation of carryover effects. The good designs fall into a class named "digram–balanced" squares by Wagenaar (1969), which have the property that each treatment is preceded and followed equally often by each other treatment. It was shown by Williams (1949) that for Latin squares of order n (n rows, n columns and n treatments), it is possible to achieve balance for the estimation of residual effects of treatments with a single Latin square for any even value of n, while for odd n, balance can be achieved with a pair of n x n squares. For even values of n, Wagenaar (1969) provided a simpler method than that advanced by Williams (1949) for generating such squares. Table 2.1 gives typical examples of such squares for even values of n up to 12, and Table 2.2 gives examples of pairs of squares for odd values of n up to 11.

Although digram–balanced squares are better than any other Latin squares insofar as the separability of treatment effects from carryover effects are concerned, the separability is not perfect. The measure of separability S defined in Section 1.2.1 for designs having a high degree of balance, based on a contingency table and Cramer's V, is given in Table 2.3 for digram–balanced squares up to order $n = 12$. Also included in this table are the more traditional

Table 2.1. Examples of n x n digram–balanced Latin squares, n even. Rows represent different subjects or sequences, columns different time periods.

$n=2$	$n=4$	$n=6$	$n=8$
A B	A B C D	A B C D E F	A B C D E F G H
B A	B D A C	B D A F C E	B D A F C H E G
	C A D B	C A E B F D	C A E B G D H F
	D C B A	D F B E A C	D F B H A G C E
		E C F A D B	E C G A H B F D
		F E D C B A	F H D G B E A C
			G E H C F A D B
			H G F E D C B A

$n=10$	$n=12$
A B C D E F G H I J	A B C D E F G H I J K L
B D F H J A C E G I	B D F H J L A C E G I K
C F I A D G J B E H	C F I L B E H K A D G J
D H A E I B F J C G	D H L C G K B F J A E I
E J D I C H B G A F	E J B G L D I A F K C H
F A G B H C I D J E	F L E K D J C I B H A G
G C J F B I E A H D	G A H B I C J D K E L F
H E B J G D A I F C	H C K F A I D L G B J E
I G E C A J H F D B	I E A J F B K G C L H D
J I H G F E D C B A	J G D A K H E B L I F C
	K I G E C A L J H F D B
	L K J I H G F E D C B A

measures of efficiency, E_d for "direct" treatment effects, and E_r for "residual" treatment effects (carryover). These latter efficiency measures were calculated from formulae presented in Patterson and Lucas (1962), which, for these Latin square designs, reduce to

$$E_d = \frac{n^2-n-2}{n^2-n-1} \cdot (100)$$

and

$$E_r = \frac{(n^2-n-2)}{n^2} \cdot (100),$$

respectively.

Except for the design for $n = 2$, where E_d and E_r are zero, S always falls between E_d and E_r, generally being closer to E_r. The fact that S is bounded by

Table 2.2.   Examples of pairs of n x n digram–balanced Latin squares, n odd. Rows represent different subjects or sequences, columns different time periods.

$n=3$

```
A B C     A C B
B C A     B A C
C A B     C B A
```

$n=5$

```
A B D E C     A C B E D
B C E A D     B D C A E
C D A B E     C E D B A
D E B C A     D A E C B
E A C D B     E B A D C
```

$n=7$

```
A B E C G F D     A G D F B C E
B C F D A G E     B A E G C D F
C D G E B A F     C B F A D E G
D E A F C B G     D C G B E F A
E F B G D C A     E D A C F G B
F G C A E D B     F E B D G A C
G A D B F E C     G F C E A B D
```

$n=9$

```
A B I C H D G E F     A I B H C G D F E
B C A D I E H F G     B A C I D H E G F
C D B E A F I G H     C B D A E I F H G
D E C F B G A H I     D C E B F A G I H
E F D G C H B I A     E D F C G B H A I
F G E H D I C A B     F E G D H C I B A
G H F I E A D B C     G F H E I D A C B
H I G A F B E C D     H G I F A E B D C
I A H B G C F D E     I H A G B F C E D
```

$n=11$

```
A K B J C I D H E G F     A B K C J D I E H F G
B A C K D J E I F H G     B C A D K E J F I G H
C B D A E K F J G I H     C D B E A F K G J H I
D C E B F A G K H J I     D E C F B G A H K I J
E D F C G B H A I K J     E F D G C H B I A J K
F E G D H C I B J A K     F G E H D I C J B K A
G F H E I D J C K B A     G H F I E J D K C A B
H GI F J E K D A C B     H I G J F K E A D B C
I H J G K F A E B D C     I J H K G A F B E C D
J I K H A G B F C E D     J K I A H B G C F D E
K J A I B H C G D F E     K A J B I C H D G E F
```

Table 2.3. Separability "efficiency", S, direct effect efficiency E_d, and residual effect efficiency E_r, for digram–balanced Latin squares of order $n = 2,3,...,12$.

n	V	S=100(1 – V)	E_d	E_r
2	0.707	29.3%	0	0
3	0.408	59.2	80.0	44.4
4	0.289	71.1	90.9	62.5
5	0.224	77.6	94.7	72.0
6	0.183	81.7	96.6	77.8
7	0.154	84.6	97.6	81.6
8	0.134	86.6	98.2	84.4
9	0.118	88.2	98.6	86.4
10	0.105	89.5	98.9	88.0
11	0.095	90.5	99.1	89.3
12	0.087	91.3	99.2	90.3

the more traditional efficiency measures lends some weight to the suggestion in Section 1.2 that efficiency of a cross–over design is closely related to the degree of separability of treatment and carryover effects.

S, E_d and E_r in Table 2.3 are rather low for $n = 2$ and $n = 3$, but attain respectable levels for $n \geq 4$. This suggests that there may be problems of interpretability associated with the designs for $n = 2$ and $n = 3$, and therefore it may not be advisable to use these designs in the form presented in Tables 2.1 and 2.2. Indeed, the design for $n = 2$, also known as the two–treatment, two–period, two–sequence cross–over design, presents problems of parameter estimability, stemming from the fact that the design has a minimum of five parameters to be estimated but only four cell means (the 2 x 2 combinations of sequences with periods). For this reason, the design cannot be recommended when there is the possibility of carryover effects being present. That design will be the subject of Chapter 3, where its statistical estimation properties will be examined in detail; further discussion will be deferred until then.

The design for $n = 3$ is the same design that was used for the illustrative example on "milk yield" in Chapter 1. Although it appears to be a perfectly good design for use where first-order carryover effects may be present, the fact that the separability of treatment and carryover effects S has a value of only

59.2% may present problems of interpretation for certain data sets. It has been demonstrated in Section 1.3 that, in extreme cases where the separability of treatment and carryover is low, it is possible to get a situation where the adjusted sums of squares for treatment and carryover, which are the appropriate ones for testing the significance of those effects, are both spuriously low, leading to the apparently sound, but nevertheless erroneous, conclusion that neither direct nor residual treatment effects are significant.

For the "milk yield" data, the P–value for carryover was 0.0597, which provides only marginal evidence of a carryover effect, if one requires that P < 0.05 or P < 0.01 be attained for significance to be declared. It is possible that the significance of this effect has been underestimated as a result of the analog of the multicollinearity phenomenon operating for qualitative variables, and that the P–value for carryover should have been lower. For treatment, the adjusted sum of squares resulted in a P–value of 0.0043, which leads to a similar inference to that obtained using the sequential sum of squares (P = 0.0065 in the ANOVA table in Section 1.1.1). Thus, although no major problems of interpretability are encountered here, for other data sets there is the possibility of the phenomenon occurring that was observed in the example in Section 1.3, with both treatment and carryover effects being apparently non–significant.

All the designs included in Tables 2.1 and 2.2 can be made "fully efficient" from the point–of–view of separating treatment from carryover effects, by use of an extra treatment period. Extra–period cross–over designs have been known for some time [Patterson (1959) and Patterson and Lucas (1962)], and construction of these designs simply involves repeating the treatments given to each subject in the last period for one extra period. Thus, for even values of n, for example, the design given in Table 2.1 for $n = 4$, the addition of the extra period results in the design given in Table 2.4.

For odd n, the addition of the extra period must be made to both squares of the pair. For example, for $n = 5$, the modification to the pair of squares in Table 2.2 is given in Table 2.5.

It is easily demonstrated that the measure of separability S developed in Section 1.2.1 is 100% for any extra–period design that is constructed by adding

Table 2.4. Fully–efficient four–treatment Latin square design with extra period.

		PERIOD				
		1	2	3	4	5
SEQUENCE	1	A	B	C	D	D
	2	B	D	A	C	C
	3	C	A	D	B	B
	4	D	C	B	A	A

Table 2.5. Fully–efficient five–treatment pair of Latin squares with extra period.

		PERIOD							PERIOD					
		1	2	3	4	5	6		1	2	3	4	5	6
	1	A	B	D	E	C	C	6	A	C	B	E	D	D
	2	B	C	E	A	D	D	7	B	D	C	A	E	E
SEQUENCE	3	C	D	A	B	E	E	8	C	E	D	B	A	A
	4	D	E	B	C	A	A	9	D	A	E	C	B	B
	5	E	A	C	D	B	B	10	E	B	A	D	C	C

an extra period to a digram–balanced Latin square, where the extra period contains the treatments that were used in the last period of the original design.

When there is more than one subject per sequence, the model for the response variable Y may be written in a form similar to Equation 1.1, but containing a term for subjects within sequences. This term is conveniently placed directly after the sequence effect term, so that all the terms that are in the Between–subjects stratum are adjacent, as are all the terms of the Within–subjects stratum. The model is written as

$$Y_{ijk} = \mu + \gamma_i + \xi_{i(k)} + \pi_j + \tau_t + \delta_j \alpha_r + \epsilon_{ijk} \qquad (2.1)$$

where μ is an overall mean, γ_i is the effect of sequence i, $i=1,2,...,4$, $\xi_{i(k)}$ is a random effect due to sampling unit k, $k=1,2,...,$ n_i, n_i being the number of subjects in sequence i, π_j is the effect of period j, $j=1,2,...,5$, τ_t is the effect of treatment t, where t is labelled A,B,C or D, α_r is the carryover effect due to treatment r having been applied in the preceding period, δ_j is an indicator variable whose value is zero in the first period and unity in subsequent periods, and ϵ_{ijk} is a random experimental error effect corresponding to sequence i, period j, and sampling unit k.

The code using PROC GLM to estimate the parameters in model (2.1) can look as follows, with SEQ used to represent sequence:

```
PROC GLM;
CLASS SEQ SUBJECT PERIOD TREAT CARRY;
MODEL Y=SEQ SUBJECT(SEQ) PERIOD TREAT CARRY/SOLUTION SS1
    SS2;
RANDOM SUBJECT(SEQ);
LSMEANS TREAT CARRY/PDIFF;
RUN;
```

Analysis of data arising from use of the design in Table 2.4 is readily handled using the coding method of Section 1.1.1. Omitting the actual observations of the response variable Y, using a period "." to indicate that the values of Y are missing, the SAS data step could look as follows, if there were, for example, two subjects per sequence:

```
DATA EXTRPRD;
INPUT SEQ SUBJECT PERIOD TREAT $ CARRY $ Y @@;
IF CARRY = '0' THEN CARRY = 'D';
CARDS;
1 1 1 A 0 .    1 2 1 A 0 .
1 1 2 B A .    1 2 2 B A .
1 1 3 C B .    1 2 3 C B .
1 1 4 D C .    1 2 4 D C .
1 1 5 D D .    1 2 5 D D .
2 1 1 B 0 .    2 2 1 B 0 .
2 1 2 D B .    2 2 2 D B .
2 1 3 A D .    2 2 3 A D .
2 1 4 C A .    2 2 4 C A .
2 1 5 C C .    2 2 5 C C .
3 1 1 C 0 .    3 2 1 C 0 .
3 1 2 A C .    3 2 2 A C .
3 1 3 D A .    3 2 3 D A .
3 1 4 B D .    3 2 4 B D .
3 1 5 B B .    3 2 5 B B .
4 1 1 D 0 .    4 2 1 D 0 .
4 1 2 C D .    4 2 2 C D .
```

```
4 1 3 B C .    4 2 3 B C .
4 1 4 A B .    4 2 4 A B .
4 1 5 A A .    4 2 5 A A .
RUN; .
```

The IF statement is needed for identifiability, as in Chapter 1. Although equal numbers of subjects per sequence is the most powerful way of allocating the total number of subjects available, there is no necessity for having equal numbers. Unequal numbers are readily handled by PROC GLM, as are missing values (see Section 9.4).

Because the inclusion of the extra period makes the design balanced for the estimation of treatment and carryover effects, the Type I and Type II sums of squares will be identical. The correlation coefficient between each component of the parameter estimates for treatment effects and those for carryover effects will be zero, as one would expect if treatment and carryover effects are truly separable. This is best shown by setting up the design matrix X for this design, which, if we confine ourselves to one subject per sequence for simplicity (the correlation coefficient remains unchanged as the sample size increases), will contain 20 rows, one for each combination of sequence and period, and 14 columns, corresponding to the 14 parameters of model (2.1) to be estimated in the following order: μ, γ_1, γ_2, γ_3, π_1, π_2, π_3, π_4, τ_1, τ_2, τ_3, λ_1, λ_2 and λ_3. The design matrix X, using the "set-to-zero" convention discussed in Chapter 1, is:

$$
X = \begin{pmatrix}
1 & 1 & 0 & 0 & 1 & 0 & 0 & 0 & 1 & 0 & 0 & 0 & 0 & 0 \\
1 & 1 & 0 & 0 & 0 & 1 & 0 & 0 & 0 & 1 & 0 & 1 & 0 & 0 \\
1 & 1 & 0 & 0 & 0 & 0 & 1 & 0 & 0 & 0 & 1 & 0 & 1 & 0 \\
1 & 1 & 0 & 0 & 0 & 0 & 0 & 1 & 0 & 0 & 0 & 0 & 0 & 1 \\
1 & 1 & 0 & 0 & 0 & 0 & 0 & 0 & 0 & 0 & 0 & 0 & 0 & 0 \\
1 & 0 & 1 & 0 & 1 & 0 & 0 & 0 & 0 & 1 & 0 & 0 & 0 & 0 \\
1 & 0 & 1 & 0 & 0 & 1 & 0 & 0 & 0 & 0 & 0 & 0 & 1 & 0 \\
1 & 0 & 1 & 0 & 0 & 0 & 1 & 0 & 1 & 0 & 0 & 0 & 0 & 0 \\
1 & 0 & 1 & 0 & 0 & 0 & 0 & 1 & 0 & 0 & 1 & 1 & 0 & 0 \\
1 & 0 & 1 & 0 & 0 & 0 & 0 & 0 & 0 & 0 & 1 & 0 & 0 & 1 \\
1 & 0 & 0 & 1 & 1 & 0 & 0 & 0 & 0 & 0 & 1 & 0 & 0 & 0 \\
1 & 0 & 0 & 1 & 0 & 1 & 0 & 0 & 1 & 0 & 0 & 0 & 0 & 1 \\
1 & 0 & 0 & 1 & 0 & 0 & 1 & 0 & 0 & 0 & 0 & 1 & 0 & 0 \\
1 & 0 & 0 & 1 & 0 & 0 & 0 & 1 & 0 & 1 & 0 & 0 & 0 & 0 \\
1 & 0 & 0 & 1 & 0 & 0 & 0 & 0 & 0 & 1 & 0 & 0 & 1 & 0 \\
1 & 0 & 0 & 0 & 1 & 0 & 0 & 0 & 0 & 0 & 1 & 0 & 0 & 0 \\
1 & 0 & 0 & 0 & 0 & 1 & 0 & 0 & 0 & 0 & 1 & 0 & 0 & 1 \\
1 & 0 & 0 & 0 & 0 & 0 & 1 & 0 & 0 & 1 & 0 & 0 & 1 & 0 \\
1 & 0 & 0 & 0 & 0 & 0 & 0 & 1 & 1 & 0 & 0 & 0 & 1 & 0 \\
1 & 0 & 0 & 0 & 0 & 0 & 0 & 0 & 1 & 0 & 0 & 1 & 0 & 0
\end{pmatrix} .
$$

The correlation coefficient is obtained from the elements of the inverse of $X'X$, which is readily computed using a matrix language such as GAUSS©. The variances and covariances of these elements are proportional to the elements of

	τ_1	τ_2	τ_3	λ_1	λ_2	λ_3
τ_1	0.4167	0.2083	0.2083	0	0	0
τ_2	0.2083	0.4167	0.2083	0	0	0
τ_3	0.2083	0.2083	0.4167	0	0	0
λ_1	0	0	0	0.50	0.25	0.25
λ_2	0	0	0	0.25	0.50	0.25
λ_3	0	0	0	0.25	0.25	0.50

Since the covariance of τ_i and λ_j is zero, for all $i,j=1,2,3$, it is clear that treatment and carryover effects are completely separable, which is consistent with the conclusion based upon the separability measure $S=100(1 - V)$, which predicted a 100% efficiency.

We must note, however, that the 100% efficiency determined above is an efficiency for <u>separating</u> treatment and carryover effects, and will not necessary agree with other definitions of efficiency, such as the ones defined by Patterson and Lucas (1962). For the above extra–period design, their definitions result in efficiencies of $E_d = 0.960$ and $E_r = 0.800$, respectively (see Design PL34, Appendix 5.A). These lower values are a result of the fact that those efficiencies also take account of the number of periods in the design, with a penalty for using the subjects in the extra period.

2.2. Estimating Period Effects in Digram–Balanced Latin Square Designs

The Latin squares designs presented in Tables 2.1 and 2.2 are referred to as digram–balanced, and are the most efficient Latin squares designs available for separating treatment from carryover effects. Experimenters may also be interested in whether there are differences among periods, or even whether there are differences among sequences. In the presence of significant carryover effects, there is, in general, no way to test whether periods are significant. There is a confounding of carryover with the component of the first period of the period effect that makes it impossible to obtain a valid estimate of the effect due to periods, if periods are a qualitative variable. As discussed in Section 1.4, it is possible to resolve this problem by making period a continuous variable.

However, even in the case of a qualitative period effect, carryover may prove to be non–significant; if so, it is possible to remove carryover from the model and re–estimate all of the other effects and sums of squares in the absence of that term. Statistically, there are problems associated with dropping terms from a model and re–estimating, as true significance levels tend to become altered by such decisions in a manner which is often impossible or difficult to quantify. Thus, we do not in this book adopt a view that encourages the reader to engage in sequential hypothesis testing, except in rare instances. The assessment of period effects is one of those rare instances where relaxing the recommendation against sequential testing may be justified. Otherwise, there can not be a valid test for period differences, unless the assumptions made when one uses continuous period effects are valid.

We use the milk yield example of Section 1.1 to illustrate the procedure. Carryover did not reach significance at the commonly–used significance levels of $\alpha=0.05$ or $\alpha=0.01$. If one re–estimates the effects without carryover in the model, using the following code for PROC GLM of SAS,

```
PROC GLM;
CLASS SQUARE COW PERIOD TREAT;
MODEL Y=SQUARE COW(SQUARE) PERIOD(SQUARE) TREAT /
     SOLUTION SS1;
RANDOM COW(SQUARE);
LSMEANS TREAT COW(SQUARE) PERIOD(SQUARE)/PDIFF;
RUN;
```

one obtains the following analysis of variance table after some editing:

Source of variation	df	Sum of squares	Mean square	F value	Pr > F
Square	1	18.0	18.0	0.13	0.7284
Cow(Square)	4	5763.1	1440.8	10.60	0.0069
Period(Square)	4	11489.1	2872.3	21.13	0.0011
Treatment	2	2276.8	1138.4	8.38	0.0183
Error	6	815.4	135.9		
Total (corrected)	17	20362.4			

An important feature of the analysis is the fact that the Type I and Type II sums of squares are identical and therefore the sums of squares presented in the above table are both sequential and adjusted values. That is, the order in which the terms are entered into the model is immaterial. The same will be true for any balanced Latin square or set of Latin squares, since, in the absence of carryover effects in the model, Latin squares are balanced for the estimation of subject, period and treatment effects. It is only the introduction of carryover effects into the model that disturbs this balance; excluding the carryover term from the statistical model restores the estimation situation to one which has full balance.

To carry out appropriate tests of significance, it is necessary to know the "expected mean squares" associated with the model that has been specified. The model also has to declare which effects are "fixed" and which are "random". In SAS, random effects are declared using the RANDOM statement, the undeclared effects being assumed to be fixed. The RANDOM statement in the above example specifies that cows within squares are allocated at random. The resulting expected mean squares table produced by SAS is as follows, the Type I and Type II tables being identical:

SOURCE	EXPECTED MEAN SQUARES
SQUARE	VAR(ERROR) + 3 VAR(COW(SQUARE)) + Q(SQUARE,PERIOD(SQUARE))
COW(SQUARE)	VAR(ERROR) + 3 VAR(COW(SQUARE))
PERIOD(SQUARE)	VAR(ERROR) + Q(PERIOD(SQUARE))
TREAT	VAR(ERROR) + Q(TREAT)

In the above table, Q(TREAT) refers to a quadratic form involving treatment (but not any other effect), and Q(PERIOD(SQUARE)) to a quadratic form involving periods within squares (but not any other effect). Therefore, treatments and periods within squares are both tested against the error mean square VAR(ERROR) in the usual way, which is shown in the above analysis of variance table. Similarly, cows within squares may be tested against VAR(ERROR), since under the null hypothesis of no differences between cows, VAR(COW(SQUARE)) = 0, and the ratio of the mean squares of "cows within squares" and "error" would be distributed as a central F–distribution. However, cows may be thought of as "blocks", such as occurs in a randomized block design

In that case, a test of blocks against the error term is often not advisable, since a block is a larger experimental unit than the plots within the blocks.

The above table also tells us that there is no valid test for differences between squares, since the quadratic form Q(SQUARE,PERIOD(SQUARE)) indicates an entangling of the effects of squares and periods within squares. Had the quadratic form involved only the effect of squares, that is, had it been Q(SQUARE) rather than as above, a valid test of squares could have been constructed by taking the ratio of the mean squares for SQUARES and for COWS(SQUARE), since that ratio would have had a central F–distribution. As it stands, there is no valid test for differences between squares, and the F–ratio of 0.13 and its associated P–value of 0.7284 in the ANOVA table is not correct, as SAS PROC GLM, by default, always produces a test against the error mean square.

The SOLUTION option of SAS provides valid estimates of the various parameters of the model using the set–to–zero restrictions. The LSMEANS statement provides appropriate least–squares means for each of the effects listed in the statement, plus probabilities of pairwise differences between each of the cell means. The treatment means,

| A | 45.167 | B | 57.500 | C | 72.667 |

the cows within squares means,

SQUARE 1		SQUARE 2	
COW 1	26.000	COW 4	69.333
COW 2	78.000	COW 5	48.000
COW 3	74.333	COW 6	55.000

and the periods within squares means,

SQUARE 1		SQUARE 2	
PERIOD 1	90.333	PERIOD 1	87.333
PERIOD 2	61.000	PERIOD 2	58.000
PERIOD 3	27.000	PERIOD 3	27.000

are precisely the same means that one would obtain by taking simple arithmetic averages of the data in Table 1.1. This is due to the fact that in the absence of carryover effects in the model, the design is orthogonal.

The question arises as to what may be validly said about period effects when carryover effects are significant. We will once again use the milk yield data set of Section 1.1 as an example. Although carryover would not have been adjudged significant at any of the commonly–used significance levels, the P–value of 0.0597 is close to $\alpha = 0.05$ and many workers would be uneasy about dropping the carryover term from the model. Retaining CARRY in the model leads to a mean square for error of 49.8125. The Type I mean square for periods is 2872.3, as in the earlier ANOVA table. Hence, the ratio of these two mean squares is 57.7, for a P–value of 0.0009. Therefore, period effects appear to be significant.

A valid estimate of the difference between periods 2 and 3 for the first square of that design is given directly by the estimate $\hat{\pi}_2 = 34.0$ from the solution vector using "set–to–zero" restrictions. Similarly, the difference between periods 2 and 3 for the second square is given directly by $\hat{\pi}_4 = 31.0$ of that same solution vector. These same estimates can be obtained from the solution vector using the sum–to–zero convention, with some computation. Thus, from the values $\hat{\pi}_1 = 30.889$ and $\hat{\pi}_2 = 1.556$ of the solution vector obtained using the sum–to–zero convention, the difference between the second and third periods in the first square is $1.556 - (-30.889 - 1.556) = 34.0$, as above. Similarly, from $\hat{\pi}_3 = 29.889$ and $\hat{\pi}_4 = 0.556$, the difference between the second and third periods in the second square is $29.889 + 2(0.556) = 31.0$, as above. Thus, although nothing can be said about period differences involving the first period, differences involving the second and subsequent periods are perfectly valid. Also, it should be noted that these differences are identical to the differences between the means of the second and third periods obtained from the raw data in each square. In fact, this will happen whenever the basic design is a Latin square design of the usual type, that is, balanced for the estimation of subject, period and treatment effects.

Least squares means may be obtained from SAS PROC GLM using the statement

LSMEANS TREAT CARRY PERIOD(SQUARE) / PDIFF

and this leads to valid contrasts for all pairwise comparisons for treatment, carryover and periods within squares with the exception of contrasts involving the first period. Hence, for treatment, the table of difference probabilities generated by the option PDIFF results in

	A	B	C
A	.	0.0403	0.0017
B	0.0403	.	0.0106
C	0.0017	0.0106	.

,

whereas for carryover, the difference probabilities are

	A	B	C
A	.	0.5605	0.0296
B	0.5605	.	0.0553
C	0.0296	0.0553	.

These latter should be used cautiously in view of the fact that the test for carryover gave a non–significant result, or, at most, a marginally significant one.

For periods, the valid difference probabilities are

SQUARE 1
2 vs. 3   0.0041

SQUARE 2
2 vs. 3   0.0058

supporting the feeling that the apparently large differences in means between the second and third periods are real.

For this data set, it was shown in Section 1.4 that if periods are considered to be a continuous variable, the explained sum of squares for periods is almost as high as when periods are considered to be qualitative. Since the effect was highly significant (P = 0.0009), one needs only in that case to look at the regression coefficients (–33.25 for Square 1 and –31.75 for Square 2) to see the extent to which the average milk yield changes between periods.

2.3. Analyzing Latin Square Designs Not Balanced for Estimation of Carryover Effects

In the present section, we consider a set of Latin squares that were not chosen to produce balance in the estimation of carryover effect. The data come from Bliss (1967) and form three 4 x 4 Latin squares. The treatments consisted of four dosages of a complex amine given to rabbits, the dosages corresponding to A = 12.5 mg, B = 50 mg, C = 200 mg, and D = 800 mg. The dosages were given to four animals in each square on four different days, the days being the same in the three squares, with the four animals being different from square to square. The response variable was increase in pupil diameter (mm). The layout of the squares, and the responses (as a subscript) were as follows, the subjects (rabbits) constituting the rows (reproduced here with the permission of McGraw–Hill, Inc.):

SQUARE

1	2	3
PERIOD	PERIOD	PERIOD
1 2 3 4	1 2 3 4	1 2 3 4

		1 2 3 4	1 2 3 4	1 2 3 4
	1	$D_7\ C_5\ B_2\ A_1$	$C_4\ D_5\ A_1\ B_2$	$B_3\ A_0\ D_5\ C_3$
RABBIT	2	$C_4\ D_6\ A_1\ B_3$	$D_6\ C_4\ B_2\ A_0$	$A_0\ D_4\ C_3\ B_2$
	3	$B_3\ A_1\ C_6\ D_7$	$A_1\ B_3\ C_4\ D_5$	$D_7\ C_3\ B_2\ A_0$
	4	$A_1\ B_3\ D_6\ C_3$	$B_2\ A_2\ D_7\ C_4$	$C_4\ B_2\ A_0\ D_6$

To examine the balance of this design using the omnibus test of separability of treatment from carryover effects described in Section 1.2, an incidence table is formed. This yields

CARRYOVER

		0	A	B	C	D	
	A	3	0	7	0	2	12
TREATMENT	B	3	4	0	5	0	12
	C	3	1	1	0	7	12
	D	3	4	1	4	0	12
		12	9	9	9	9	48

and indicates some extreme imbalances such as the fact that treatment A follows treatment B seven times and never follows treatment C, that C follows D seven times, that B follows C five times, etc. The chi–square value for this contingency table is 43.111, Cramer's V is 0.547, and the measure of separability efficiency is S = 100(1 – V) = 45.3%. This contrasts with an efficiency of 71.1% (see Table 2.3) that could have been obtained had three identical squares of the form for n = 4 in Table 2.1 been chosen, or had a set of three mutually orthogonal Latin squares (see Section 2.4) been used.

The analysis of the data set from Bliss (1967) can be carried out using SAS PROC GLM. The statistical model is similar to that of Equation 2.1 with some slight modification of interpretation. Writing the model as before,

$$Y_{ijk} = \mu + \gamma_i + \xi_{i(k)} + \pi_j + \tau_t + \delta_j \alpha_r + \epsilon_{ijk}$$

where the term γ_i now refers to "squares" rather than sequences, i=1,...,3, $\xi_{i(k)}$ now refers to rabbits within squares, k=1,...,4, π_j still refers to periods, j=1,...,4, τ_t still refers to treatments, t=1,...,4, α_r is still the carryover effect from the rth treatment having been applied in the previous period, δ_j is still an indicator variable whose value is zero in the first period and unity in subsequent periods, and ϵ_{ijk} is a random experimental error corresponding to square i, period j, and rabbit k.

The coding of the data step might look as follows.

```
DATA AMINE;
INPUT Y SQUARE RABBIT PERIOD TREAT $ CARRY $ @@;
IF CARRY = '0' THEN CARRY = 'D';
CARDS;
7 1  1 1 D 0    5 1  1 2 C D    2 1  1 3 B C    1 1  1 4 A B
4 1  2 1 C 0    6 1  2 2 D C    1 1  2 3 A D    3 1  2 4 B A
3 1  3 1 B 0    1 1  3 2 A B    6 1  3 3 C A    7 1  3 4 D C
1 1  4 1 A 0    3 1  4 2 B A    6 1  4 3 D B    3 1  4 4 C D
4 2  5 1 C 0    5 2  5 2 D C    1 2  5 3 A D    2 2  5 4 B A
6 2  6 1 D 0    4 2  6 2 C D    2 2  6 3 B C    0 2  6 4 A B
1 2  7 1 B 0    3 2  7 2 B A    4 2  7 3 C B    5 2  7 4 D C
2 2  8 1 B 0    2 2  8 2 A B    7 2  8 3 D A    4 2  8 4 C D
3 3  9 1 B 0    0 3  9 2 A B    5 3  9 3 D A    3 3  9 4 C D
0 3 10 1 A 0    4 3 10 2 D A    3 3 10 3 C D    2 3 10 4 B C
7 3 11 1 D 0    3 3 11 2 C D    2 3 11 3 B C    0 3 11 4 A B
4 3 12 1 C 0    2 3 12 2 B C    0 3 12 3 A B    6 3 12 4 D A
RUN;
```

The code for PROC GLM might read

```
PROC GLM;
CLASS SQUARE RABBIT PERIOD TREAT CARRY;
MODEL Y=SQUARE RABBIT(SQUARE) PERIOD TREAT CARRY /
     SOLUTION SS1 SS2;
RANDOM RABBIT(SQUARE);
LSMEANS TREAT CARRY / PDIFF;
MEANS PERIOD;
RUN;
```

The SAS output from the above run gave an error sum of squares of 12.2391 with 27 degrees of freedom, for a mean square of 0.4533. The relevant portions of the ANOVA tables are as follows:

SOURCE	DF	TYPE I SS	MEAN SQ.	F VALUE	PR > F
SQUARE	2	7.0417	3.5208	7.77	0.0022
RABBIT(SQ.)	9	5.1875	0.5764	1.27	0.2966
PERIOD	3	1.5625	0.5208	1.15	0.3473
TREAT	3	179.0625	59.6875	131.67	0.0001
CARRY	3	1.3859	0.4620	1.02	0.3996

SOURCE	DF	TYPE II SS	MEAN SQ.	F VALUE	PR > F
TREAT	3	135.4882	45.1627	99.63	0.0001
CARRY	3	1.3859	0.4620	1.02	0.3996

The P–value associated with the carryover effect, 0.3996, indicates that there are no significant residual effects between pairs of treatments. Although the Type II sum of squares is the appropriate one to use to test whether there are significant treatment effects, we have seen in Section 1.3 that, in a design that separates treatment from carryover effects poorly, it is possible that the adjusted treatment sum of squares may be underestimated. This is not the case here, for although its value of 135.4882 is less than the Type I SS of 179.0625, the treatment effect is still highly significant ($P < 0.0001$). Thus, one can confidently conclude that there are significant treatment differences. The LSMEANS statement produces the following table of treatment means and associated difference probabilities:

TREAT	Y LSMEAN	PROB > |T| H0: LSMEAN(I)=LSMEAN(J) I/J	1	2	3	4
A	0.7747	1	.	0.0007	0.0001	0.0001
B	2.3762	2	0.0007	.	0.0006	0.0001
C	3.9371	3	0.0001	0.0006	.	0.0001
D	5.8748	4	0.0001	0.0001	0.0001	.

From this table, it is clear that the pupil diameter increases almost linearly as the dosage of amine increases logarithmically, and that each successive pairwise difference is highly significant (P < 0.001). It might also be of interest to the investigator to test whether there are significant period differences. Since carryover was non–significant, period effects can be assessed using the Type I sum of squares. The P–value of 0.3473 can be read directly from the ANOVA table above, which indicates that periods are not significant. This P–value was based on the error mean square of 0.4533, computed with CARRY in the model. Removing carryover from the model and re–calculating the error mean square yields 0.4542, which not surprisingly leads to the same conclusion about the non–significance of period effects. Had period effects proven to be significant, then the LSMEANS statement would have provided relevant difference probabilities for contrasts involving the means of the second and subsequent periods only. Contrasts involving the mean of the first period are not estimable with carryover effects in the model. Alternatively, if period means appear to change linearly, one might consider the use of continuous period effects, as in Section 1.4.

2.4. Mutually Orthogonal Latin Squares versus Digram–balanced Latin Squares

An alternative to the digram–balanced form of Latin square design is a set of n–1 mutually orthogonal Latin squares, where n is the number of rows, columns and treatments in each square. Mutually orthogonal squares were first suggested for use in cross–over trials by Cochran *et al.* (1941), and have the property that each treatments is preceded and followed equally often by each other treatment not only in successive periods, but in periods separated by two, three, etc., units of time. A consequence of this is that not only are treatment effects orthogonal to first–order carryover, as they are in digram–balanced Latin squares, but they are also orthogonal to second– and higher–order carryover effects.

For $n = 3$, the 2 ($= n$–1) squares are identical to the digram–balanced squares. For $n = 4$, the following is a set of three mutually orthogonal squares, copied directly from Appendix 5.C of Chapter 5.

Design PL6. $t = k = 4$, $p = 4$, $b = 3$, $n = 12$, VIF = 1.100

A set of t-1 mutually orthogonal Latin squares. An example is

		Block	
	1	2	3
Subject →	1 2 3 4	1 2 3 4	1 2 3 4

Period				
1	A B C D	A B C D	A B C D
2	B A D C	C D A B	D C B A
3	C D A B	D C B A	B A D C
4	D C B A	B A D C	C D A B

Readers can readily check that each treatment is followed by each other treatment an equivalent number of times for each time difference. Since such a design has 12 distinct sequences, it may be more difficult to use it in practice compared with the use of three replicates of a single digram–balanced 4 x 4 Latin square having but four sequences. Unless it is thought that second– and higher–order carryover effects are likely, there seems little reason to use such a design.

We also remark elsewhere that if second–order carryover is really likely to be observed, then a cross–over design may not be the appropriate experimental design to use. Another factor militating against the use of a set of mutually orthogonal Latin squares is the fact that no set of such squares exists for $n = 6$. In Appendix 5.B of Chapter 5, Design PL11 is a set of four mutually orthogonal Latin squares for $n = 5$, but no similar set of squares is available for $n = 6$, although Design PL15 is balanced both for first– and second–order carryover effects, but not higher.

Suppose one wished to use a 4 x 4 Latin square with three replications. It is not necessary to use the same square for each replication. For example, the following set of three squares, each individual square having digram–balance, can be employed (see top of next page).

Each square retains variance–balance and orthogonality of treatment and carryover effects, and consequently, so does the entire set. However, since there are 12 distinct sequences, this design is more prone to error than simply using three exact replicates of the same square.

SQUARE

		1 PERIOD				2 PERIOD				3 PERIOD		
		1	2	3	4	1	2	3	4	1	2	3	4
	1	A	D	C	B	C	B	A	D	A	C	B	D
SUBJECT	2	D	B	A	C	B	D	C	A	C	D	A	B
	3	C	A	B	D	A	C	D	B	B	A	D	C
	4	B	C	D	A	D	A	B	C	D	B	C	A

Even more complicated conditions than mutually orthogonal Latin squares have been studied. Williams (1950) considered designs such that each treatment is preceded by each *pair* of other treatments, so that the designs would be balanced for pairs of carryover effects. For n treatments, $n (n - 1)$ replications are required, so that even for a modest four treatments, 12 Latin squares are needed to achieve the desired condition of balance.

Exercises

2.1. Consider only one of the two squares in a digram–balanced set of Latin squares for $n = 5$, viz.

PERIOD

		1	2	3	4	5
	1	A	B	D	E	C
	2	B	C	E	A	D
SEQUENCE	3	C	D	A	B	E
	4	D	E	B	C	A
	5	E	A	C	D	B

(a) Using the omnibus test of separability of treatment and carryover effects described in Section 1.2, calculate the efficiency of separation S.

(b) Compare the value of S calculated in part (a) with the corresponding value in Table 2.3 for a digram–balanced pair of squares for $n=5$.

2.2. For the design in Exercise 2.1, set up the design matrix and determine the (a) variance inflation factor VIF, (b) efficiency E_d of the design.

[Procedure: Find the inverse of the 25 x 17 matrix, assuming one subject per sequence, and then find the inverse of a similar matrix, of size 25 x 13, having the carryover parameters omitted. The ratio of diagonal elements corresponding to treatment parameters, from the two matrices, is the VIF and its reciprocal is the required efficiency.]

2.3. Hedayat and Afsarinejad (1978) reported a design, attributed to K. B. Mertz, for $n = 9$ which has the same property of balance as the digram–balanced Latin squares of Williams (1949), where each treatment is preceded and followed equally often by each other treatment. This is one of several known designs for odd n in which balance can be achieved with a single square, but it is the only one for a small number of treatments. Hedayat and Afsarinejad (1978) also showed plans for $n = 15$ (due to E. Sonnemann) and $n = 27$. Previously, Hedayat and Afsarinejad (1975) gave an example of a balanced design for $n = 21$. The balanced design for $n = 9$ is given below.

```
                         Periods
                  1 2 3 4 5 6 7 8 9

                  A I G F D H E C B
                  B G H D E I F A C
                  C H I E F G D B A
   Subjects       D F B C I A H G E
  (Sequences)     E D C A G B I H F
                  F E A B H C G I D
                  G A D I B F C E H
                  H B E G C D A F I
                  I C F H A E B D G
```

(a) Calculate the omnibus measure of efficiency S for this design and compare it to the value of S in Table 2.3 for the corresponding digram–balanced Latin square of order $n = 9$.

(b) What advantages can you enumerate for using the above square in preference to the digram–balanced pair of squares if $n = 9$?

2.4. The following set of data came from a study of the effect of the order of administration of a set of six treatments on testicular diffusing factor. Six rabbits were employed in this study, using six positions on the backs of each of these rabbits. The responses were as follows, with the treatment listed as a letter in parentheses (data from Gill, 1978, Exercise 8.6, reproduced here with the permission of the Iowa State University Press):

		PERIOD (POSITION)					
		1	2	3	4	5	6
	1	7.9(C)	6.1(D)	7.5(A)	6.9(F)	6.7(B)	7.3(E)
	2	8.7(E)	8.2(B)	8.1(C)	8.5(A)	9.9(D)	8.3(F)
RABBIT	3	7.4(D)	7.7(F)	6.0(E)	6.8(C)	7.3(A)	7.3(B)
	4	7.4(A)	7.1(E)	6.4(F)	7.7(B)	6.4(C)	5.8(D)
	5	7.1(F)	8.1(C)	6.2(B)	8.5(D)	6.4(E)	6.4(A)
	6	8.2(B)	5.9(A)	7.5(D)	8.5(E)	7.3(F)	7.7(C)

(a) Using the omnibus test of separability of treatment and carryover effects described in Section 1.2, calculate the efficiency of separation S.

(b) As this is not the best 6 x 6 design that could have been used, give a digram–balanced Latin square having the same first sequence as the above design.

(c) Analyze the above design and determine whether there are significant treatment and carryover effects.

3
The 2-Treatment, 2-Period, 2-Sequence Design

Probably the most widely used and most abused cross-over design in current employ is the 2-treatment, 2-period, 2-sequence design. It is sometimes referred to as the "Grizzle" design or the "Hills-Armitage" design after authors (Grizzle, 1965, 1974; Hills and Armitage, 1979) who were among the first to describe the statistical analysis of the design. In the past decade, there have been literally dozens of papers written about various aspects of this design, and there has been much controversy about the analysis of this design. The salient feature of this design that dominates above any other question is the fact that, with a carryover parameter present in the model, the design is not analyzable. This has not stopped an almost incessant flow of literature on the subject, much of which is reviewed in Jones and Kenward (1989).

It is inescapable that a design that yields only four cells means (the responses for each of the two sequences in each of the two periods) cannot be used to estimate more than four parameters. As one of these parameters is the overall grand mean, another represents differences between periods, and a third differences between treatments, one can get an estimate of differential carryover effects as the fourth parameter only by making some strong assumption, such as, for example, that (1) there is no sequence effect, or (2) there is no period-by-treatment interaction. Assumptions such as these cannot be tested using the data sets to hand, and unless there is prior information about the process, it is not possible to proceed. This "intrinsic aliasing" of the various alternative parameterizations cannot be resolved by statistical techniques.

The purpose of including a carryover parameter in the model is not necessarily because one believes that carryover will be present, but because one wishes to be able to test whether or not it is significant. Not being able to make such a test without making other, possibly unwarranted, assumptions is an important restriction which militates against use of this design in practice. An analogy would be a factorial experiment with two factors A and B. A term representing the A · B interaction is often included in the analysis of variance model to be tested, not necessarily because the investigator believes in the existence of that interaction, but to test whether a significant interaction exists.

In this chapter, we will examine various approaches that are possible to the analysis of the 2 x 2 cross–over trial, and look closely at some modifications that have been suggested to resolve the difficulty, such as the use of baseline measurements and "wash–out" periods. It will be shown in Section 3.6 that the use of a baseline period plus two wash–out periods, one following each treatment period, results in an efficient design for the separation of treatment and first–order carryover effects. The barrier to its adoption may be the fact that the experimental subjects must remain in the trial for five measurements sessions.

3.1. The Basic Layout of the 2 x 2 Cross–over Trial

The 2 x 2 design is identical to the digram–balanced Latin square design of Table 2.1 for $n = 2$, when equal numbers of subjects are allocated to the two sequences. Since equal allocation leads to the most powerful test statistically, viewing this design as a set of replicated Latin squares is a good way of looking at these trials. In any event, all estimable functions of interest can be written in terms of the cell means of the four period–by–sequence combinations, so that even when there are unequal numbers in the two sequences, the estimation approach to be discussed here still applies. In clinical trials and other experiments which have used this trial, subjects are allocated at random to one of two sequences. Those subjects falling into the first sequence receive treatment A during the first treatment period, and then treatment B during the second treatment period. Subjects allocated to the second sequence receive the treatments in the reverse order, that is, B in the first treatment period and A in the second treatment period. Thus, the design is simply represented as follows:

$$
\begin{array}{c c}
& \text{Period} \\
& 1 \qquad 2
\end{array}
$$

		Period 1	Period 2
Sequence (subject group)	1	A	B
	2	B	A

For the moment we exclude the situations where there is a baseline reading on each subject prior to Period 1, or where a measurement is taken between Periods 1 and 2 after allowing a reasonable time to elapse (the wash–out period) so that the effect of the treatment applied in the first period should no longer persist. These modifications will be considered later in this chapter. We saw in Chapter 2, Table 2.3, that the separation efficiency of treatment and carryover effect is very low, and that the treatment and carryover efficiencies E_d and E_r are zero. That probably explains why Patterson and Lucas (1962) ignored the 2 x 2 design and its derivatives, in their catalog of change–over designs. Indeed, their calculations showed that *all* two–period designs were of low efficiency.

The model that one would like to apply to the above design is the same as that used in Chapter 2, viz.

$$
Y_{ijk} = \mu + \gamma_i + \xi_{i(k)} + \pi_j + \tau_t + \delta_j \lambda_r + \epsilon_{ijk}, \tag{3.1}
$$

where here $i,j,t,r = 1,2$, and $k=1,2,...,n_i$. The true cell means may be denoted by μ_{ij}, the subscripts identifying the ith sequence and the jth period. From a set of data, the observed cell means can be denoted by $\overline{Y}_{ij}$, as in the following table:

		Period 1	Period 2
Sequence (subject group)	1	$\overline{Y}_{11}$	$\overline{Y}_{12}$
	2	$\overline{Y}_{21}$	$\overline{Y}_{22}$

Taking expectations, one can write down four equations in terms of model parameters as follows:

$$\mu_{11} = \mu + \gamma_1 + \pi_1 + \tau_1$$
$$\mu_{12} = \mu + \gamma_1 + \pi_2 + \tau_2 + \lambda_1$$
$$\mu_{21} = \mu + \gamma_2 + \pi_1 + \tau_2$$
$$\mu_{22} = \mu + \gamma_2 + \pi_2 + \tau_1 + \lambda_2.$$

Adopting the "sum–to–zero" convention (although use of the "set–to–zero" convention is also adequate), the relations

$$\gamma_2 = -\gamma_1, \quad \pi_2 = -\pi_1, \quad \tau_2 = -\tau_1, \text{ and } \lambda_2 = -\lambda_1$$

transform the above four equations into the following:

$$\mu_{11} = \mu + \gamma_1 + \pi_1 + \tau_1$$
$$\mu_{12} = \mu + \gamma_1 - \pi_1 - \tau_1 + \lambda_1$$
$$\mu_{21} = \mu - \gamma_1 + \pi_1 - \tau_1$$
$$\mu_{22} = \mu - \gamma_1 - \pi_1 + \tau_1 - \lambda_1.$$

Writing these equations in matrix form as

$$\mathbf{X}\beta = \begin{pmatrix} 1 & 1 & 1 & 1 & 0 \\ 1 & 1 & -1 & -1 & 1 \\ 1 & -1 & 1 & -1 & 0 \\ 1 & -1 & -1 & 1 & -1 \end{pmatrix} \begin{pmatrix} \mu \\ \gamma_1 \\ \pi_1 \\ \tau_1 \\ \lambda_1 \end{pmatrix} = \begin{pmatrix} \mu_{11} \\ \mu_{12} \\ \mu_{21} \\ \mu_{22} \end{pmatrix},$$

and finding the echelon form of $\mathbf{X}$ as described in Section 1.5 leads to

$$\begin{pmatrix} 1 & 0 & 0 & 0 & 0 \\ 0 & 1 & 0 & 0 & 1 \\ 0 & 0 & 1 & 0 & 0 \\ 0 & 0 & 0 & 1 & -1 \end{pmatrix}.$$

Since the echelon form is not an identity matrix, some parameters are confounded with others. From the non–zero off–diagonal elements, one can conclude that carryover is confounded with sequence and treatment. Thus, the least squares estimator of the parameter vector, $\hat{\beta}$, does not exist. This result is no surprise; as already mentioned, one cannot expect to be able to estimate five parameters, μ, γ_1, π_1, τ_1, λ_1, from just four pieces of information, the cell means. One way out of the dilemma is simply to recognize that estimation of the carryover parameter, λ_1, is not possible with this design, and to eliminate that parameter from the model. The reformulated model excluding λ_1 then gives rise to four equations in four unknown parameters and readily can be solved. However, because of the great interest in wishing to be able to make some kind of statement about the effect of carryover, many workers, including Grizzle (1965, 1974), Hills and Armitage (1979), and others, have persevered and have made one of a set of possible assumptions enabling one parameter of the model to be eliminated.

The various assumptions that are made are not equivalent, and lead to different parameterizations of the basic model having vastly different interpretations. Many workers have glossed over the difference in meanings of the various parameterizations, and have treated the results as being equivalent. That is not the case, as the following treatment will show. Let us consider each parameterization in turn.

Parameterization 1:

This model contains period, treatment and carryover effects, but assumes that there are no sequence effects. A model without sequence effects can be justified if one believes that subjects have been allocated to the two sequences in an appropriately random fashion and that there are no other factors operating which might cause the mean of subjects in Sequence 1 to differ from the mean of subjects in Sequence 2 other than as a result of random effects. This assumption is, of course, not testable using the same data set that is being analyzed. Making this assumption relies on the belief that, if λ_1 proves to be significantly different from zero, this effect is indeed a measure of a late response (delayed effect) of one or other of the treatments.

This parameterization should not be applied to a clinical trial in which there is a possibility of a "psychological" carryover effect where, say, the absence of a positive effect of the treatment given in the first period may condition the subjects in that sequence to expect a poor result from the treatment to be applied in the second period. The subjects in the other sequence receive the second treatment first, and they may experience a different psychological response to that treatment. A situation where psychological carryover may be present is best handled by a model which includes a treatment–by–period interaction parameter (see Parameterization 2).

The absence of a sequence effect profoundly changes the model structure from one which previously had a between–subjects stratum with a sequence effect and a subjects within sequence term to one which has only between–subjects variability. Thus, instead of Equation (3.1), the model becomes

$$Y_{ijk} = \mu + \xi_{i(k)} + \pi_j + \tau_t + \delta_j \lambda_r + \epsilon_{ijk}$$

with the parameter γ_i deleted. When expectations are taken, the random effects have expectation zero, enabling the cell means to be written as

$$E(Y_{ijk}) = \mu_{ij} = \mu + \pi_j + \tau_t + \lambda_r \tag{3.2}$$

where μ_{ij}, despite the absence of a sequence parameter in this model, continues to represent the true mean of the response in the ith sequence and jth period, as before.

The set of equations (3.2) may be written in matrix form as

$$\mathbf{X}\beta \;=\; \begin{pmatrix} 1 & 1 & 1 & 0 \\ 1 & -1 & -1 & 1 \\ 1 & 1 & -1 & 0 \\ 1 & -1 & 1 & -1 \end{pmatrix} \begin{pmatrix} \mu \\ \pi_1 \\ \tau_1 \\ \lambda_1 \end{pmatrix} \;=\; \begin{pmatrix} \mu_{11} \\ \mu_{12} \\ \mu_{21} \\ \mu_{22} \end{pmatrix}.$$

The portion of the inverse of the $\mathbf{X}'\mathbf{X}$ matrix relating to the treatment and carryover parameters is

$$(\mathbf{X}'\mathbf{X})^{-1} \;=\; \begin{pmatrix} 0.5 & 0.5 \\ 0.5 & 1 \end{pmatrix}$$

and the least–squares estimators of the parameters of Equation (3.2) are

$$
\hat{\beta} = \begin{pmatrix} \hat{\mu} \\ \hat{\tau}_1 \\ \hat{\tau}_1 \\ \hat{\lambda}_1 \end{pmatrix} = (X'X)^{-1}X'Y = \begin{pmatrix} (\overline{Y}_{11}+\overline{Y}_{12}+\overline{Y}_{21}+\overline{Y}_{22})/4 \\ (\overline{Y}_{11}-\overline{Y}_{12}+\overline{Y}_{21}-\overline{Y}_{22})/4 \\ (\overline{Y}_{11}-\overline{Y}_{21})/2 \\ (\overline{Y}_{11}+\overline{Y}_{12}-\overline{Y}_{21}-\overline{Y}_{22})/2 \end{pmatrix} .
$$

These results show that the carryover parameter λ_1 is estimated from all four observed cell means and is a contrast between the average response in the two time periods in Sequence 1 and the average response in the two time periods of Sequence 2. The direct treatment parameter τ_1, on the other hand, is estimated using results only from the first period, and is simply one–half the difference between the responses to Treatment A and Treatment B in that period. Various authors (for example, Grizzle, 1965) have advocated a two–stage testing procedure, in which one first tests whether carryover λ_1 is significant. In that approach, if carryover is found to be significant (Grizzle recommends that a conservative significance level of $\alpha=0.15$ be used), then one should estimate the treatment effect τ_1 using results from only the first period. It is important to realize that the estimator for τ_1 obtained above is the solution of the ordinary least–squares equations resulting from model (3.2). The estimator does not depend upon any prior test procedure or subjective evaluations derived from the data set itself. It depends only upon the specification of the model (3.2), which, of course, rested upon the non–testable assumption that sequence effects are absent. That was a rather important assumption, since the treatment estimator

$$
\hat{\tau}_1 = (\overline{Y}_{11}-\overline{Y}_{21})/2
$$

is confounded with differences between sequences. That is, any difference between sequences in Period 1 cannot be distinguished from a genuine difference between treatments. This illustrates the weakness of having an over–parameterized model such as (3.1); one can not proceed without making a non–testable and perhaps unjustifiable assumption.

The covariance matrix of the parameter estimates is proportional to

$(\mathbf{X}'\mathbf{X})^{-1}$, which is given above for τ_1 and λ_1. From this matrix, we see that the estimator of λ_1 is twice as variable as the estimator of τ_1, and that the correlation between the estimators of the treatment and carryover parameters is

$$\text{Corr}(\hat{\tau}_1, \hat{\lambda}_1) = 0.5/(0.5 \cdot 1)^{1/2} = 0.707.$$

This correlation of 0.707 is identical to the value of Cramèr's V in Table 2.3 for this design, a Latin square of order $n = 2$, helping to justify the use of V as part of a measure of separability of treatment and carryover effects. Its large value should serve as a caution to users intending to employ the 2–treatment, 2–period, 2–sequence cross–over design. That is, even after making non–testable and probably unwarranted assumptions, the resulting model has low separability of the parameters describing the direct and residual effects of treatment. In Section 1.3, we showed that, under such conditions, the effects of true treatment differences may be grossly underestimated.

Parameterization 2:

This model is similar to Parameterization 1 in that it assumes that there is no sequence effect. It contains period, treatment and the treatment–by–period interaction terms, and, after the taking of expectations, can be written as follows,

$$E(Y_{ijk}) = \mu_{ij} = \mu + \pi_j + \tau_t + (\tau\pi)_{tj} \tag{3.3}$$

where $(\tau\pi)_{tj}$ is the parameter representing treatment–by–period interaction and μ_{ij} is used to represent the true mean of the response in the ith sequence and the jth period for consistency with previous usage. This parameterization may be more pertinent than Parameterization 1 when there is the possibility of a psychological carryover effect, which results in the subjects in one of the sequences responding differently to the treatment given in the second period than they would if that treatment had been given in the first period. This is one interpretation of a departure from a model with simple period and treatment effects, and there may be other circumstances in which a treatment–by–period interaction is a sensible way of viewing or interpreting reality. This model differs from Parameterization 1, as that parameterization has a carryover parameter λ_1 which represents a genuine persistent effect, or late response, of a treatment.

The set of equations (3.3) may be written in matrix form as follows, the elements of the interaction term $(\tau\pi)_{11}$, similar to other experimental designs, being determined by multiplying the elements of τ and those of π:

$$
X\beta = \begin{pmatrix} 1 & 1 & 1 & 1 \\ 1 & -1 & -1 & 1 \\ 1 & 1 & -1 & -1 \\ 1 & -1 & 1 & -1 \end{pmatrix} \begin{pmatrix} \mu \\ \pi_1 \\ \tau_1 \\ (\tau\pi)_{11} \end{pmatrix} = \begin{pmatrix} \mu_{11} \\ \mu_{12} \\ \mu_{21} \\ \mu_{22} \end{pmatrix}
$$

from which the portion of the inverse of $X'X$ relating to the treatment and treatment–by–period interaction parameters is

$$
(X'X)^{-1} = \begin{pmatrix} 0.25 & 0 \\ 0 & 0.25 \end{pmatrix} .
$$

Thus, in this parameterization, treatment and the treatment–by–period interaction parameters are orthogonal to each other.

The least–squares estimators of the parameters of Equation (3.3) are given by

$$
\hat{\beta} = \begin{pmatrix} \hat{\mu} \\ \hat{\pi}_1 \\ \hat{\tau}_1 \\ \widehat{(\tau\pi)}_{11} \end{pmatrix} = (X'X)^{-1}X'Y = \begin{pmatrix} (\overline{Y}_{11}+\overline{Y}_{12}+\overline{Y}_{21}+\overline{Y}_{22})/4 \\ (\overline{Y}_{11}-\overline{Y}_{12}+\overline{Y}_{21}-\overline{Y}_{22})/4 \\ (\overline{Y}_{11}-\overline{Y}_{12}-\overline{Y}_{21}+\overline{Y}_{22})/4 \\ (\overline{Y}_{11}+\overline{Y}_{12}-\overline{Y}_{21}-\overline{Y}_{22})/4 \end{pmatrix} .
$$

The estimators of the overall mean and period parameters μ and π_1, respectively, are the same as in Parameterization 1 and have the same interpretation. The estimator of $(\tau\pi)_{11}$, except for a factor of 2, is the same as the estimator of the carryover parameter λ_1 in Parameterization 1, but their interpretations differ greatly. One might ask how is it possible for a treatment–by–period interaction to equal a late response to treatment? Obviously, these effects can be equal only by chance, and although the same values for their sums of squares may be obtained from the different models, it is necessary to accept the fact that there is an "intrinsic aliasing" [see McCullagh and Nelder (1983), Jones and Kenward (1989)] between parameters in these alternative parameterizations.

Parameterization 3:

This model contains sequence, period and treatment effects, eliminating the carryover parameter λ_1. It would be the model that one would use to analyze the 2 x 2 design if it were felt that carryover was impossible to occur. In that case, the cell means can be written

$$E(Y_{ijk}) = \mu_{ij} = \mu + \gamma_i + \pi_j + \tau_t \ . \tag{3.4}$$

This model assumes that there is no carryover effect of the type that is traditionally viewed as a persistent or "late response" to treatment, nor is there a treatment–by–period interaction such as might be termed a "psychological" carryover effect. A model such as (3.4) can be applicable to the situation, arising as a result of the randomization process by which subjects are allocated to sequences, where one sequence may get a preponderance of individuals that have a characteristic (e.g., age, sex, general health condition) that predisposes them to react to the sequence of treatments in such a way as to give rise to a significant "sequence effect" when tested statistically. The least–squares equations for this model can be written as

$$\mathbf{X}\beta \ = \ \begin{pmatrix} 1 & 1 & 1 & 1 \\ 1 & 1 & -1 & -1 \\ 1 & -1 & 1 & -1 \\ 1 & -1 & -1 & 1 \end{pmatrix} \begin{pmatrix} \mu \\ \gamma_1 \\ \pi_1 \\ \tau_1 \end{pmatrix} = \begin{pmatrix} \mu_{11} \\ \mu_{12} \\ \mu_{21} \\ \mu_{22} \end{pmatrix} .$$

The portion of the inverse of the $\mathbf{X}'\mathbf{X}$ matrix relating to treatment is

$$(\mathbf{X}'\mathbf{X})^{-1} = 0.25$$

which is the same as for τ_1 in Parameterization 2. The estimators of the parameters are given by

$$
\hat{\beta} = \begin{pmatrix} \hat{\mu} \\ \hat{\gamma}_1 \\ \hat{\pi}_1 \\ \hat{\tau}_1 \end{pmatrix} = (X'X)^{-1}X'Y = \begin{pmatrix} (\overline{Y}_{11}+\overline{Y}_{12}+\overline{Y}_{21}+\overline{Y}_{22})/4 \\ (\overline{Y}_{11}+\overline{Y}_{12}-\overline{Y}_{21}-\overline{Y}_{22})/4 \\ (\overline{Y}_{11}-\overline{Y}_{12}+\overline{Y}_{21}-\overline{Y}_{22})/4 \\ (\overline{Y}_{11}-\overline{Y}_{12}-\overline{Y}_{21}+\overline{Y}_{22})/4 \end{pmatrix} ,
$$

so that the estimator for the treatment parameter τ_1, unlike that for the treatment parameter in Parameterization 1, involves all four cell means and has twice the precision of the estimator in that parameterization. All estimators are intuitive ones, the one for μ being the average of all four cell means, the one for γ_1 being a contrast of the average response in Sequence 1 with that in Sequence 2, the one for π_1 being a contrast of the average response in Period 1 with that in Period 2, and the one for τ_1 being a contrast of the average response to Treatment A with that to Treatment B. The test for treatment does not depend upon the result of the test for sequences, the treatment test being conducted against the within–subjects error, while the sequence test involves the between–subjects error.

Parameterization 4:

This parameterization was used by Milliken and Johnson (1984) and contains sequence, treatment and sequence–by–treatment interaction terms as follows,

$$
E(Y_{ijk}) = \mu_{ij} = \mu + \gamma_i + \tau_t + (\gamma\tau)_{it} \tag{3.5}
$$

where $(\gamma\tau)_{it}$ is the parameter representing the interaction between sequence and treatment. The notation μ_{ij}, representing the true mean response in the ith sequence and jth period, is retained for consistency with previous models.

The set of equations (3.5) may be written in matrix form as

$$
X\beta = \begin{pmatrix} 1 & 1 & 1 & 1 \\ 1 & 1 & -1 & -1 \\ 1 & -1 & -1 & 1 \\ 1 & -1 & 1 & -1 \end{pmatrix} \begin{pmatrix} \mu \\ \gamma_1 \\ \tau_1 \\ (\gamma\tau)_{11} \end{pmatrix} = \begin{pmatrix} \mu_{11} \\ \mu_{12} \\ \mu_{21} \\ \mu_{22} \end{pmatrix} .
$$

from which the portion of the inverse of $X'X$ relating to the treatment and sequence–by–treatment parameters is

$$
(X'X)^{-1} = \begin{pmatrix} 0.25 & 0 \\ 0 & 0.25 \end{pmatrix} .
$$

The least–squares estimators of the parameters of Equation (3.5) are

$$
\hat{\beta} = \begin{pmatrix} \hat{\mu} \\ \hat{\gamma}_1 \\ \hat{\tau}_1 \\ (\widehat{\gamma\tau})_{11} \end{pmatrix} = (X'X)^{-1}X'Y = \begin{pmatrix} (\overline{Y}_{11}+\overline{Y}_{12}+\overline{Y}_{21}+\overline{Y}_{22})/4 \\ (\overline{Y}_{11}+\overline{Y}_{12}-\overline{Y}_{21}-\overline{Y}_{22})/4 \\ (\overline{Y}_{11}-\overline{Y}_{21}-\overline{Y}_{12}+\overline{Y}_{22})/4 \\ (\overline{Y}_{11}-\overline{Y}_{12}+\overline{Y}_{21}-\overline{Y}_{22})/4 \end{pmatrix} .
$$

The estimator of the sequence parameter γ_1 is the same as in Parameterization 3, and the estimator of the treatment parameter τ_1 is the same as in Parameterizations 2 and 3. The estimator of the sequence–by–treatment interaction parameter $(\gamma\tau)_{11}$ is the same as the estimator of the period parameter π_1 in Parameterizations 1, 2 and 3. If the sequence–by–treatment interaction can be considered to be synonymous with a period effect, this parameterization becomes identical to Parameterization 3. Otherwise, the two parameterizations lead to different interpretations and statistical tests, as we shall see in the next section.

3.2. Tests of Significance in the 2 x 2 Cross–over Design

It might be thought that once having detailed the model terms in each of the four parameterizations, the statistical analyses that follow would be straight–forward. Although this should be the case, many published examples of the analysis of variance of one or other of these models suffers from errors of commission or omission. Consider the analysis of variance reported by Jones and Kenward (1989, Table 2.5). A portion of that ANOVA table is reproduced here as Table 3.1; the formulae for the sums of squares in their original table are

Table 3.1. Part of ANOVA table, Table 2.5, from Jones and Kenward (1989).

Source of Variation	df	Expected Mean Square
Between–subjects:		
Carryover	1	$\dfrac{2n_1 n_2}{(n_1+n_2)}(\lambda_1-\lambda_2)^2 + 2\sigma_s^2 + \sigma^2$
Between–subjects residual	(n_1+n_2-2)	$2\sigma_s^2 + \sigma^2$
Within–subjects:		
Direct Treatments (adjusted for periods)	1	$\dfrac{2n_1 n_2}{(n_1+n_2)}[(\tau_1-\tau_2)^2 - \dfrac{(\lambda_1-\lambda_2)^2}{2}] + \sigma^2$
Periods (adjusted for treatments)	1	$\dfrac{2n_1 n_2}{(n_1+n_2)}(\pi_1-\pi_2)^2 + \sigma^2$
Within–subjects residual	(n_1+n_2-2)	σ^2

omitted for brevity. Also, as they do not assume equal numbers of subjects in each sequence, the sequences have n_1 and n_2 subjects, respectively. Their carryover effect, with one degree of freedom, is situated in the between–subjects stratum of Table 3.1. The expected mean square indicates that a test of the null hypothesis H_0: $\lambda_1 = \lambda_2$ should be made by taking the ratio of the mean square for carryover to the mean square between–subjects residual, that ratio having a central F–distribution. But the question immediately arises as to how the parameter difference $\lambda_1-\lambda_2$, which relates to the differential effect of a late response or a persistent effect of treatment and is derived from information *within* subjects, can possibly be present in an expected mean square in a *between* subjects stratum? Obviously, true carryover, parameterized using λ_1 and λ_2, should not appear in a quadratic form involving information between subjects. The mystery is solved when one realizes that the term labelled "Carryover" is mislabelled, and should really read "Sequences". Replacing the first term in that expected mean square by

$$2n_1 n_2 (\gamma_1-\gamma_2)^2/(n_1+n_2),$$

where γ_1 is the sequence parameter (with $\gamma_2 = -\gamma_1$), leads to an appropriate test of sequence differences in Parameterization 3. Similarly, the term involving $(\lambda_1 - \lambda_2)^2$ also should not appear in the Direct Treatments term in the within–subjects stratum, since this parameter is not part of Parameterization 3. The correct tests for whether there are significant period effects and treatment effects are made by taking the ratio of the mean squares for these terms against the within–subjects residual mean square. The validity of these tests does not depend upon whether the sequence test is significant or not, contrary to the beliefs of many workers in this field who argue that an unbiased test for treatments can only be made if carryover is negligible. That would follow if the expected mean squares in Table 3.1 were correct, but they cannot be correct if they contain parameters that are not in Parameterization 3.

Suppose Parameterization 1 is the parameterization of interest. This parameterization has no sequence term in the model and the carryover term appears in the within–subjects stratum. The ordinary least squares estimators for τ_1 and λ_1 are given in the equations below the definition of model (3.2). Associated with these estimators are adjusted sums of squares, leading to mean squares which can be used to form central F–distributions to test relevant null hypotheses. The test for treatments uses only the means from the first period, and although this test is apparently valid, one must bear in mind that any sampling variation due to sequences, plus any other variations in the sequence sum of squares that are not accounted for in the model, must inflate the within–subjects residual sum of squares. Thus, a different residual mean square will be used for testing effects for Parameterization 1, than was used for Parameterization 3. Also, the poor separability of the treatment and carryover parameters, indicated by the correlation coefficient of 0.707, may lead to both treatment and carryover effects being non–significant (see Section 1.3).

Parameterization 2 is similar to Parameterization 1 in that there is no sequence term in the model. The treatment, period, and treatment–by–period interaction terms all appear in the within–subjects stratum. The residual sum of squares and mean squares are the same as in Parameterization 1, but the adjusted treatment mean squares differ, as does the interpretation of the effects.

Parameterization 4, like Parameterization 3, contains a sequence parameter, but when it is tested against the between–subjects residual mean square, the resulting F–ratio is not an unadulterated test of sequence differences. As we will see in the next paragraph, the quadratic form being tested in the null hypothesis contains elements of the sequence–by–treatment interaction as well as elements of the sequence effects. Furthermore, the supposed test for "treatments" is also not an unadulterated test for treatment differences.

Hills and Armitage (1979) presented an analysis of variance table which is similar to that of Table 2.5 of Jones and Kenward (1989), which is our Table 3.1 but with one major exception. Their carryover term, which also appears in the between–subjects stratum, is labelled T x P, designating the treatment–by–period interaction that we have denoted $(\tau\pi)_{ij}$. For reasons already discussed, that interaction properly belongs in the within–subjects stratum. What is really being tested in the Hills-Armitage table is a *sequence* effect, and it should be labelled as such. Another reference to an analysis of variance table for the 2 x 2 cross–over design is that of Milliken and Johnson (1984, p. 437). Their Table 32.2 is formulated as follows, except for some differences in notation:

Source	df	EMS
Sequence	1	$\sigma_e^2 + 2\sigma_\epsilon^2 + Q(\text{Sequence})$
Error(Between subjects)	$n_1 + n_2 - 2$	$\sigma_e^2 + 2\sigma_\epsilon^2$
Treatment	1	$\sigma_e^2 + Q(\text{Treatment})$
Treatment · Sequence	1	$\sigma_e^2 + Q(\text{Treatment} \cdot \text{Sequence})$
Error(Within subjects)	$n_1 + n_2 - 2$	σ_e^2

The term Q(Sequence) refers to a quadratic form containing sequence parameters but not other model parameters; similarly, Q(Treatment) refers to a quadratic form involving only treatment parameters, etc. These expected mean squares are different from the expected mean squares produced by SAS where the term Q(Sequence) is replaced by Q(Sequence, Sequence · Treatment) and the term Q(Treatment) is Q(Treatment, Sequence · Treatment). These quadratic forms do not provide an appropriate test for sequence or treatment effects. Hocking (1973) discusses the choice between competing models which give rise to different expected mean squares. We choose to present expected mean squares produced by SAS.

A problem with Milliken and Johnson's analysis is that they purport to be conducting a test for sequence differences without having any sequence parameters in their model on pp. 435–436. Instead they write the sequence effect as a difference between the carryover parameters λ_1 and λ_2, leading to the same potential confusion as we have decried earlier. To avoid such confusion, one should write down a model, such as one of the models given in Parameterizations 1, 2, 3 or 4, and present expected mean squares which contain only parameters that appear in that model. Mixing of parameters should be seen to be a major source of confusion to users attempting to model experimental results.

3.3. An Example of Data Analysis in the 2 x 2 Cross-over Design

We will now present a numerical illustration of the various problems associated with the 2 x 2 design using the data set of Patel (1983). These data were also used for illustrative purposes by Jones and Kenward (1989), and have the advantage over the data set of Grizzle (1965), for example, in that they also contain readings for baseline (run-in) and wash-out periods. Although those values will not be needed at present, they will be considered in later sections of this chapter. Table 3.2 presents the data of Patel (1983), which represent values of forced expiratory volume in one second (FEV_1) obtained using subjects suffering from various degrees of asthma, and given two active drugs, A and B, in the two treatment periods (reproduced here with the permission of Marcel Dekker, Inc.).

Table 3.2.   Data of Patel (1983), values of FEV_1.

Sequence 1:

Drug → Subject ↓	– Run–in period	A First treatment	– Wash–out period	B Second treatment
1	1.09	1.28	1.24	1.33
2	1.38	1.60	1.90	2.21
3	2.27	2.46	2.19	2.43
4	1.34	1.41	1.47	1.81
5	1.31	1.40	0.85	0.85
6	0.96	1.12	1.12	1.20
7	0.66	0.90	0.78	0.90
8	1.69	2.41	1.90	2.79

(continued on following page)

Sequence 2:

Subject ↓	Drug → – Run–in period	B First treatment	– Wash–out period	A Second treatment
1	1.74	3.06	1.54	1.38
2	2.41	2.68	2.13	2.10
3	3.05	2.60	2.18	2.32
4	1.20	1.48	1.41	1.30
5	1.70	2.08	2.21	2.34
6	1.89	2.72	2.05	2.48
7	0.89	1.94	0.72	1.11
8	2.41	3.35	2.83	3.23
9	0.96	1.16	1.01	1.25

Ignoring the run–in and wash–out periods, a 2 x 2 table of sequence–by–period means is given in the following table, with the treatment label (A or B) also indicated.

		Period 1	Period 2
Sequence	1	1.5725 (A)	1.6900 (B)
	2	2.3411 (B)	1.9456 (A)

Superficially, it appears that treatment B is more effective than treatment A, but that it may be less so when given after treatment A than when given first. Formal testing is now made considering each parameterization in turn.

We first fit Parameterization 3, which contains an explicit sequence term, thereby producing an analysis of variance table close in appearance to that of Table 3.1. Table 3.3 contains the numerical results, which agree with those presented in Kenward and Jones, p. 32. Because there is no carryover parameter in the model, all effects are orthogonal and the sequential and adjusted sums of squares are identical. The only difference in the two approaches is that we use the label "Sequence", whereas Jones and Kenward use the label "Carry–over". Since the correct test for the sequence parameter γ_1 leads to a non–significant result (P–value = 0.125), one may conclude that there is no sequence effect. A further assumption is necessary to go from here to the conclusion that there is no differential carryover effect, that is, $\lambda_1 = \lambda_2$, a conclusion often desired by users of the 2 x 2 cross–over design. In other words, because of the intrinsic aliasing that is inherent in this design, workers wish to interpret the significance test of the null hypothesis of no sequence differences (H_0: $\gamma_1 = \gamma_2$) as being indicative of whether or not there are any carryover effects.

Testing whether there are significant treatment effects and period effects using the adjusted sum of squares in Table 3.3 indicates no differential period effect (P–value = 0.260) but a marginally significant differential treatment effect (P–value = 0.047). Jones and Kenward (1989) examined the residuals from the fitted model and searched for potential outliers. As one of the data points appeared to be suspect, they omitted this data point and analyzed the results again, this time getting a P–value of 0.06 for the test of treatment differences. We shall also examine the residuals, using methods described in Section 1.6, but we postpone that until later in this chapter, when we have considered the full data set including baseline and wash–out measurements.

Table 3.3. ANOVA of Patel's Data Using Parameterization 3, [model (3.4)]

Source of variation	df	Sum of Squares	Mean Squares	F	Pr > F
Between–subjects:					
Sequence	1	2.2212	2.2212	2.636	0.125
Residual	15	12.6415	0.8428		
Within–subjects:					
Periods	1	0.1637	0.1637	1.373	0.260
Treatments	1	0.5574	0.5574	4.676	0.047
Residual	15	1.7882	0.1192		

We now fit Parameterization 1, which contains period, treatment, and carryover effects but no sequence effects. The ANOVA table is given in Table 3.4. The Type II or adjusted sums of squares are the appropriate ones for testing the significance of treatment and carryover effects. Because a carryover parameter is in the model, no contrast involving the first period is estimable. Thus, the adjusted period sum of squares, mean square, etc., produced by SAS and other statistical packages are likely to be spurious. That term is omitted

Table 3.4. ANOVA of Patel's Data Using Parameterization 1, [model (3.2)]

Source of variation	df	Type I SS	Mean Squares	F	Pr > F
Periods	1	0.2019	0.2019	0.420	0.522
Treatments	1	0.5574	0.5574	1.159	0.290
Carryover	1	2.2212	2.2212	4.618	0.040
	df	Type II SS	Mean Squares	F	Pr > F
Treatments	1	2.5021	2.5021	5.202	0.030
Carryover	1	2.2212	2.2212	4.618	0.040
Residual	30	14.4297	0.4810		

from the Type II portion of Table 3.4. The tests of significance give a P–value of 0.040 for carryover and a P–value of 0.030 for treatment. The conclusion with respect to treatment is similar to that obtained from Parameterization 3, the P–value for treatment in that parameterization having been 0.047. In both analyses there is marginal evidence for a treatment difference. For carryover, the P–value of 0.040 provides marginal evidence of a differential effect, whereas in Parameterization 3 there was less evidence (P–value = 0.125) for such an effect if the sequence parameter is interpreted as indicative of carryover. One must remember that the meanings of the parameterizations are different, and in Parameterization 1 any differences due to sequences are included in the "residual" term which has 30 degrees of freedom.

We now look at the results of fitting Parameterization 2, which contains period and treatment effects and a period–by–treatment interaction. The analysis of variance table is given in Table 3.5. The Type II or adjusted sums of squares are the appropriate ones for testing the significance of treatment effects and the period–by–treatment interaction. The period–by–treatment interaction leads to exactly the same result as for Parameterization 1 (Table 3.4), providing marginal evidence for "carryover", if one wishes to interpret the interaction as indicative of such, but the test for a differential treatment effect gives a result (P–value = 0.290) which is quite different from that of Parameterizations 1 and 3. The reason is that, although Parameterizations 2 and 3 have the same adjusted

Table 3.5. ANOVA of Patel's Data Using Parameterization 2, [model (3.3)]

Source of variation	df	Type I SS	Mean Squares	F	Pr > F
Periods (P)	1	0.2019	0.2019	0.420	0.522
Treatments (T)	1	0.5574	0.5574	1.159	0.290
P · T	1	2.2212	2.2212	4.618	0.040
	df	Type II SS	Mean Squares	F	Pr > F
Treatments	1	0.5574	0.5574	1.159	0.290
P · T	1	2.2212	2.2212	4.618	0.040
Residual	30	14.4297	0.4810		

treatment mean squares of 0.5574, the latter model tests this against a much larger residual mean square (0.4810) than the former model does (0.1192), producing a smaller and less significant F–value. It should never be forgotten that the models correspond to different assumptions of the nature of the underlying physical process. From the numerical results alone, one cannot say which interpretation is more likely to prove the more correct.

Finally, we look at the numerical results for the fit of Parameterization 4, which contains sequence and treatment parameters and the sequence–by–treatment interaction. Results are given in Table 3.6 (see top of following page). The F–ratio of the "Sequence" mean square against the between–subjects residual mean square, as mentioned in the last paragraph of Section 3.2, does not have a central F–distribution and is not a test of sequence effects. Similarly, the test of "treatments" in the within–subjects portion of the table does not provide an unadulterated test of a differential treatment effect, as the quadratic form being tested contains some elements of the sequence–by–treatment interaction. The only test that leads to an interpretable result is that for the sequence–by–treatment interaction, which gives an identical result to that for the period effect in Parameterization 3.

3.4. Adding Baseline Measurements to the 2 x 2 Cross–over Design

As we have seen in the previous sections, the 2 x 2 cross–over design without baseline measurements leads to an overparameterized model that requires, if

Table 3.6. ANOVA of Patel's Data Using Parameterization 4, [model (3.5)]

Source of variation	df	Sum of Squares	Mean Squares	F	Pr > F
Between–subjects:					
Sequence	1	2.2212	2.2212	2.636	0.125
Residual	15	12.6415	0.8428		
Within–subjects:					
Treatments	1	0.5956	0.5956	4.996	0.041
Seq. · Treat.	1	0.1637	0.1637	1.373	0.260
Residual	15	1.7882	0.1192		

meaningful solutions are to be obtained, additional assumptions to be made to remove one of the parameters. Although removal of a parameter does lead to a solution of the estimation equations, the interpretations may differ greatly depending upon how the model is parameterized. This was shown in Section 3.3. Freeman (1989) and others have suggested that use of baseline measurements, that is, readings taken on each subject *before* any treatments are given to them, leads to a unique resolution of the estimation difficulties.

In this section, we examine the consequences of the use of baseline measurements, and in Section 3.5, we look at what happens if a wash–out period is added as well. [Some authors, such as Kershner and Federer (1981), use the name *baseline* to mean any measurement taken before a treatment is given to the subject. In this work, we use baseline to mean only those readings taken before the subject is given the *first* treatment. Readings taken after the effect of a treatment is presumed to have died away are referred to as wash–out readings.] In the altered design, there are three periods, with the first period being used for recording the baseline measurements. The design can thus be represented as follows:

```
                    Period
                1    2    3
            1 | –    A    B |
Sequence      |            |
            2 | –    B    A |
```

The hyphen (–) serves to indicate that there is no treatment applied, and follows a notation used by Kershner and Federer (1981). The model to be used is the same as Equation (3.1), which, after taking expectations, can be written as

$$E(Y_{ijk}) = \mu_{ij} = \mu + \gamma_i + \pi_j + \tau_t + \delta_j \lambda_r \tag{3.6}$$

where μ_{ij} represents the true cell mean in the ith sequence and jth period, as before. From the six cells of the sequence–by–period combination, one can write down six equations in terms of the model parameters as follows:

$$\mu_{11} = \mu + \gamma_1 + \pi_1$$
$$\mu_{12} = \mu + \gamma_1 + \pi_2 + \tau_1$$
$$\mu_{13} = \mu + \gamma_1 + \pi_3 + \tau_2 + \lambda_1$$
$$\mu_{21} = \mu + \gamma_2 + \pi_1$$
$$\mu_{22} = \mu + \gamma_2 + \pi_2 + \tau_2$$
$$\mu_{23} = \mu + \gamma_2 + \pi_3 + \tau_1 + \lambda_2 .$$

Employing sum–to–zero restrictions on the parameters leads to the relations

$$\gamma_2 = -\gamma_1, \quad \pi_3 = -\pi_1 - \pi_2, \quad \tau_2 = -\tau_1, \text{ and } \lambda_2 = -\lambda_1,$$

which transform the above equations into the following form, written in matrix notation:

$$
X\beta \;=\;
\begin{pmatrix}
1 & 1 & 1 & 0 & 0 & 0 \\
1 & 1 & 0 & 1 & 1 & 0 \\
1 & 1 & -1 & -1 & -1 & 1 \\
1 & -1 & 1 & 0 & 0 & 0 \\
1 & -1 & 0 & 1 & -1 & 0 \\
1 & -1 & -1 & -1 & 1 & -1
\end{pmatrix}
\begin{pmatrix}
\mu \\ \gamma_1 \\ \pi_1 \\ \pi_2 \\ \tau_1 \\ \lambda_1
\end{pmatrix}
\;=\;
\begin{pmatrix}
\mu_{11} \\ \mu_{12} \\ \mu_{13} \\ \mu_{21} \\ \mu_{22} \\ \mu_{23}
\end{pmatrix} .
$$

Unlike the situation which prevailed with this same design but without baseline information, the $X'X$ matrix is not singular and a unique least–squares solution to the above equations is obtained as follows:

$$\hat{\beta} = \begin{pmatrix} \hat{\mu} \\ \hat{\gamma} \\ \hat{\pi}_1 \\ \hat{\pi}_2 \\ \hat{\tau}_1 \\ \hat{\lambda}_1 \end{pmatrix} = \begin{pmatrix} (\overline{Y}_{11}+\overline{Y}_{12}+\overline{Y}_{13}+\overline{Y}_{21}+\overline{Y}_{22}+\overline{Y}_{23})/6 \\ (\overline{Y}_{11}-\overline{Y}_{21})/2 \\ (\overline{Y}_{11}+\overline{Y}_{21})/3 - (\overline{Y}_{12}+\overline{Y}_{13}+\overline{Y}_{22}+\overline{Y}_{23})/6 \\ (\overline{Y}_{12}+\overline{Y}_{22})/3 - (\overline{Y}_{11}+\overline{Y}_{13}+\overline{Y}_{21}+\overline{Y}_{23})/6 \\ (\overline{Y}_{12}-\overline{Y}_{11})/2 - (\overline{Y}_{22}-\overline{Y}_{21})/2 \\ [(\overline{Y}_{12}+\overline{Y}_{13}-2\overline{Y}_{11}) - (\overline{Y}_{22}+\overline{Y}_{23}-2\overline{Y}_{21})]/2 \end{pmatrix} .$$

The resulting estimators bear much resemblance to those obtained for the model where no baseline data are present, and the following points can be noted:

(1) The estimator of the population parameter μ is the average of the six observed cell means,

(2) The difference in sequences is determined wholly by the difference in the baseline observations of the two groups,

(3) The "baseline" period parameter π_1 is estimated by the difference between the baseline cell means and the average of the cell means of the other two periods, averaged over the groups,

(4) The first "treatment" period parameter π_2 is estimated by the difference between the cell means of that period and the average of the cell means of the baseline and second treatment period, averaged over the groups,

(5) The treatment effect parameter τ_1 involves the cell means of the first treatment and baseline periods, but not those of the second treatment period,

(6) The estimator of the carryover parameter λ_1 is a difference between contrasts, one for each sequence, between the average of the two active treatment periods and the baseline.

Some of these estimators are intuitive ones, such as those for μ, π_1 and π_2. So too is the estimator for γ_1, which establishes that a sequence effect is nothing more than a difference between the means of the two groups before any treatments have been applied. Clearly, the strength of having baseline readings is the fact that one can estimate true carryover, as embodied in the λ_1 parameter, separately from the sequence parameter γ_1, which will reflect imperfections in the randomization of individuals to groups, and which will be influenced by imbalances in various resulting factors such as age, gender, health, and so on. The sequence parameter is also aliased with treatment–by–period interactions such as might be caused by "psychological" carryover (see discussion under Parameterization 1 of Section 3.1).

Perhaps it is surprising that the treatment effect does not involve the measurement taken in Period 3, the last of the "treatment" periods. In that respect it mirrors the result obtained for the treatment parameters of Parameterization 1 without baseline information. A difference, however, is that with baselines there is no confounding of carryover with sequence differences. The carryover effect also is analogous to the result for Parameterization 1 without baselines, the difference being that the baseline readings provide a correction for differences between sequences.

The $(X'X)^{-1}$ matrix is proportional to the covariances of the parameter estimates, the multiplying factor being $\hat{\sigma}^2/n$ for the period, treatment and carryover effects. The portion of the matrix for treatment and carryover is

$$(X'X)^{-1} = \begin{pmatrix} 1.0 & 1.5 \\ 1.5 & 3.0 \end{pmatrix}$$

which tells potential users of this design that the variance of λ_1 is three times that of τ_1, indicating a much higher degree of variation in the estimation of carryover than in the estimation of "direct" treatment effects. Secondly, the correlation between the estimators of τ_1 and λ_1 is given by

$$\text{Corr}(\hat{\tau}_1, \hat{\lambda}_1) = 1.5/(3 \cdot 1)^{1/2} = 0.866.$$

This very high correlation, which was also reported by Freeman (1989), points to a difficulty with the practical use of the 2 x 2 design, even when baseline observations are present. The high correlation between direct and residual treatments effects should lead to difficulties in interpreting the results of experiments obtained using this design. The experimenter may not be able to distinguish between a direct effect due to one of the treatments being superior to the other and a residual effect due to the other treatment being superior to the first. Furthermore, as discussed in Section 1.3, the phenomenon akin to multicollinearity may lead to apparently non–significant treatment and carryover effects when in fact those effects are significant.

3.5. Adding Baseline and Wash–out Measurements to the 2 x 2 Cross–over Design

We now examine the consequences of adding wash–out measurements as well as a baseline measurement to the 2 x 2 cross–over design. There will now be four periods, with the first used for recording the baseline measurements, the third for recording the wash–out period measurements, and the second and fourth periods for administering the treatments and recording their effects, as indicated by the following design:

		Period			
		1	2	3	4
Sequence	1	–	A	–	B
	2	–	B	–	A

As in Section 3.4, the hyphen (–) serves to indicate that there is no treatment applied in that period. The model to be used is a slight extension of Equation (3.6), as follows,

$$E(Y_{ijk}) = \mu_{ij} = \mu + \gamma_i + \pi_j + \tau_t + \lambda_r + \theta_s \, , \qquad (3.7)$$

where μ_{ij} represents the cell mean in the ith sequence and jth period, as before. The parameter λ_r represents a first–order carryover effect, which would be manifest in the third period from a treatment given in the second period. The parameter θ_s represents a second–order carryover effect, which would be observed in the fourth period from a treatment given in the second period. It is labelled as a second–order effect because of its spacing two periods after the application of the treatment. The addition of the extra period requires an additional period parameter π_3, and with the inclusion of θ_s, this makes a total of eight parameters to be estimated. As there are eight cell means, there are just enough equations to estimate the required parameters. Such a model is referred to as "saturated". The equations are:

$$\mu_{11} = \mu + \gamma_1 + \pi_1$$
$$\mu_{12} = \mu + \gamma_1 + \pi_2 + \tau_1$$
$$\mu_{13} = \mu + \gamma_1 + \pi_3 + \lambda_1$$
$$\mu_{14} = \mu + \gamma_1 + \pi_4 + \tau_2 + \theta_1$$
$$\mu_{21} = \mu + \gamma_2 + \pi_1$$
$$\mu_{22} = \mu + \gamma_2 + \pi_2 + \tau_2$$
$$\mu_{23} = \mu + \gamma_2 + \pi_3 + \lambda_2$$
$$\mu_{24} = \mu + \gamma_2 + \pi_4 + \tau_1 + \theta_2 .$$

Employing sum–to–zero restrictions on the parameters leads to the relations

$$\gamma_2 = -\gamma_1, \quad \pi_4 = -\pi_1-\pi_2-\pi_3, \quad \tau_2=-\tau_1, \quad \lambda_2=-\lambda_1, \text{ and } \theta_2=-\theta_1,$$

which transform the above equations into the following set of equations, written in matrix notation:

$$X\beta = \begin{pmatrix} 1 & 1 & 1 & 0 & 0 & 0 & 0 & 0 \\ 1 & 1 & 0 & 1 & 0 & 1 & 0 & 0 \\ 1 & 1 & 0 & 0 & 1 & 0 & 1 & 0 \\ 1 & 1 & -1 & -1 & -1 & -1 & 0 & 1 \\ 1 & -1 & 1 & 0 & 0 & 0 & 0 & 0 \\ 1 & -1 & 0 & 1 & 0 & -1 & 0 & 0 \\ 1 & -1 & 0 & 0 & 1 & 0 & -1 & 0 \\ 1 & -1 & -1 & -1 & -1 & 1 & 0 & -1 \end{pmatrix} \begin{pmatrix} \mu \\ \gamma_1 \\ \pi_1 \\ \pi_2 \\ \pi_3 \\ \tau_1 \\ \lambda_1 \\ \theta_1 \end{pmatrix} = \begin{pmatrix} \mu_{11} \\ \mu_{12} \\ \mu_{13} \\ \mu_{14} \\ \mu_{21} \\ \mu_{22} \\ \mu_{23} \\ \mu_{24} \end{pmatrix} .$$

The solution to the above set of least–squares equations is given by the following matrix equations:

$$
\hat{\beta} = \begin{pmatrix} \hat{\mu} \\ \hat{\gamma} \\ \hat{\pi}_1 \\ \hat{\pi}_2 \\ \hat{\pi}_3 \\ \hat{\tau}_1 \\ \hat{\lambda}_1 \\ \hat{\theta}_1 \end{pmatrix} = \begin{pmatrix} (\bar{Y}_{11}+\bar{Y}_{12}+\bar{Y}_{13}+\bar{Y}_{14}+\bar{Y}_{21}+\bar{Y}_{22}+\bar{Y}_{23}+\bar{Y}_{24})/8 \\ (\bar{Y}_{11}-\bar{Y}_{21})/2 \\ 3(\bar{Y}_{11}+\bar{Y}_{21})/8 - (\bar{Y}_{12}+\bar{Y}_{13}+\bar{Y}_{14}+\bar{Y}_{22}+\bar{Y}_{23}+\bar{Y}_{24})/8 \\ 3(\bar{Y}_{12}+\bar{Y}_{22})/8 - (\bar{Y}_{11}+\bar{Y}_{13}+\bar{Y}_{14}+\bar{Y}_{21}+\bar{Y}_{23}+\bar{Y}_{24})/8 \\ 3(\bar{Y}_{13}+\bar{Y}_{23})/8 - (\bar{Y}_{11}+\bar{Y}_{12}+\bar{Y}_{14}+\bar{Y}_{21}+\bar{Y}_{22}+\bar{Y}_{24})/8 \\ (\bar{Y}_{12}-\bar{Y}_{11})/2 - (\bar{Y}_{22}-\bar{Y}_{21})/2 \\ (\bar{Y}_{13}-\bar{Y}_{11})/2 - (\bar{Y}_{23}-\bar{Y}_{21})/2 \\ (\bar{Y}_{12}+\bar{Y}_{14}-2\bar{Y}_{11})/2 - (\bar{Y}_{22}+\bar{Y}_{24}-2\bar{Y}_{21})/2 \end{pmatrix}.
$$

There is a strong similarity between the above estimators and the estimators obtained in Section 3.4 which involved baseline measurements but not wash-out measurements. The following points can be noted:

(1) The difference in sequences is determined wholly by the difference in the baseline observations of the two groups,

(2) The estimators of the period parameters π_j are contrasts between the means for the jth period and the average of the means of the other periods, averaged over the groups,

(3) The treatment effect parameter τ_1 is a contrast involving the differences of the cell means of the first treatment and baseline periods,

(4) The estimator of the first-order carryover parameter λ_1 is a contrast involving the differences of the cell means of the wash-out and baseline periods,

(5) The estimator of the second-order carryover parameter θ_1 involves a contrast between the average of the two active treatment periods and the baseline.

The $(X'X)^{-1}$ matrix is proportional to the covariances of the parameter estimates, the multiplier being σ^2/n for the period, treatment and carryover parameters. The portion of the matrix for treatment, first-order carryover and second-order carryover is as follows:

$$(\mathbf{X}'\mathbf{X})^{-1} = \begin{pmatrix} 1.0 & 0.5 & 1.5 \\ 0.5 & 1.0 & 1.0 \\ 1.5 & 1.0 & 3.0 \end{pmatrix}.$$

This indicates that the variance of the estimator of θ_1, the parameter associated with second–order carryover, is three times that of the treatment parameter τ_1 and three times that of the first–order carryover effect parameter λ_1. The correlation of $\hat{\theta}_1$ and $\hat{\tau}_1$ is a rather high 0.866, but this may not present any real difficulties to users of this design, since if the wash–out period is long enough in time to be effective, θ_1 should be small compared with τ_1. More important is the relationship between the estimators of first–order carryover λ_1 and treatment τ_1. The variances of $\hat{\lambda}_1$ and $\hat{\tau}_1$ are equal and their correlation is 0.5. Although this is still a rather high correlation, it is much less than the correlation of 0.866 which applies to the design situation, as in Section 3.4, where baseline measurements but no wash–out measurements are present.

We now present a numerical illustration of fitting the 2 x 2 design with baseline and wash–out measurements using the data set of Patel (1983) in Table 3.2. A table of sequence–by–period means in presented in the following table, with treatment labels (A, B or –) also indicated, where appropriate.

		Period			
		1	2	3	4
	1	1.3375 –	1.5725 (A)	1.4313 –	1.6900 (B)
Sequence					
	2	1.8056 –	2.3411 (B)	1.7867 –	1.9456 (A)

The following code shows how the analysis can be carried out using SAS.

```
DATA PATEL;
INPUT SEQ SUBJ PERIOD TREAT $ CARRY1 $ CARRY2 $ Y @@;
IF CARRY1 = '0' THEN CARRY1='B'; /* These IF statements are */
IF CARRY2 = '0' THEN CARRY2='B'; /* included here to make */
IF TREAT = '0' THEN TREAT='B';   /* various LSMEANS estimable */
```

(continued on following page)

```
CARDS;
1 1 1 0 0 0 1.09   1 1 2 A 0 0 1.28   1 1 2 0 A 0 1.24   1 1 4 B 0 A 1.33
1 2 1 0 0 0 1.38   1 2 2 A 0 0 1.60   1 2 3 0 A 0 1.90   1 2 4 B 0 A 2.21
1 3 1 0 0 0 2.27   1 3 2 A 0 0 2.46   1 3 3 0 A 0 2.19   1 3 4 B 0 A 2.43
1 4 1 0 0 0 1.34   1 4 2 A 0 0 1.41   1 4 3 0 A 0 1.47   1 4 4 B 0 A 1.81
1 5 1 0 0 0 1.31   1 5 2 A 0 0 1.40   1 5 3 0 A 0 0.85   1 5 4 B 0 A 0.85
1 6 1 0 0 0 0.96   1 6 2 A 0 0 1.12   1 6 3 0 A 0 1.12   1 6 4 B 0 A 1.20
1 7 1 0 0 0 0.66   1 7 2 A 0 0 0.90   1 7 3 0 A 0 0.78   1 7 4 B 0 A 0.90
1 8 1 0 0 0 1.69   1 8 2 A 0 0 2.41   1 8 3 0 A 0 1.90   1 8 4 B 0 A 2.79
2 1 1 0 0 0 1.74   2 1 2 B 0 0 3.06   2 1 3 0 B 0 1.54   2 1 4 A 0 B 1.38
2 2 1 0 0 0 2.41   2 2 2 B 0 0 2.68   2 2 3 0 B 0 2.13   2 2 4 A 0 B 2.10
2 3 1 0 0 0 3.05   2 3 2 B 0 0 2.60   2 3 3 0 B 0 2.18   2 3 4 A 0 B 2.32
2 4 1 0 0 0 1.20   2 4 2 B 0 0 1.48   2 4 3 0 B 0 1.41   2 4 4 A 0 B 1.30
2 5 1 0 0 0 1.70   2 5 2 B 0 0 2.08   2 5 3 0 B 0 2.21   2 5 4 A 0 B 2.34
2 6 1 0 0 0 1.89   2 6 2 B 0 0 2.72   2 6 3 0 B 0 2.05   2 6 4 A 0 B 2.48
2 7 1 0 0 0 0.89   2 7 2 B 0 0 1.94   2 7 3 0 B 0 0.72   2 7 4 A 0 B 1.11
2 8 1 0 0 0 2.41   2 8 2 B 0 0 3.35   2 8 3 0 B 0 2.83   2 8 4 A 0 B 3.23
2 9 1 0 0 0 0.96   2 9 2 B 0 0 1.16   2 9 3 0 B 0 1.01   2 9 4 A 0 B 1.25
RUN;

PROC GLM;
CLASS SEQ SUBJ PERIOD TREAT CARRY1 CARRY2;
MODEL Y = SEQ SUBJ(SEQ) PERIOD TREAT CARRY1
     CARRY2 / SOLUTION SS1 SS2;
RANDOM SUBJ(SEQ);
LSMEANS TREAT CARRY1 CARRY2;
RUN;
```

An analysis of variance table is given in Table 3.7, which has been adapted from the output of PROC GLM, and which contains various Type I (sequential) sums of squares and relevant Type II (adjusted) sums of squares needed for making tests of significance. The appropriate tests for treatment, first–order carryover and second–order carryover effects are carried out using the adjusted sums of squares. Their P–values in Table 3.7 (0.1456, 0.5817 and 0.8034, respectively) lead to the conclusion that none of these effects is significant. The non–significance of both the first- and second-order carryover effects indicates that either the wash–out period was effective, or that the spacing between treatment periods (Periods 2 and 4 in this design) was sufficiently long to prevent the treatments from persisting, or a combination of both processes was operating. The absence of treatment effects contradicts the result obtained from the sequential sum of squares, which, if tested against the residual mean square after removal of the carryover contributions, gives a P–value of 0.0151, indicating at least marginal evidence of a treatment effect.

Table 3.7. ANOVA of Patel's Data with Baseline and Wash–out Measurements, model used is Equation (3.7).

Source of variation	df	Type I SS	Mean Squares	F	Pr > F
Between–subjects:					
Sequence	1	3.6146	3.6146	2.48	0.1361
Residual	15	21.8503	1.4567		
Within–subjects:					
Periods	3	1.7418	0.5806	6.66	0.0008
Treatments	1	0.5574	0.5574	6.39	0.0151
1st–order carry	1	0.0640	0.0640	0.73	0.3961
2nd–order carry	1	0.0055	0.0055	0.06	0.8034
Residual	45	3.9258	0.0872		

Source of variation	df	Type II SS	Mean Squares	F	Pr > F
Treatments	1	0.1913	0.1913	2.19	0.1456
1st–order carry	1	0.0269	0.0269	0.31	0.5817
2nd–order carry	1	0.0055	0.0055	0.06	0.8034

It was seen in Section 1.3 that if treatment and carryover effects were poorly separable, it was possible for the adjusted treatment sum of squares to be vastly smaller than the sequential treatment sum of squares. It is likely that a similar phenomenon is occurring here, in view of the high correlation between treatment and carryover parameters. Since the evidence for carryover effects is low, one course of action is to declare that both first– and second–order carryover effects are negligible, and to reanalyze the design with parameters for treatment effects, but not carryover effects, in the model. Generally we caution against this "two–stage" approach, because this procedure affects the significance levels of the subsequent tests in a difficult–to–calculate manner. This procedure leads to a P–value for the test of treatment effects of 0.0137, which is very close to the value based on the Type I SS when carryover effects were in the model. This is to be expected, as the residual mean square changes only slightly as a result of removing carryover effects from the model.

When carryover effects are in the model, there is no valid test for sequence effects. Although it might appear to be plausible to test the sequence mean square against the residual of the between–subjects stratum, the quadratic form being tested in the hypothesis involves carryover effects as well as sequence effects. However, for this data set the P–value of 0.1361 is not significant, so although the mean square may be inflated by containing carryover effects as well as sequence effects, there is no hint of sequence effects. A sequence effect, had it been significant, might reflect chance differences resulting from the randomization procedure by which subjects were allocated to the groups. Another difficulty with the model which includes carryover effects is the fact that period effects cannot be tested, because of the inherent confounding of carryover and period effects in the first treatment period.

Removing the carryover parameters from the model, if it is permissible, can lead to a valid test for periods as well as sequences. The resulting simplified model becomes

$$E(Y_{ijk}) = \mu_{ij} = \mu + \gamma_i + \pi_j + \tau_t , \tag{3.8}$$

and the analysis of variance is given in Table 3.8.

Table 3.8. ANOVA of Patel's Data with Baseline and Wash–out Measurements, Carryover Effects Excluded, model used is given by Equation (3.8).

Source of variation	df	Type I SS	Mean Squares	F	Pr > F
Between–subjects:					
Sequence	1	3.6146	3.6146	2.48	0.1361
Residual	15	21.8503	1.4567		
Within–subjects:					
Periods	3	1.7418	0.5806	6.83	0.0006
Treatments	1	0.5574	0.5574	6.56	0.0137
Residual	47	3.9953	0.0850		

With carryover no longer in the model, the sequential sums of squares are appropriate. The results indicate strong evidence for period effects, as well as at least marginal evidence for treatment differences. The F–ratio of 2.48 obtained from the ratio of the sequence to the between–subjects residual mean square is truly a test of sequence effects, since the quadratic form in the null hypothesis is now free from carryover effects.

We now examine the residuals, that is, the differences between the observed readings and the observations, after fitting the model given by Equation (3.8). As shown in Section 1.6, the residuals may be saved by inserting the following statement in the PROC GLM code after the MODEL statement:

OUTPUT OUT=DRES STUDENT=STDRES R=RESID P=FITVAL;   .

The following code will then produce a rough normal probability plot and some descriptive statistics, including the Wilk–Shapiro statistic:

```
PROC UNIVARIATE DATA=DRES PLOT NORMAL;
   VAR RESID STDRES;
RUN;     .
```

The unstandardized residuals RESID are probably the more appropriate to use with the Wilk–Shapiro statistic, which, for these data came out to be 0.9149 with a P–value of 0.3028, indicating that the residuals are sufficiently close enough to being considered to be normally distributed. A plot of these residuals against the fitted values, which can be achieved using

```
PROC PLOT DATA=DRES;
   PLOT RESID*FITVAL;
RUN;
```

shows a couple of potential outliers, values which have high positive residuals and are conspicuous on that graph. Despite these two points, the Wilk–Shapiro statistic did not indicate lack of normality, but it needs to be remembered that that test is not a very powerful one.

The studentized residuals STDRES are more appropriate than the "raw" residuals RESID for use with the normal probability plot. As described in Section 1.6 of Chapter 1, if high quality graphics are available, one can obtain a better quality graph than that produced by the PLOT option of PROC UNIVARIATE by calculating the expected normal order statistics and plotting the residuals against them using a good quality plotter. For Patel's data, the resulting graph is shown at the top of the following page. From that graph, it can be seen that the residuals conform closely to a straight line, except for the two points with the largest positive residuals. These two residuals can be

Normal Probability Plot

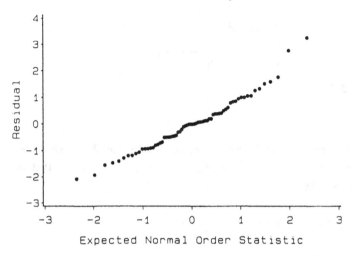

identified by printing out the data set DRES containing the variables SEQ, SUBJ, PERIOD, TREAT and STDRES. Since that data set has been sorted, an extreme residual is easy to identify, as well as the experimental unit and period associated with it. The two highest residuals come from Subject 1 of the second sequence (the observed value of 3.06 in the second period appears high) and from Subject 3 of the second sequence (the observed value of 3.05 in the first period appears high). As these do not appear to be extreme outliers, no further action need be taken. In more extreme cases, some users would eliminate those subjects containing the unusual observations.

3.6. Adding Baseline and Two Wash-out Measurements to the 2 x 2 Cross-over Design

We now examine the effect of adding two wash-out periods to the 2 x 2 crossover design, one after the first treatment period (Period 2) and one after the second treatment period (Period 4). The initial baseline reading in Period 1 is retained. That is, the subjects are asked to appear a total of five times. This results in five periods, giving the following design:

		Period			
	1	2	3	4	5
Sequence 1	–	A	–	B	–
Sequence 2	–	B	–	A	–

As in Sections 3.4 and 3.5, the hyphen (–) serves to indicate that there is no treatment applied in that period. The model to be used is identical to that of Equation (3.7), as follows,

$$E(Y_{ijk}) = \mu_{ij} = \mu + \gamma_i + \pi_j + \tau_t + \lambda_r + \theta_s \, ,$$

where μ_{ij} represents the cell mean in the ith sequence and jth period, as before. The parameter λ_r represents a first-order carryover effect, which would be manifest in the third period from a treatment given in the second period, or in the fifth period from a treatment given in the fourth period. The parameter θ_s represents a second-order carryover effect, which would be observed in the fourth period from a treatment given in the second period. The addition of an extra period requires an extra period parameter in the model, making a total of nine parameters to be estimated. As there are ten cell means, there are more than enough equations to estimate the required parameters. The equations are:

$$\mu_{11} = \mu + \gamma_1 + \pi_1$$
$$\mu_{12} = \mu + \gamma_1 + \pi_2 + \tau_1$$
$$\mu_{13} = \mu + \gamma_1 + \pi_3 + \lambda_1$$
$$\mu_{14} = \mu + \gamma_1 + \pi_4 + \tau_2 + \theta_1$$
$$\mu_{15} = \mu + \gamma_1 + \pi_5 + \lambda_2$$
$$\mu_{21} = \mu + \gamma_2 + \pi_1$$
$$\mu_{22} = \mu + \gamma_2 + \pi_2 + \tau_2$$
$$\mu_{23} = \mu + \gamma_2 + \pi_3 + \lambda_2$$
$$\mu_{24} = \mu + \gamma_2 + \pi_4 + \tau_1 + \theta_2$$
$$\mu_{25} = \mu + \gamma_2 + \pi_5 + \lambda_1 .$$

Employing sum–to–zero restrictions on the parameters leads to the relations

$$\gamma_2 = -\gamma_1, \quad \pi_5 = -\pi_1-\pi_2-\pi_3-\pi_4, \quad \tau_2=-\tau_1, \quad \lambda_2=-\lambda_1, \text{ and } \theta_2=-\theta_1,$$

which transform the above equations into the following set of equations, written in matrix notation:

$$
\mathbf{X\beta} =
\begin{pmatrix}
1 & 1 & 1 & 0 & 0 & 0 & 0 & 0 & 0 \\
1 & 1 & 0 & 1 & 0 & 0 & 1 & 0 & 0 \\
1 & 1 & 0 & 0 & 1 & 0 & 0 & 1 & 0 \\
1 & 1 & 0 & 0 & 0 & 1 & -1 & 0 & 1 \\
1 & 1 & -1 & -1 & -1 & -1 & 0 & -1 & 0 \\
1 & -1 & 1 & 0 & 0 & 0 & 0 & 0 & 0 \\
1 & -1 & 0 & 1 & 0 & 0 & -1 & 0 & 0 \\
1 & -1 & 0 & 0 & 1 & 0 & 0 & -1 & 0 \\
1 & -1 & 0 & 0 & 0 & 1 & 1 & 0 & -1 \\
1 & -1 & -1 & -1 & -1 & -1 & 0 & 1 & 0
\end{pmatrix}
\begin{pmatrix}
\mu \\ \gamma_1 \\ \pi_1 \\ \pi_2 \\ \pi_3 \\ \pi_4 \\ \tau_1 \\ \lambda_1 \\ \theta_1
\end{pmatrix}
=
\begin{pmatrix}
\mu_{11} \\ \mu_{12} \\ \mu_{13} \\ \mu_{14} \\ \mu_{15} \\ \mu_{21} \\ \mu_{22} \\ \mu_{23} \\ \mu_{24} \\ \mu_{25}
\end{pmatrix}.
$$

The solution to the above set of least–squares equations is as follows:

$$
\begin{pmatrix}
\hat{\mu} \\ \hat{\gamma} \\ \hat{\pi}_1 \\ \hat{\pi}_2 \\ \hat{\pi}_3 \\ \hat{\pi}_4 \\ \hat{\tau}_1 \\ \hat{\lambda}_1 \\ \hat{\theta}_1
\end{pmatrix}
=
\begin{pmatrix}
(\overline{Y}_{11}+\overline{Y}_{12}+\overline{Y}_{13}+\overline{Y}_{14}+\overline{Y}_{15}+\overline{Y}_{21}+\overline{Y}_{22}+\overline{Y}_{23}+\overline{Y}_{24}+\overline{Y}_{25})/10 \\
(\overline{Y}_{11}+\overline{Y}_{13}+\overline{Y}_{15}-\overline{Y}_{21}-\overline{Y}_{23}-\overline{Y}_{25})/6 \\
[4(\overline{Y}_{11}+\overline{Y}_{21}) - (\overline{Y}_{12}+\overline{Y}_{13}+\overline{Y}_{14}+\overline{Y}_{15}+\overline{Y}_{22}+\overline{Y}_{23}+\overline{Y}_{24}+\overline{Y}_{25})/10 \\
[4(\overline{Y}_{12}+\overline{Y}_{22}) - (\overline{Y}_{11}+\overline{Y}_{13}+\overline{Y}_{14}+\overline{Y}_{15}+\overline{Y}_{21}+\overline{Y}_{23}+\overline{Y}_{24}+\overline{Y}_{25})/10 \\
[4(\overline{Y}_{13}+\overline{Y}_{23}) - (\overline{Y}_{11}+\overline{Y}_{12}+\overline{Y}_{14}+\overline{Y}_{15}+\overline{Y}_{21}+\overline{Y}_{22}+\overline{Y}_{24}+\overline{Y}_{25})/10 \\
[4(\overline{Y}_{14}+\overline{Y}_{24}) - (\overline{Y}_{11}+\overline{Y}_{12}+\overline{Y}_{13}+\overline{Y}_{15}+\overline{Y}_{21}+\overline{Y}_{22}+\overline{Y}_{23}+\overline{Y}_{25})/10 \\
(\overline{Y}_{12}-\overline{Y}_{22})/2 - (\overline{Y}_{11}+\overline{Y}_{13}+\overline{Y}_{15}-\overline{Y}_{21}-\overline{Y}_{23}-\overline{Y}_{25})/6 \\
(\overline{Y}_{13}-\overline{Y}_{15}-\overline{Y}_{23}+\overline{Y}_{25})/4 \\
(\overline{Y}_{12}+\overline{Y}_{14}-\overline{Y}_{22}-\overline{Y}_{24})/2 - (\overline{Y}_{11}+\overline{Y}_{13}+\overline{Y}_{15}-\overline{Y}_{21}-\overline{Y}_{23}-\overline{Y}_{25})/3
\end{pmatrix}.
$$

Some of the parameter estimators differ considerably from the estimators obtained in Section 3.5 which involved baseline measurements and a single wash–out period. Others bear close similarities to those in Section 3.5. The following points can be noted:

(1) The difference in sequences is a contrast between the means of the measurements in the three "non–treatment" periods of the two groups,

(2) The estimators of the period parameters π_j are contrasts between the means for the jth period and the average of the means of the other periods, averaged over the groups,

(3) The treatment effect parameter τ_1 is estimated by a contrast between the mean of the first treatment period and the average of the three "non–treatment" periods, contrasted over the two groups,

(4) The estimator of the first–order carryover parameter λ_1 involves means of the first and second wash–out periods of the two groups,

(5) The estimator of the second–order carryover parameter θ_1 involves a contrast between the average of the two active treatment periods and the average of the baseline and both wash–out periods, contrasted over the two groups.

The $(\mathbf{X}'\mathbf{X})^{-1}$ matrix is proportional to the covariances of the parameter estimates, the multiplier being $\hat{\sigma}^2/n$ for period, treatment and first– and second–order carryover. The portion of the matrix relating to treatment, first– and second–order carryover is given by

$$(\mathbf{X}'\mathbf{X})^{-1} = \begin{pmatrix} 2/3 & 0 & 5/6 \\ 0 & 1/4 & 0 \\ 5/6 & 0 & 5/3 \end{pmatrix} .$$

which reveals an interesting and important practical result, namely that the estimators of first–order carryover λ_1 and treatment τ_1 are uncorrelated. Not only that, their variances are lower than the variances of the corresponding estimators in Section 3.5 where only a single wash–out period was used. The variance of $\hat{\tau}_1$ is two–thirds of its previous value, and the variance of $\hat{\lambda}_1$ is only a quarter of what it was before. In addition, $\hat{\lambda}_1$ and $\hat{\theta}_1$ are also uncorrelated, although there is still a rather high correlation of 0.791 between $\hat{\tau}_1$ and $\hat{\theta}_1$. The implications of these results for the continued use of the two–treatment, two–sequence cross–over design are important. Many of the objections to the use of that design are now overcome, provided that the subjects, in a clinical trial, for example, are willing or able to visit the facility where the research is being carried out for a total of five occasions.

The first visit is for the purpose of having the baseline measurement recorded, and for receiving the first treatment. The length of time on this medication will depend upon the ailment or disease, and is related to the time that it may take for the treatment to have an effect. At the end of that time, the second measurement is taken. The length of the wash–out period again is related to the time necessary for the effect of the treatment to wear off. When the wash–out measurement is made on this third visit, the subject is then given the second active treatment. On the next visit, after a length of time deemed sufficient for the treatment to have had its effect, the fourth measurement is made. The subject must return once again, after a sufficient length of time to have allowed the treatment effect to wear off, to have the fifth measurement taken. Although the above is written for a human subject attending a clinical trial, the same number of measurement periods will apply if the cross–over trial involves feeding animals with different diets, for example.

Let us now examine the consequences of omitting the second–order carryover parameter θ_1. This may be justified on the grounds that second–order carryover is less likely than first–order carryover. Provided that the distance between periods is sufficiently long, the treatment may not persist more than one extra time period, if at all. The resulting set of least–squares equations, with the second–order carryover parameter omitted, is, in matrix form:

$$
\mathbf{X}\beta \;=\;
\begin{pmatrix}
1 & 1 & 1 & 0 & 0 & 0 & 0 & 0 \\
1 & 1 & 0 & 1 & 0 & 0 & 1 & 0 \\
1 & 1 & 0 & 0 & 1 & 0 & 0 & 1 \\
1 & 1 & 0 & 0 & 0 & 1 & -1 & 0 \\
1 & 1 & -1 & -1 & -1 & -1 & 0 & -1 \\
1 & -1 & 1 & 0 & 0 & 0 & 0 & 0 \\
1 & -1 & 0 & 1 & 0 & 0 & -1 & 0 \\
1 & -1 & 0 & 0 & 1 & 0 & 0 & -1 \\
1 & -1 & 0 & 0 & 0 & 1 & 1 & 0 \\
1 & -1 & -1 & -1 & -1 & -1 & 0 & 1
\end{pmatrix}
\begin{pmatrix}
\mu \\
\gamma_1 \\
\pi_1 \\
\pi_2 \\
\pi_3 \\
\pi_4 \\
\tau_1 \\
\theta_1
\end{pmatrix}
\;=\;
\begin{pmatrix}
\mu_{11} \\
\mu_{12} \\
\mu_{13} \\
\mu_{14} \\
\mu_{15} \\
\mu_{21} \\
\mu_{22} \\
\mu_{23} \\
\mu_{24} \\
\mu_{25}
\end{pmatrix} .
$$

The solution to the above set of least–squares equations is:

$$
\begin{pmatrix}
\hat{\mu} \\
\hat{\gamma} \\
\hat{\pi}_1 \\
\hat{\pi}_2 \\
\hat{\pi}_3 \\
\hat{\pi}_4 \\
\hat{\tau}_1 \\
\hat{\lambda}_1
\end{pmatrix}
=
\begin{pmatrix}
(\overline{Y}_{11}+\overline{Y}_{12}+\overline{Y}_{13}+\overline{Y}_{14}+\overline{Y}_{15}+\overline{Y}_{21}+\overline{Y}_{22}+\overline{Y}_{23}+\overline{Y}_{24}+\overline{Y}_{25})/10 \\
(\overline{Y}_{11}+\overline{Y}_{12}+\overline{Y}_{13}+\overline{Y}_{14}+\overline{Y}_{15}-\overline{Y}_{21}-\overline{Y}_{22}-\overline{Y}_{23}-\overline{Y}_{24}-\overline{Y}_{25})/10 \\
[4(\overline{Y}_{11}+\overline{Y}_{21}) - (\overline{Y}_{12}+\overline{Y}_{13}+\overline{Y}_{14}+\overline{Y}_{15}+\overline{Y}_{22}+\overline{Y}_{23}+\overline{Y}_{24}+\overline{Y}_{25})]/10 \\
[4(\overline{Y}_{12}+\overline{Y}_{22}) - (\overline{Y}_{11}+\overline{Y}_{13}+\overline{Y}_{14}+\overline{Y}_{15}+\overline{Y}_{21}+\overline{Y}_{23}+\overline{Y}_{24}+\overline{Y}_{25})]/10 \\
[4(\overline{Y}_{13}+\overline{Y}_{23}) - (\overline{Y}_{11}+\overline{Y}_{12}+\overline{Y}_{14}+\overline{Y}_{15}+\overline{Y}_{21}+\overline{Y}_{22}+\overline{Y}_{24}+\overline{Y}_{25})]/10 \\
[4(\overline{Y}_{14}+\overline{Y}_{24}) - (\overline{Y}_{11}+\overline{Y}_{12}+\overline{Y}_{13}+\overline{Y}_{15}+\overline{Y}_{21}+\overline{Y}_{22}+\overline{Y}_{23}+\overline{Y}_{25})]/10 \\
(\overline{Y}_{12}-\overline{Y}_{14}-\overline{Y}_{22}+\overline{Y}_{24})/4 \\
(\overline{Y}_{13}-\overline{Y}_{15}-\overline{Y}_{23}+\overline{Y}_{25})/4
\end{pmatrix}.
$$

All these estimators are intuitive ones. For example, the estimator of the sequence parameter is just a contrast between the average response over the five periods of each sequence. The estimator of the treatment parameter is simply a contrast of Treatment A with Treatment B in the two treatment periods. The estimator of first–order carryover remains unchanged from what it was in the model that also contained second–order carryover. The portion of the $(X'X)^{-1}$ matrix relating to treatment and carryover is:

$$
(X'X)^{-1} =
\begin{pmatrix}
0.25 & 0 \\
0 & 0.25
\end{pmatrix}.
$$

The variances of τ_1 and λ_1 are each 0.25, a considerable reduction for τ_1 compared to the model that also included θ_1.

Let us now consider a set of data from a feeding trial conducted by the Department of Primary Industry, Tasmania, the purpose of which was to test whether dairy cows could be fed, and thrive upon, a waste product from cheese–making known as delactose permeate. The design used was the 2 x 2 design with an initial baseline reading and wash–out readings after each of the two treatment periods. There were only four cows available for each of the two sequences, and one cow in the second sequence (Cow No. 6) had to be withdrawn from the trial before completion because it was showing signs of ketosis (a condition unrelated to the diet the animal was receiving). The following data on average milk yield during the week following the commencement of the treatments (which includes baseline and wash-out periods as well as active treatments) is reproduced here with the permission of Dr. G.T. Stevenson and the Department of Primary Industry, Tasmania, Australia :

				Period		
		1	2	3	4	5
Cow↓						
1	1	14.9	15.3(A)	14.8	14.7(B)	12.7
	2	25.4	24.1(A)	24.6	24.4(B)	21.0
	3	22.8	22.2(A)	21.2	23.9(B)	19.3
	4	19.5	17.6(A)	18.6	17.6(B)	15.4
Seq.						
	5	22.5	20.6(B)	17.9	17.6(A)	18.2
2	6	24.2	21.6(B)	20.4	—	—
	7	18.2	15.0(B)	16.4	16.0(A)	15.5
	8	16.9	17.3(B)	15.6	14.7(A)	13.5

Although use can be made of the information that is available on Cow 6 during the first three periods (see Section 9.4), for simplicity the data for that cow will be ignored in the current analysis. The coding using SAS might be as follows, where carryover is restricted to being of first order.

```
DATA STEVO;
INPUT SEQ MLKYLD COW PERIOD TREAT $ CARRY $ @@;
IF TREAT='0' THEN TREAT='B';
IF CARRY='0' THEN CARRY='B';
CARDS;
1 14.9 1 1 0 0   1 15.3 1 2 A 0   1 14.8 1 3 0 A   1 14.7 1 4 B 0   1 12.7 1
1 25.4 2 1 0 0   1 24.1 2 2 A 0   1 24.6 2 3 0 A   1 24.4 2 4 B 0   1 21.0 2
1 22.8 3 1 0 0   1 22.2 3 2 A 0   1 21.2 3 3 0 A   1 23.9 3 4 B 0   1 19.3 3
1 19.5 4 1 0 0   1 17.6 4 2 A 0   1 18.6 4 3 0 A   1 17.6 4 4 B 0   1 15.4 4
2 22.5 5 1 0 0   2 20.6 5 2 B 0   2 17.9 5 3 0 B   2 17.6 5 4 A 0   2 18.2 5
2 18.2 7 1 0 0   2 15.0 7 2 B 0   2 16.4 7 3 0 B   2 16.0 7 4 A 0   2 15.5 7
2 16.9 8 1 0 0   2 17.3 8 2 B 0   2 15.6 8 3 0 B   2 14.7 8 4 A 0   2 13.5 8
RUN;

PROC GLM;
CLASS SEQ COW PERIOD TREAT CARRY;
MODEL MLKYLD=SEQ COW(SEQ) PERIOD TREAT CARRY /
    SOLUTION SS1 SS2;
RANDOM COW(SEQ);
LSMEANS TREAT CARRY;
RUN;
```

The portion of the output pertaining to the treatment and carryover effects is as follows:

Source	df	Type II SS	Mean Squares	F	Pr > F
TREAT	1	3.04024	3.04024	3.31992	0.08207
CARRY	1	2.77714	2.77714	3.03263	0.09557

Hence, there is marginal evidence of treatment and carryover differences.

3.7. Recommendations About the Use of the 2 x 2 Cross–over Design

We have seen in this chapter that the use of the 2–treatment, 2–period, 2–sequence cross–over design is very problematic. In the absence of baseline and wash–out observations, the basic model, Equation (3.1), has too many parameters and does not lend itself to a unique solution. Attempts have been made by workers to overcome this problem by making assumptions that result in one of the parameters being dropped from the model, so that a solution is possible. However, it is difficult to justify these assumptions, none of which can be tested using the actual data themselves. Different assumptions lead to different tests of significance for treatment effects, as well as to different tests for the various manifestations of "carryover" effects, some of which are best interpreted as a sequence effect or a treatment–by–period interaction.

Other authors have recommended using baseline measurements to solve the problem of non–estimability of the basic model. It is true that the inclusion of baseline measurements does lead to estimability, but it does not solve the problem of poor separability of treatment from carryover effects. The correlation coefficient between the treatment and carryover parameters is 0.866, and furthermore, the variance of the carryover parameter is three times that of the treatment parameter. Thus, it may be difficult to detect a carryover effect even when it is present, and a significant treatment effect, apparently pointing to one treatment being better than another, may really be due to the persistent effect (carryover) of the other treatment.

Using a single wash–out period as well as a baseline measurement hardly changes anything, although it does result in a lower correlation between the treatment and carryover parameters. In the numerical example using the data set of Patel (1983), we saw how the treatment sum of squares, when adjusted for carryover effects, came out to be non–significant, although it was significant in a model that did not include carryover effects. Contradictions like these result from poor separability of direct and residual treatment effects in the 2 x 2 design.

Addition of a second wash–out period to the above, giving a design that has five measurement periods, remarkably improves the efficiency of the 2 x 2 design. The variances of the estimators of treatment and first–order carryover drop dramatically, and the estimators are uncorrelated. Omitting a second–order carryover parameter improves the design still further. The main problem with this design is the requirement that the experimental subjects be available for a total of five measurement periods. This may preclude its use in some clinical trials where the "dropout" rate is high. There does not appear to be any literature on this design, despite its obvious nature.

We shall see in the next chapter that there is a relatively simple modification to the basic 2 x 2 design involving only three measurement periods that leads to efficient estimation of the parameters of interest to investigators in clinical trials, agricultural experimentation, and psychological studies. Designers of clinical trials, agricultural feeding trials, and other experiments for which cross–over designs are suitable, should give serious consideration towards adopting a design such as (4.1.1) in the next chapter.

It is necessary, though, to recognize that the 2 x 2 trial, despite its limitations, is the most widely–used trial in current use. Despite a recommendation from the U.S. Food and Drug Administration that this trial not be used for drug evaluation, preferring a "parallel groups" trial where a subject is given only a single medication, its use is likely to continue unless the users recognize the inherent limitations of a design producing only four cell means with which to estimate a minimum of five parameters. Patterson and Lucas (1962) recognized at an early stage the low efficiencies associated with two–period designs. Although dropout rates in clinical trials tend to be higher as the number of periods increase, the benefits of the 2–sequence, 3–period trial with two treatments, Design (4.1.1), should become apparent when one considers the fact that the treatment and carryover parameters are orthogonal in that design, guaranteeing complete separability of the two manifestations of treatment.

Exercises

3.1. For the data set of Patel (1983) in Table 3.2, the table of sequence–by–period means was as follows:

		Period 1	Period 2
	1	1.5725 (A)	1.6900 (B)
Sequence			
	2	2.3411 (B)	1.9456 (A)

(a) Using the parameter estimators given in Section 3.1 for each of the four parameterizations, calculate $\hat{\tau}_1$ and show that the estimates of τ_1 for Parameterizations 2, 3, and 4 are much smaller in absolute magnitude than the estimate of τ_1 in Parameterization 1.

(b) Calculate the estimates of λ_1, $(\tau\pi)_{11}$, γ_1, and γ_1 in Parameterizations 1–4, respectively, and show that they are equal (except for a factor of 2 for λ_1 in Parameterization 1). What does this tell us about our ability to differentiate between a late response to treatment, a psychological carryover, and a sequence effect?

3.2. Consider the following set of data for comparing a new drug (B) with a standard formulation (A). Patients were randomly allocated to two sequences AB and BA, respectively. Baseline readings were taken before application of the first drug, and during a wash–out period between the two treatment periods. The followings results were obtained on a scale with a maximum value of 15, there being 10 patients in each of the two sequences (see top of next page).

(a) If the baseline and wash–out readings are to be used, what statements must be added to the DATA step, to ensure estimability of treatment and first–order carryover effects?

(b) Carry out an analysis of variance of this data set, using both the baseline and wash–out readings, to test whether there are significant treatment and first–order carryover effects.

Sequence 1 (AB):

Drug →	–	A	–	B
Subject ↓	Baseline reading	First treatment	Wash–out reading	Second treatment
1	3	2	2	4
2	4	4	5	8
3	5	6	4	8
4	8	8	8	8
5	8	9	8	7
6	13	12	10	11
7	8	9	10	11
8	10	10	9	12
9	4	3	3	6
10	7	8	9	14

Sequence 2:

Drug →	–	B	–	A
Subject ↓	Baseline reading	First treatment	Wash–out reading	Second treatment
1	1	2	2	0
2	3	3	4	5
3	5	7	5	6
4	6	7	8	9
5	6	10	8	6
6	9	11	9	6
7	10	13	12	12
8	10	14	12	10
9	11	13	11	8
10	3	6	3	2

3.3. Consider a two–sequence, two–treatment design with two wash–out periods but with no initial baseline reading, viz.

		Period			
		1	2	3	4
Sequence	1	A	–	B	–
	2	B	–	A	–

(a) Examine the properties of this design using matrix operations to determine the elements of the $(X'X)^{-1}$ matrix. Do this both with a second–order carryover parameter in the model and for the case where there is only first–order carryover.

(b) Compare your results with those obtained in Section 3.6 where there was a baseline reading as well as two wash–out periods in the model. What does this tell us about the relative importance of wash-out periods compared to baseline readings?

3.4. Consider the example of a two-sequence, two-treatment design with an initial baseline reading and two wash-out periods that was analyzed in Section 3.6, involving two diets fed to dairy cows with average weekly milk production readings as the response variable. Show that the same analysis of variance would be obtained irrespective of whether the cows were fed a third diet, Treatment C, say, during the baseline and wash-out periods, or whether one of the two "treatment" diets (A or B) was given during those periods. In fact, in order to accustom cows to a diet supplemented with delactose permeate (Treatment A), that diet was given to all cows, irrespective of sequence, during the baseline and wash-out periods.

4

Modifications of the
2-Treatment, 2-Period,
2-Sequence Design

We have seen in Chapter 3 that, despite the fact that the 2-treatment, 2-period, 2-sequence cross-over design is widely used in practice, there is a major problem with this design deriving from the fact that its basic model is over-parameterized. Assumptions are usually made which allow the model equations to produce a unique solution, but the assumptions are seldom verifiable. The use of baseline measurements alleviates the problem somewhat, leading to unique estimators of the parameters, but the correlation between the estimators of the treatment and carryover parameters is very high and the power of the test for carryover effects is so low that the design remains problematic. Incorporation of a wash-out period between the administration of the two active treatments does little to help solve the basic problem, although the use of a second wash-out period after the last treatment may be beneficial. However, this latter modification appears to be almost untried.

In this chapter, we consider other modifications to the standard design, some of which result in complete separability of treatment and carryover effects. Usually the modifications involve the addition of extra periods or extra sequences, or both.

4.1. Designs With Two Sequences and Three Periods

We will restrict ourselves to "dual balanced" designs, that is, ones containing dual sequences with equal replication within each sequence. A dual sequence is a pair of sequences in which the second sequence is obtained from the first sequence by interchanging the letters A and B in the treatment labels. This guarantees that each treatment will be applied the same number of times in each period. Thus, the following are the four possible dual–balanced designs having two sequences and three periods :

Design 4.1.1

		Period	
Seq.	1	2	3
↓			
1	A	B	B
2	B	A	A

Design 4.1.2

		Period	
Seq.	1	2	3
↓			
1	A	B	A
2	B	A	B

Design 4.1.3

		Period	
Seq.	1	2	3
↓			
1	A	A	B
2	B	B	A

Design 4.1.4

		Period	
Seq.	1	2	3
↓			
1	A	A	A
2	B	B	B

In these four designs, Treatments A and B do not appear an equal number of times in each sequence. Hence, it is not possible to assess the separability of treatment and carryover effects using the omnibus measure described in Section 1.2. Instead, we have to use the more difficult and time–consuming procedure of examining the estimation equations and determining the variances and covariances of the estimators of the parameters.

We look first at Design 4.1.1. The model is the same as that used in Chapter 2, but now there are three periods, resulting in the following,

$$Y_{ijk} = \mu + \gamma_i + \xi_{i(k)} + \pi_j + \tau_t + \delta_j\lambda_r + \epsilon_{ijk} \qquad (4.1)$$

where $i,t,r = 1,2$, $j = 1,2,3$ and $k = 1,2,...,n_i$. The true cell means and the observed cell means can be denoted by μ_{ij} and $\overline{Y}_{ij}$ respectively, the subscripts identifying the ith sequence and the jth period. The observed cell means are represented as follows:

		Period		
		1	2	3
	1	$\overline{Y}_{11}$	$\overline{Y}_{12}$	$\overline{Y}_{13}$
Sequence				
	2	$\overline{Y}_{21}$	$\overline{Y}_{22}$	$\overline{Y}_{23}$

This leads to six equations in terms of model parameters as follows:

$$\mu_{11} = \mu + \gamma_1 + \pi_1 + \tau_1$$
$$\mu_{12} = \mu + \gamma_1 + \pi_2 + \tau_2 + \lambda_1$$
$$\mu_{13} = \mu + \gamma_1 + \pi_3 + \tau_2 + \lambda_2$$
$$\mu_{21} = \mu + \gamma_2 + \pi_1 + \tau_2$$
$$\mu_{22} = \mu + \gamma_2 + \pi_2 + \tau_1 + \lambda_2$$
$$\mu_{23} = \mu + \gamma_2 + \pi_3 + \tau_1 + \lambda_1.$$

Adopting the "sum–to–zero" convention, the relations

$$\gamma_2 = -\gamma_1, \quad \pi_3 = -\pi_1 - \pi_2, \quad \tau_2 = -\tau_1, \quad \text{and } \lambda_2 = -\lambda_1$$

transform the above six equations into the following set of equations, written in matrix form:

$$\mathbf{X}\beta = \begin{pmatrix} 1 & 1 & 1 & 0 & 1 & 0 \\ 1 & 1 & 0 & 1 & -1 & 1 \\ 1 & 1 & -1 & -1 & -1 & -1 \\ 1 & -1 & 1 & 0 & -1 & 0 \\ 1 & -1 & 0 & 1 & 1 & -1 \\ 1 & -1 & -1 & -1 & 1 & 1 \end{pmatrix} \begin{pmatrix} \mu \\ \gamma_1 \\ \pi_1 \\ \pi_2 \\ \tau_1 \\ \lambda_1 \end{pmatrix} = \begin{pmatrix} \mu_{11} \\ \mu_{12} \\ \mu_{13} \\ \mu_{21} \\ \mu_{22} \\ \mu_{23} \end{pmatrix}.$$

The portion of the inverse of the $X'X$ matrix pertaining to the treatment and carryover parameters τ_1 and λ_1 is

$$(X'X)^{-1} = \begin{pmatrix} 3/16 & 0 \\ \\ 0 & 1/4 \end{pmatrix}.$$

and the least–squares estimators of the parameters are

$$\begin{pmatrix} \hat{\mu} \\ \hat{\gamma}_1 \\ \\ \hat{\pi}_1 \\ \hat{\pi}_2 \\ \hat{\tau}_1 \\ \\ \hat{\lambda}_1 \end{pmatrix} = \begin{pmatrix} (\overline{Y}_{11}+\overline{Y}_{12}+\overline{Y}_{13}+\overline{Y}_{21}+\overline{Y}_{22}+\overline{Y}_{23})/6 \\ [3(\overline{Y}_{11}+\overline{Y}_{12}+\overline{Y}_{13}-\overline{Y}_{21}-\overline{Y}_{22}-\overline{Y}_{23}) + \\ \quad (\overline{Y}_{11}+\overline{Y}_{22}+\overline{Y}_{23}-\overline{Y}_{12}-\overline{Y}_{13}-\overline{Y}_{21})]/16 \\ (\overline{Y}_{11}+\overline{Y}_{21})/3 - (\overline{Y}_{12}+\overline{Y}_{13}+\overline{Y}_{22}+\overline{Y}_{23})/6 \\ (\overline{Y}_{12}+\overline{Y}_{22})/3 - (\overline{Y}_{11}+\overline{Y}_{13}+\overline{Y}_{21}+\overline{Y}_{23})/6 \\ [(\overline{Y}_{11}+\overline{Y}_{12}+\overline{Y}_{13}-\overline{Y}_{21}-\overline{Y}_{22}-\overline{Y}_{23}) + \\ \quad 3(\overline{Y}_{11}+\overline{Y}_{22}+\overline{Y}_{23}-\overline{Y}_{12}-\overline{Y}_{13}-\overline{Y}_{21})]/16 \\ (\overline{Y}_{12}-\overline{Y}_{13}-\overline{Y}_{22}+\overline{Y}_{23})/4 \end{pmatrix}.$$

The interpretation of the estimators and their covariances is interesting. It is seen that the estimator of the carryover parameter λ_1 is a rather simple contrast involving the cell means of readings in the second and third periods. The estimator of the treatment parameter, however, is not simply a contrast between cell means that received Treatment A versus those that received Treatment B. Although that contrast is part of the expression for $\hat{\tau}_1$, there is another term that accounts for the imbalance between sequence and treatment, a reflection of the fact that the two treatments do not appear equally often in the two sequences.

The estimators of the period parameters π_1 and π_2 are what might be expected from the fact that they are orthogonal to the other parameters of the design. The estimator of π_1 is a contrast of the cell means of the first period with those of the other periods, and the estimator of π_2 is a contrast of the cell means of the second period with those of the other periods. The estimator of the sequence parameter γ_1 involves not only the contrast between cell means in Sequence 1 versus those in Sequence 2, but there is also an additional term to reflect the imbalance of sequences with treatment. The estimator of μ is simply the mean of the six cell means.

One can set up the estimation equations for the other designs in this section in a similar fashion to the way it was done for Design 4.1.1. Results deriving from some of the key elements of the $(X'X)^{-1}$ matrix are shown in the following table for Designs 4.1.1, 4.1.2 and 4.1.3. Results for 4.1.4 are omitted, since the complete confounding of sequences and treatments in that design leads to a singular $X'X$ matrix.

	Design		
	4.1.1	4.1.2	4.1.3
$\text{Var}(\hat{\tau}_1)$	0.1875	0.750	0.250
$\text{Var}(\hat{\lambda}_1)$	0.2500	1.000	1.000
$\text{Corr}(\hat{\tau}_1,\hat{\lambda}_1)$	0.0	0.866	0.500

The numbers in the above table point to the conclusion that Design 4.1.1 is superior to the other three–period designs having two sequences. Not only are the variances of $\hat{\tau}_1$ and $\hat{\lambda}_1$ less for that design than for the other designs, the correlation between them is zero, making these effects orthogonal to each other. Thus, there is little doubt that Design 4.1.1 is the best among the class of designs with two treatments, three periods and two sequences. Indeed, Design 4.1.1 is the fully–efficient "extra treatment period" design for $n = 2$ in the class of extra–period designs proposed by Patterson and Lucas (1959). Further information on such designs is given in Section 2.1 and in Chapters 5 and 6.

We now consider a set of data by which means the analysis of Design 4.1.1 may be illustrated. The data set is from part of a clinical trial reported by Ebbutt (1984); that author used only the data for 10 subjects in each group, but the full data set was reported by Jones and Kenward (1989; pp. 156–7). The full data set contains results for four sequence groups, viz. ABB, BAA, ABA and BAB, but we will use only the data for the two sequence groups ABB and BAA.

These data are reproduced in Table 4.1. Baseline readings were also available for this data set, but as we shall see, their use scarcely improves the efficiency of estimation of the differential treatment parameter.

Table 4.1. Systolic blood pressures for a 2–treatment, 3–period, 2–sequence design. [Data of Ebbutt (1984), reproduced here with the permission of Dr. Alan Ebbutt and of The Biometric Society.]

Sequence ABB				Sequence BAA			
	Period				Period		
Subject	1	2	3	Subject	1	2	3
1	159	140	137	1	165	154	173
2	153	172	155	2	160	165	140
3	160	156	140	3	140	150	180
4	160	200	132	4	140	125	130
5	170	170	160	5	158	160	180
6	174	132	130	6	180	165	160
7	175	155	155	7	170	160	160
8	154	138	150	8	140	158	148
9	160	170	168	9	126	170	200
10	160	160	170	10	130	125	150
11	145	140	140	11	144	140	120
12	148	154	138	12	140	160	140
13	170	170	150	13	120	145	120
14	125	130	130	14	145	150	150
15	140	112	95	15	155	130	140
16	125	140	125	16	168	168	168
17	150	150	145	17	150	160	180
18	136	130	140	18	120	120	140
19	150	140	160	19	150	150	160
20	150	140	150	20	150	140	130
21	202	181	170	21	175	180	160
22	190	150	170	22	140	170	150
				23	150	160	130
				24	150	130	125
				25	140	150	160
				26	140	140	130
				27	126	140	138

We illustrate the analysis of the data in Table 4.1 using PROC GLM of SAS. Suggested code for the data step is as follows:

```
DATA ABBBAA;  /* Data of Ebbutt (1984) */
INPUT SEQUENCE SUBJECT PERIOD TREAT $ CARRY $ Y @@;
IF CARRY = '0' THEN CARRY = 'B';
CARDS;
1 1 1 A 0 159    1 1 2 B A 140    1 1 3 B B 137
1 2 1 A 0 153    1 2 2 B A 172    1 2 3 B B 155
1 3 1 A 0 160    1 3 2 B A 156    1 3 3 B B 140
```

/* Data from other subjects in sequence 1 appear here */

```
1 21 1 A 0 202      1 21 2 B A 181      1 21 3 B B 170
1 22 1 A 0 190      1 22 2 B A 150      1 22 3 B B 170
2  1 1 B 0 165      2  1 2 A B 154      2  1 3 A A 173
2  2 1 B 0 160      2  2 2 A B 165      2  2 3 A A 140
2  3 1 B 0 140      2  3 2 A B 150      2  3 3 A A 180
```

/* Data from other subjects in sequence 2 appear here */

```
2  26 1 B 0 140     2  26 1 A B 140     2  26 3 A A 130
2  27 1 B 0 126     2  27 2 A B 140     2  27 3 A A 138
RUN;
```

The statement

IF CARRY = '0' THEN CARRY = 'B';

is needed to ensure estimability of the effects in the LSMEANS statement in the following code.

```
PROC GLM;
CLASS SEQUENCE SUBJECT PERIOD TREAT CARRY;
MODEL Y = SEQUENCE SUBJECT(SEQUENCE) PERIOD TREAT
        CARRY / SOLUTION SS1 SS2 E1 E2;
RANDOM SUBJECT(SEQUENCE);
LSMEANS TREAT CARRY / PDIFF;
RUN;  .
```

The output from PROC GLM contains various information from which the following analysis of variance table can be constructed:

Table 4.2. ANOVA of data in Table 4.1 on systolic blood pressure.

Source of variation	df	Type I SS	Mean Squares	F	Pr > F
Between–subjects:					
Sequences	1	157.972	157.972	0.244	0.624
Residual	47	30490.586	648.736		
Within–subjects:					
Periods	2	275.878	137.939	0.801	0.452
Treatments	1	1133.586	1133.586	6.581	0.012
Carryover	1	173.062	173.062	1.005	0.319
Residual	94	16192.141	172.257		
Source of variation	df	Type II SS	Mean Squares	F	Pr > F
Treatments	1	1133.586	1133.586	6.581	0.012
Carryover	1	173.062	173.062	1.005	0.319

Note that the Type I and Type II sums of squares for treatments are identical. This is because $\hat{\tau}_1$ and $\hat{\lambda}_1$ are uncorrelated; in other designs, such as 4.1.2 or 4.1.3, the sequential (Type I) and adjusted (Type II) treatment sums of squares would be different. The probability 0.319, which is the chance of obtaining an F–value as extreme as or more extreme than the one observed, under the null hypothesis of no differential carryover effects, does not suggest that there are any significant carryover effects. The probability 0.012 associated with Treatments does suggest that there are treatment differences. The LSMEANS statement produced the following information on treatment means and differences:

TREAT	Y LSMEAN	PROB > |T| H0: LSMEAN1=LSMEAN2
A	153.8186	0.0119
B	147.8969	

Thus, the least–squares mean for treatment A is greater than that of treatment B by 5.922 units. The probability associated with the difference, 0.0119, is identical, except for the number of digits reported, with the probability associated with the test for treatment effects in the ANOVA table. This will always be the case when there are but two treatments.

Generally speaking, there is no valid test for period differences when period is modeled as a qualitative variable. The Type II Period sum of squares is always spurious in designs in which carryover effects are being estimated, due to the irresolvable confounding of carryover with first period effects. For that reason, Type II period sum of squares are not reported in the above table. However, when carryover effects prove to be non–significant, as is the case with the present data set, it is possible to use the Type I sum of squares to produce a statistical test that is only slightly inexact. From the above table, the F–value of 0.801 leads to a probability of 0.452, providing no evidence of significant period effects. Some readers may prefer to pool the non–significant carryover term with the error term to obtain a sum of squares of 16365.2 with 95 degrees of freedom, for a mean square of 172.265. The F–ratio for periods is then 0.801, which is indistinguishable from its value without pooling.

The ANOVA table shows an F-value of 0.244, and an associated probability of 0.624, for the test of sequence differences. This test is, in fact, not correct, as the ratio of the sequence mean square to the residual mean square between subjects does not have a central F–distribution under the null hypothesis of no sequence differences. The expected mean squares produced by PROC GLM for sequences is as follows:

Source	Type I Expected Mean Square
SEQUENCE	$\sigma^2 + 3\,\sigma_s^2 + Q(SEQUENCE,TREAT)$
SUBJECT(SEQUENCE)	$\sigma^2 + 3\,\sigma_s^2$

where σ_s^2 is the variance of subjects within sequences and $Q(SEQUENCE, TREAT)$ is a quadratic form involving both sequences and treatments. Had this test produced a significant value of F, it would not have been certain whether there really was a difference due to sequences, or whether it was due to the contribution from treatment differences. Since the test in this example is clearly non–significant, one can safely assume that the randomization of subjects to sequences was carried out in an appropriate fashion and that there is no preponderance of subjects in one sequence with characteristics that influence the effect of treatment. In general, however, readers should realize that it may not be possible to obtain an exact test for sequence differences.

We now look at the residuals from the fitted model (4.1) to see if they are normally distributed, using methods described in Section 1.6. First, one saves both the "raw" and the "studentized" residuals by inserting the following statement in the PROC GLM code after the model statement:

```
OUTPUT OUT=DRES STUDENT=STDRES R=RESID P=FITVAL;  .
```

Then a rough normal probability plot and some descriptive statistics, including a box plot and the Wilk–Shapiro statistic, may be obtained using the following code:

```
PROC UNIVARIATE DATA=DRES PLOT NORMAL;
   VAR RESID STDRES;
RUN;
```

The reason for saving both kinds of residuals is that the raw residuals RESID are probably the more appropriate ones to use with the Wilk–Shapiro statistic, which, for this data set, is 0.983 with a P–value of 0.594, indicating that the residuals are sufficiently close to being normally distributed. One can plot these residuals against the fitted values, using code such as

```
PROC PLOT DATA=DRES;
   PLOT RESID*FITVAL;
RUN;
```

from which one then can note that there appears to be a total of four potential outliers, two large positive ones and two large negative ones. If high quality graphics are available, one can obtain a quality plot of the standardized residuals STDRES against their expected normal order statistics, as described in Section 1.6. For the data set under consideration here on systolic blood pressures, the following normal probability plot was obtained (see top of next page). The majority of the 147 points fall on a good straight line. The two largest positive standardized residuals of 3.447 and 3.194, and the two largest negative standardized residuals of –3.526 and –2.527, are conspicuous on that graph. By printing out the variables in data set DRES, including the sequence, subject and period numbers as well as STDRES, one can determine that these four most outlying points come from just two experimental units, namely Subject 4 in Sequence 1 and Subject 9 in Sequence 2. Examining the data in Table 4.1 shows the inconsistency in having such discrepant values as 200 in Period 2 and 132 in Period 3 for Subject 4, Sequence 1, although Treatment B was applied in both cases. For Subject 9, Sequence 1, the large residuals result from the values of 126 in Period 1 and 200 in Period 3. The user has to make a decision whether to ignore this information, or to re–analyze the data after removal of these two experimental units. Since the Wilk–Shapiro statistic had a P–value of 0.594, far from such decision–making critical values as 0.10, 0.05 or 0.01, we choose not to delete those two units.

It was mentioned earlier that baseline measurements, that is, readings taken on each subject prior to the administration of treatment, were available for the data in Table 4.1. The use of baseline readings would make Design 4.1.1 appear

Normal Probability Plot

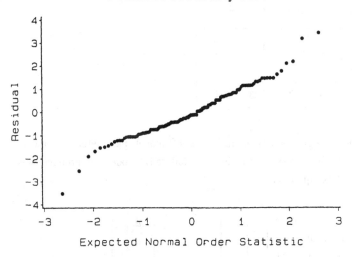

as follows, the hyphen (−) indicating that no treatment is given in the first period, but that a measurement of the response variable is made:

Design 4.1.1 With Baseline Measurements

	Period			
Seq.	1	2	3	4
↓				
1	−	A	B	B
2	−	B	A	A

Using model 4.1, and denoting true cell means by μ_{ij} and observed cell means by $\overline{Y}_{ij}$ as in the following table,

	Period		
	1	2	3	4

Sequence

1 | $\overline{Y}_{11}$ $\overline{Y}_{12}$ $\overline{Y}_{13}$ $\overline{Y}_{14}$

2 | $\overline{Y}_{21}$ $\overline{Y}_{22}$ $\overline{Y}_{23}$ $\overline{Y}_{24}$

one can write down eight equations in terms of model parameters, adopt a convention such as "sum–to–zero" to obtain a full–rank model, to produce the following set of matrix equations:

$$
X\beta = \begin{pmatrix}
1 & 1 & 1 & 0 & 0 & 0 & 0 \\
1 & 1 & 0 & 1 & 0 & 1 & 0 \\
1 & 1 & 0 & 0 & 1 & -1 & 1 \\
1 & 1 & -1 & -1 & -1 & -1 & -1 \\
1 & -1 & 1 & 0 & 0 & 0 & 0 \\
1 & -1 & 0 & 1 & 0 & -1 & 0 \\
1 & -1 & 0 & 0 & 1 & 1 & -1 \\
1 & -1 & -1 & -1 & -1 & 1 & 1
\end{pmatrix}
\begin{pmatrix}
\mu \\
\gamma_1 \\
\pi_1 \\
\pi_2 \\
\pi_3 \\
\tau_1 \\
\lambda_1
\end{pmatrix}
=
\begin{pmatrix}
\mu_{11} \\
\mu_{12} \\
\mu_{13} \\
\mu_{14} \\
\mu_{21} \\
\mu_{22} \\
\mu_{23} \\
\mu_{24}
\end{pmatrix} .
$$

The portion of the $(X'X)^{-1}$ matrix for the estimators of the parameters τ_1 and λ_1 enables one to write down the covariance of $\hat{\tau}_1$ and $\hat{\lambda}_1$, omitting the multiplying factor $\hat{\sigma}^2/n$:

$$
(X'X)^{-1} = \begin{pmatrix}
2/11 & 0 \\
0 & 1/4
\end{pmatrix} .
$$

This may be compared with the corresponding portion of the $(X'X)^{-1}$ matrix for those same parameter estimators without baseline readings, obtained earlier in this section:

$$(\mathbf{X}'\mathbf{X})^{-1} = \begin{pmatrix} 3/16 & 0 \\ 0 & 1/4 \end{pmatrix}.$$

The variances of $\hat{\lambda}_1$ are identical, so there is no increase in efficiency in the estimation of carryover effects. In both cases, the estimator of the treatment parameter is uncorrelated with that of the carryover parameter. Only the variances of the estimators of $\hat{\tau}_1$ differ, and that difference is only −0.00568 [= 2/11 − 3/16], an improvement in efficiency of about 3%. In view of this, there would be little reason to incorporate baseline readings into the analysis of the data. Laska *et al.* (1983) proved a general result showing that baseline readings [Note: they used the term "baseline" to mean readings taken before *each* treatment period] make little improvement in the precision of the estimators when three or more periods are used. Therefore, there seems little or no reason for utilizing baseline readings in the statistical analysis of data from clinical trials with more than two periods, although such readings can have other uses. For example, they can help stratify the subjects prior to randomization, to ensure that each sequence will contain persons of approximately the same average values of the response prior to the administration of the treatments.

4.2. Designs With Four Sequences and Two Periods

We will now consider a modification of the 2–treatment, 2–period, 2–sequence cross–over design which is a special case of two–period designs having t^2 sequences proposed by Balaam (1968), where t is the number of treatments. Thus, in the present case where $t = 2$, there are four sequences, the extra sequences being AA and BB, that is, in these two groups, the subjects receive the same treatment in Period 2 that they received in Period 1. The design therefore looks as follows:

Design 4.2

		Period	
		1	2
Sequence	1	A	B
	2	B	A
	3	A	A
	4	B	B

Using model (4.1) and representing the true cell means by μ_{ij} and observed cell means by $\overline{Y}_{ij}$ with the subscripts identifying the ith sequence and the jth period, one obtains the following observed cell means for the four sequences combined with the two periods:

$$
\begin{array}{c}
\text{Period} \\
\begin{array}{cc}
1 \qquad 2
\end{array}
\end{array}
$$

$$
\text{Sequence}
\begin{array}{c}
1 \\
2 \\
3 \\
4
\end{array}
\left[
\begin{array}{cc}
\overline{Y}_{11} & \overline{Y}_{12} \\
\overline{Y}_{21} & \overline{Y}_{22} \\
\overline{Y}_{31} & \overline{Y}_{32} \\
\overline{Y}_{41} & \overline{Y}_{42}
\end{array}
\right]
$$

This leads to eight equations, one each corresponding to each of the cell means. Although a maximum of eight parameters can be estimated, we restrict ourselves here to seven, viz. an overall grand mean μ, three sequence parameters γ_1, γ_2, and γ_3, a period parameter π_1, a treatment parameter τ_1, and a first–order carryover parameter λ_1. The eight linear equations, in terms of model parameters, are as follows:

$$
\begin{aligned}
\mu_{11} &= \mu + \gamma_1 + \pi_1 + \tau_1 \\
\mu_{12} &= \mu + \gamma_1 + \pi_2 + \tau_2 + \lambda_1 \\
\mu_{21} &= \mu + \gamma_2 + \pi_1 + \tau_2 \\
\mu_{22} &= \mu + \gamma_2 + \pi_2 + \tau_1 + \lambda_2 \\
\mu_{31} &= \mu + \gamma_3 + \pi_1 + \tau_1 \\
\mu_{32} &= \mu + \gamma_3 + \pi_2 + \tau_1 + \lambda_1 \\
\mu_{41} &= \mu + \gamma_4 + \pi_1 + \tau_2 \\
\mu_{42} &= \mu + \gamma_4 + \pi_2 + \tau_2 + \lambda_2.
\end{aligned}
$$

Adopting the "sum–to–zero" convention, the relations

$$
\gamma_4 = -\gamma_1 - \gamma_2 - \gamma_3, \quad \pi_2 = -\pi_1, \quad \tau_2 = -\tau_1, \text{ and } \lambda_2 = -\lambda_1
$$

transform the above eight equations into the following set of equations, written in matrix form:

$$\mathbf{X}\beta = \begin{pmatrix} 1 & 1 & 0 & 0 & 1 & 1 & 0 \\ 1 & 1 & 0 & 0 & -1 & -1 & 1 \\ 1 & 0 & 1 & 0 & 1 & -1 & 0 \\ 1 & 0 & 1 & 0 & -1 & 1 & -1 \\ 1 & 0 & 0 & 1 & 1 & 1 & 0 \\ 1 & 0 & 0 & 1 & -1 & 1 & 1 \\ 1 & -1 & -1 & -1 & 1 & -1 & 0 \\ 1 & -1 & -1 & -1 & -1 & -1 & -1 \end{pmatrix} \begin{pmatrix} \mu \\ \gamma_1 \\ \gamma_2 \\ \gamma_3 \\ \pi_1 \\ \tau_1 \\ \lambda_1 \end{pmatrix} = \begin{pmatrix} \mu_{11} \\ \mu_{12} \\ \mu_{21} \\ \mu_{22} \\ \mu_{31} \\ \mu_{32} \\ \mu_{41} \\ \mu_{42} \end{pmatrix}.$$

The inverse of the portion of the $\mathbf{X}'\mathbf{X}$ matrix pertaining to the treatment parameter τ_1 and the carryover parameter λ_1 is

$$(\mathbf{X}'\mathbf{X})^{-1} = \begin{pmatrix} 0.5 & 0.5 \\ 0.5 & 1.0 \end{pmatrix}.$$

and the least–square estimators of the parameters are

$$\begin{pmatrix} \hat{\mu} \\ \hat{\gamma}_1 \\ \hat{\gamma}_2 \\ \hat{\gamma}_3 \\ \hat{\pi}_1 \\ \hat{\tau}_1 \\ \hat{\lambda}_1 \end{pmatrix} = \begin{pmatrix} (\overline{Y}_{11}+\overline{Y}_{12}+\overline{Y}_{21}+\overline{Y}_{22}+\overline{Y}_{31}+\overline{Y}_{32}+\overline{Y}_{41}+\overline{Y}_{42})/8 \\ [3(\overline{Y}_{11}+\overline{Y}_{12}-\overline{Y}_{32}-\overline{Y}_{41}) - (\overline{Y}_{21}+\overline{Y}_{22}-\overline{Y}_{31}-\overline{Y}_{42})]/8 \\ [3(\overline{Y}_{21}+\overline{Y}_{22}-\overline{Y}_{31}-\overline{Y}_{42}) - (\overline{Y}_{11}+\overline{Y}_{12}-\overline{Y}_{32}-\overline{Y}_{41})]/8 \\ [7\overline{Y}_{31} - 5\overline{Y}_{41} + 3(\overline{Y}_{42}-\overline{Y}_{11}-\overline{Y}_{22}) + (\overline{Y}_{12}+\overline{Y}_{21}-\overline{Y}_{32})]/8 \\ [(\overline{Y}_{11}+\overline{Y}_{21}+\overline{Y}_{31}+\overline{Y}_{41}) - (\overline{Y}_{12}+\overline{Y}_{22}+\overline{Y}_{32}+\overline{Y}_{42})]/8 \\ [(\overline{Y}_{11}+\overline{Y}_{22}+\overline{Y}_{32}+\overline{Y}_{41}) - (\overline{Y}_{12}+\overline{Y}_{21}+\overline{Y}_{31}+\overline{Y}_{42})]/4 \\ (\overline{Y}_{32}+\overline{Y}_{41}-\overline{Y}_{31}-\overline{Y}_{42})/2 \end{pmatrix}.$$

Interestingly, the estimator of the carryover parameter λ_1 involves only cell means from the third and fourth sequences. Still more intriguing is the fact that the estimator of the treatment parameter τ_1 is not a simple contrast of means from cells having had Treatment A against means from cells having had Treatment B. Instead the means $\overline{Y}_{31}$ and $\overline{Y}_{41}$ seem to be in the "wrong" places. The reason for these apparently surprising results is due to the fact that the estimators of the treatment and carryover parameters are correlated with the estimators of the sequence parameters. On the other hand, the estimator of the period parameter π_1, which is uncorrelated with the estimators of any other parameter, is a simple contrast of first period results with second period results. The estimators of the sequence parameters, as might be expected from the above discussion, are rather complicated contrasts. The estimator of μ is uncorrelated with the estimators of the other parameters and is simply the mean of all of the eight observed cell means.

From the elements of the $(X'X)^{-1}$ matrix, the correlation of $\hat{\tau}_1$ with $\hat{\lambda}_1$ is $0.5/(0.5 \cdot 1)^{1/2} = 0.707$, a rather high value despite the apparent balance in this design resulting from the fact that Treatment A carries over into itself as frequently as it carries over into Treatment B (and similarly B carries over into itself as frequently as it carries over into A). This correlation of 0.707 is identical to that between $\hat{\tau}_1$ and $\hat{\lambda}_1$ in Parameterization 1 (Section 3.1) of the 2–treatment, 2–period, 2–sequence design, that is, the present design without the addition of the two extra sequences AA and BB. It appears that little benefit has actually been derived from the use of the extra sequences AA and BB. It is true that in comparison to the basic "Grizzle/ Hills–Armitage" design, the Balaam design is estimable without the necessity of making untestable assumptions, but the basic problem of poor separability of "direct" from "residual" treatment effects still remains.

Comparisons of efficiency can now be made between Balaam's Design 4.2 of this section and the optimum extra–period Design 4.1.1 of Section 4.1 using information from the $(X'X)^{-1}$ matrices for the two designs. It must be remembered that the elements of these matrices need to be multiplied by a factor $\hat{\sigma}^2/n$ to obtain variances and covariances of the estimators. Since Design 4.2 has twice as many sequences as Design 4.1.1, a comparison between the two designs, based on the total number of experimental subjects used, would require that each sequence of Design 4.2 use half as many subjects as each sequence of Design 4.1.1. Thus, to compare the efficiencies of the two designs, it is necessary to multiply the variances and covariances of Design 4.2 by a factor of 2. The following table makes this comparison. The variances and covariances for Design 4.1.1 are taken directly from results in Section 4.1; the corresponding values for Design 4.2 are obtained from the results in this section, after multiplication by a factor of 2.

	Design 4.1.1	Design 4.2
$\text{Var}(\hat{\tau}_1)$	0.1875	1.0
$\text{Var}(\hat{\lambda}_1)$	0.25	2.0
$\text{Cov}(\hat{\tau}_1, \hat{\lambda}_1)$	0.0	1.0

The conclusions from this comparison are striking. The extra–period Design 4.1.1 is much more efficient than the extra–sequences Design 4.2.

However, the efficiency of Design 4.2 can be improved considerably by the inclusion of baseline observations prior to the first treatment period. Although Laska *et al.* (1983) proved that baseline measurements scarcely improved the efficiency of cross–over designs with three or more periods, a two–period design has an opportunity to increase its efficiency significantly.

The inclusion of baseline readings would make the altered design appear as follows, the hyphen (–) indicating that no treatment is given in the first period, but that a measurement of the response variable is made:

Design 4.2 With Baseline Measurements

		Period 1	2	3
	1	–	A	B
Sequence	2	–	B	A
	3	–	A	A
	4	–	B	B

Using model 4.1, and denoting the observed cell means as in the following table,

Period

	1	2	3
1	$\overline{Y}_{11}$	$\overline{Y}_{12}$	$\overline{Y}_{13}$
Sequence 2	$\overline{Y}_{21}$	$\overline{Y}_{22}$	$\overline{Y}_{23}$
3	$\overline{Y}_{31}$	$\overline{Y}_{32}$	$\overline{Y}_{33}$
4	$\overline{Y}_{41}$	$\overline{Y}_{42}$	$\overline{Y}_{43}$

one can write down twelve equations in terms of model parameters, adopt a convention such as "sum–to–zero" to obtain a full–rank model, to produce the following set of matrix equations:

$$
\mathbf{X\beta} =
\begin{pmatrix}
1 & 1 & 0 & 0 & 1 & 0 & 0 & 0 \\
1 & 1 & 0 & 0 & 0 & 1 & 1 & 0 \\
1 & 1 & 0 & 0 & -1 & -1 & -1 & 1 \\
1 & 0 & 1 & 0 & 1 & 0 & 0 & 0 \\
1 & 0 & 1 & 0 & 0 & 1 & -1 & 0 \\
1 & 0 & 1 & 0 & -1 & -1 & 1 & -1 \\
1 & 0 & 0 & 1 & 1 & 0 & 0 & 0 \\
1 & 0 & 0 & 1 & 0 & 1 & 1 & 0 \\
1 & 0 & 0 & 1 & -1 & -1 & 1 & 1 \\
1 & -1 & -1 & -1 & 1 & 0 & 0 & 0 \\
1 & -1 & -1 & -1 & 0 & 1 & -1 & 0 \\
1 & -1 & -1 & -1 & -1 & -1 & -1 & -1
\end{pmatrix}
\begin{pmatrix}
\mu \\ \gamma_1 \\ \gamma_2 \\ \gamma_3 \\ \pi_1 \\ \pi_2 \\ \tau_1 \\ \lambda_1
\end{pmatrix}
=
\begin{pmatrix}
\mu_{11} \\ \mu_{12} \\ \mu_{13} \\ \mu_{21} \\ \mu_{22} \\ \mu_{23} \\ \mu_{31} \\ \mu_{32} \\ \mu_{33} \\ \mu_{41} \\ \mu_{42} \\ \mu_{43}
\end{pmatrix}.
$$

The portion of the $(\mathbf{X'X})^{-1}$ matrix for the estimators of the parameters τ_1 and λ_1 enables one to write down the covariance of $\hat{\tau}_1$ and $\hat{\lambda}_1$, omitting the multiplying factor $\hat{\sigma}^2/n$:

$$
\mathrm{Cov}(\hat{\tau}_1, \hat{\lambda}_1) =
\begin{pmatrix}
3/14 & 3/28 \\
3/28 & 3/7
\end{pmatrix}.
$$

This may be compared with the corresponding portion of the $(X'X)^{-1}$ matrix for those same parameter estimators without baseline readings, again omitting the factor $\hat{\sigma}^2/n$ (see material earlier in this section):

$$\text{Cov}(\hat{\tau}_1,\hat{\lambda}_1) = \begin{pmatrix} 1/2 & 1/2 \\ \\ 1/2 & 1 \end{pmatrix}.$$

Thus, the use of baseline readings reduced the variance of $\hat{\tau}_1$ from 0.5 to 0.214, a reduction of 57%. Similarly, the variance of $\hat{\lambda}_1$ was also reduced by 57%. The covariance of $\hat{\tau}_1$ and $\hat{\lambda}_1$ experienced a 78.6% reduction, from 0.5 to 0.107. The correlation between $\hat{\tau}_1$ and $\hat{\lambda}_1$ is now 0.354, or half its value of 0.707 when baseline measurements are not used. The efficiencies of Design 4.2 with baselines, compared with Design 4.1.1, is shown in the following table, where the variances and covariances of the estimators in Design 4.2 are multiplied by a factor of 2 to take account of the extra sequences, each of which contain half as many subjects as the sequences of Design 4.1.1.

	Design 4.1.1	Design 4.2 (with baselines)
$\text{Var}(\hat{\tau}_1)$	0.1875	0.429
$\text{Var}(\hat{\lambda}_1)$	0.25	0.857
$\text{Cov}(\hat{\tau}_1,\hat{\lambda}_1)$	0.0	0.214

Therefore Design 4.1.1 is considerably more efficient than Design 4.2, even when the latter design uses baselines. Since both designs require the subjects to have a response variable measured at three time periods, there seems to be little role for Design 4.2 in cross-over experimentation.

We now present a data set and illustration of how one can analyze data from Design 4.2. The data set is taken from Taka and Armitage (1983). The trial involved comparing the effect of amantadine hydrochloride (Treatment A) with a placebo (Treatment B) on subjects suffering from Parkinsonism. The

Table 4.3. Average scores for subjects in amantadine trial. [Data from Taka and Armitage (1983), reproduced here with the permission of Marcel Dekker, Inc.]

Sequence	Subject	Period		
		1	2	3
–AB	1	9	8.75	8.75
	2	12	10.50	9.75
	3	17	15.00	18.50
	4	21	21.00	21.50
–BA	1	23	22.00	18.00
	2	15	15.00	13.00
	3	13	14.00	13.75
	4	24	22.75	21.50
	5	18	17.75	16.75
–AA	1	14	12.50	14.00
	2	27	24.25	22.50
	3	19	17.25	16.25
	4	30	28.25	29.75
–BB	1	21	20.00	19.51
	2	11	10.50	10.00
	3	20	19.50	20.75
	4	25	22.50	23.50

data, which also contain baseline readings, are given in Table 4.3. We illustrate the analysis of these data by using PROC GLM of SAS©. Suggested code for the data step is as follows:

```
DATA AMANTAD; /* Data from Taka and Armitage (1983) */
INPUT SEQUENCE SUBJECT PERIOD TREAT $ CARRY $ RESP @@;
IF TREAT = '0' THEN TREAT = 'B';
IF CARRY = '0' THEN CARRY = 'B';
CARDS;
3 1 1 0 0 9      3 1 2 A 0  8.75     3 1 3 B A  8.75
3 2 1 0 0 12     3 2 2 A 0 10.50     3 2 3 B A  9.75
3 3 1 0 0 17     3 3 2 A 0 15.00     3 3 3 B A 18.50
3 4 1 0 0 21     3 4 2 A 0 21.00     3 4 3 B A 21.50
4 1 1 0 0 23     4 1 2 B 0 22.00     4 1 3 A B 18.00
4 2 1 0 0 15     4 2 2 B 0 15.00     4 2 3 A B 13.00
4 3 1 0 0 13     4 3 2 B 0 14.00     4 3 3 A B 13.75
4 4 1 0 0 24     4 4 2 B 0 22.75     4 4 3 A B 21.50
4 5 1 0 0 18     4 5 2 B 0 17.75     4 5 3 A B 16.75
1 1 1 0 0 14     1 1 2 A 0 12.50     1 1 3 A A 14.00
1 2 1 0 0 27     1 2 2 A 0 24.25     1 2 3 A A 22.50
```

```
1 3 1 0 0 19      1 3 2 A 0 17.25      1 3 3 A A 16.25
1 4 1 0 0 30      1 4 2 A 0 28.25      1 4 3 A A 29.75
2 1 1 0 0 21      2 1 2 B 0 20.00      2 1 3 B B 19.51
2 2 1 0 0 11      2 2 2 B 0 10.50      2 2 3 B B 10.00
2 3 1 0 0 20      2 3 2 B 0 19.50      2 3 3 B B 20.75
2 4 1 0 0 25      2 4 2 B 0 22.50      2 4 3 B B 23.50
```

The statement

IF TREAT = '0' THEN TREAT = 'B';

is needed to ensure estimability of the treatment effects in the LSMEANS statement in the following code, and the statement

IF CARRY = '0' THEN CARRY = 'B';

is needed for estimability of the carryover effects. The PDIFF option produces probabilities associated with the differences between the treatment levels.

```
PROC GLM;
CLASS SEQUENCE SUBJECT PERIOD TREAT CARRY;
MODEL RESP=SEQUENCE SUBJECT(SEQUENCE) PERIOD TREAT
      CARRY / SOLUTION SS1 SS2;
RANDOM SUBJECT(SEQUENCE);
TEST H=SEQUENCE E=SUBJECT(SEQUENCE)/HTYPE=1 ETYPE=1;
LSMEANS TREAT CARRY / PDIFF;
RUN;
```

The output from PROC GLM contains various information from which a analysis of variance table can be constructed as follows:

Table 4.4. ANOVA of data in Table 4.3 on amantadine trial.

Source of variation	df	Type I SS	Mean Squares	F	Pr > F
Between–subjects:					
Sequences	3	285.818	95.273	1.014	0.418
Residual	13	1221.495	93.961		
Within–subjects:					
Periods	2	15.125	7.563	7.676	0.002
Treatments	1	8.483	8.483	8.610	0.006
Carryover	1	0.114	0.114	0.115	0.737
Residual	30	29.557	0.985		

Source of variation	df	Type II SS	Mean Squares	F	Pr > F
Treatments	1	6.525	6.525	6.623	0.015
Carryover	1	0.114	0.114	0.115	0.737

The probability 0.737, which is the chance of obtaining an F–value as extreme as or more extreme than the one observed, under the null hypothesis of no differential carryover effects, provides no evidence of significant carryover effects. The probability 0.015 does suggest, however, that there are treatment differences. The LSMEANS statement produced the following information on treatment means and differences:

```
TREAT    RESP    PROB > |T|   H0:
         LSMEAN  LSMEAN1=LSMEAN2

A        17.3216    0.0153
B        18.4717
```

The least–squares mean for Treatment A is less than that for Treatment B by 1.150 units. Because there are only two treatment levels, the probability 0.0153 associated with this difference is identical, except for the number of digits reported, with the probability associated with the test for treatment effects in the ANOVA table.

As has been mentioned previously, there is no valid test for period differences, the Type II sum of squares, mean squares, and F–ratio being spurious as a result of carryover being in the model. However, in this example, since carryover is not significant, it is possible to use the Type I sum of squares to perform a test which is not far removed from exactness. From the ANOVA table, the F–value of 7.676 leads to a probability of 0.002, suggesting that there are significant period effects. One will obtain the same inference if one pools the non–significant carryover term with the error term to obtain a sum of squares of 29.671 with 31 degrees of freedom, for a mean square of 0.957. The F–ratio is then 7.902, which is not greatly different from 7.676. To interpret the significant period effects, it is sufficient to look at a table of means for the three periods, as follows:

Period	Sample Size	Mean response
1	17	18.765
2	17	17.735
3	17	17.515

It appears that the period effect is being caused by a reduction in average response score in Periods 2 and 3 with respect to the baseline readings.

The ANOVA table shows an F–value of 1.014 and an associated probability of 0.418 for the test of sequence differences. This test is not valid as the ratio of the sequence mean square to the residual mean square between subjects is not distributed as a central F–distribution, under the null hypothesis of no difference. The expected mean squares produced by PROC GLM for sequences is as follows:

Source   Type I Expected Mean Square

SEQUENCE $\sigma^2 + 3\,\sigma_s^2 + Q(\text{SEQUENCE},\text{TREAT},\text{CARRY})$

SUBJECT(SEQUENCE) $\sigma^2 + 3\,\sigma_s^2$

where σ_s^2 is the variance of subjects within sequences and Q(SEQUENCE, TREAT, CARRY) is a quadratic form involving sequence, treatment and carryover. Had this test produced a significant value of F, it would not have been certain whether there really was a difference due to sequences, or whether it was due to contributions from treatment and carryover differences, especially as the treatment difference was significant. Since the test in this example leads to a clearly non–significant result, one can safely assume that the randomization of subjects to sequences was carried out in an appropriate fashion. It is important for readers to realize that, in general, it may not be possible to obtain an exact test for sequence differences.

4.3. Designs With Four Sequences and Three Periods

The two previous sections of this chapter have considered modifications to the 2–treatment, 2–period, 2–sequence design that improve the estimability and efficiency of that design while retaining just two treatments. The design with two sequences of three periods, viz. ABB and BAA, considered in Section 4.1, was shown to result in complete separation of the treatment and carryover parameters, that is, $\text{Corr}(\tau_1,\lambda_1) = 0$. Complete orthogonality in that design was prevented only by the imbalance between sequence and treatment occasioned by the fact that with an odd number of periods it is impossible for each of the two treatments to appear an equal number of times within a sequence. Section 4.2 considered a design with four sequences of two periods, but this design is decidedly inferior to ABB and BAA, the best 2–sequence, 3–period design of

Section 4.1. Unless it is impossible or unreasonable to have the subjects in the trial participate for more than two treatment periods, that design, Design 4.1.1 of Section 4.1, is recommended. We now examine whether there is any advantage to increasing the number of sequences from two to four, or from two to six, etc., by combining some of the designs of Section 4.1.

Dual–balanced designs of four sequences and three periods can be constructed by combining the designs that appear in Section 4.1. Since Design 4.1.4, which consists of sequences having the same treatment for all three periods, is highly inefficient, this design need not be considered, but Designs 4.1.1, 4.1.2, and 4.1.3 may be joined up to produce the following 4–sequence designs for consideration:

Design 4.3.1

Seq. ↓	Period 1	2	3
1	A	B	B
2	B	A	A
3	A	B	A
4	B	A	B

Design 4.3.2

Seq. ↓	Period 1	2	3
1	A	B	B
2	B	A	A
3	A	A	B
4	B	B	A

Design 4.3.3

Seq. ↓	Period 1	2	3
1	A	B	A
2	B	A	B
3	A	A	B
4	B	B	A

Because the treatments do not appear the same number of times within each sequence, these designs cannot be assessed by use of the omnibus measure of separability described in Section 1.2. Therefore, to determine the efficiencies of these designs requires examination of the variances and covariances of the estimators. The statistical model for these designs is the same as has been used throughout these chapters, viz.

$$Y_{ijk} = \mu + \gamma_i + \xi_{i(k)} + \pi_j + \tau_t + \delta_j\lambda_r + \epsilon_{ijk}$$

where $i=1,...,4$, $j=1,...,3$, $t,r=1,2$, and $k=1,2,...,n_i$. Representing the observed cell means by $\overline{Y}_{ij}$, the subscripts identifying the ith sequence and the jth period, one has, from a set of data for any of Designs 4.3.1 – 4.3.3,

		Period		
		1	2	3
Sequence	1	$\overline{Y}_{11}$	$\overline{Y}_{12}$	$\overline{Y}_{13}$
	2	$\overline{Y}_{21}$	$\overline{Y}_{22}$	$\overline{Y}_{23}$
	3	$\overline{Y}_{31}$	$\overline{Y}_{32}$	$\overline{Y}_{33}$
	4	$\overline{Y}_{41}$	$\overline{Y}_{42}$	$\overline{Y}_{43}$

For Design 4.3.1, use of the usual representation leads to the system of 12 linear equations which may be expressed in matrix form as

$$\mathbf{X}\beta = \begin{pmatrix} 1 & 1 & 0 & 0 & 1 & 0 & 1 & 0 \\ 1 & 1 & 0 & 0 & 0 & 1 & -1 & 1 \\ 1 & 1 & 0 & 0 & -1 & -1 & -1 & -1 \\ 1 & 0 & 1 & 0 & 1 & 0 & -1 & 0 \\ 1 & 0 & 1 & 0 & 0 & 1 & 1 & -1 \\ 1 & 0 & 1 & 0 & -1 & -1 & 1 & 1 \\ 1 & 0 & 0 & 1 & 1 & 0 & 1 & 0 \\ 1 & 0 & 0 & 1 & 0 & 1 & -1 & 1 \\ 1 & 0 & 0 & 1 & -1 & -1 & 1 & -1 \\ 1 & -1 & -1 & -1 & 1 & 0 & -1 & 0 \\ 1 & -1 & -1 & -1 & 0 & 1 & 1 & -1 \\ 1 & -1 & -1 & -1 & -1 & -1 & -1 & 1 \end{pmatrix} \begin{pmatrix} \mu \\ \gamma_1 \\ \gamma_2 \\ \gamma_3 \\ \pi_1 \\ \pi_2 \\ \tau_1 \\ \lambda_1 \end{pmatrix} = \begin{pmatrix} \mu_{11} \\ \mu_{12} \\ \mu_{13} \\ \mu_{21} \\ \mu_{22} \\ \mu_{23} \\ \mu_{31} \\ \mu_{32} \\ \mu_{33} \\ \mu_{41} \\ \mu_{42} \\ \mu_{43} \end{pmatrix}.$$

The elements of the inverse of the $(\mathbf{X'X})^{-1}$ matrix, except for the multiplying factor $\hat{\sigma}^2/n$, relating to the parameters τ_1 and λ_1 are given by

$$\text{Cov}(\hat{\tau}_1,\hat{\lambda}_1) = \begin{pmatrix} 0.1154 & 0.0577 \\ 0.0577 & 0.1538 \end{pmatrix}$$

which leads to a correlation coefficient between $\hat{\tau}_1$ and $\hat{\lambda}_1$ of $0.0577/(0.1154 \cdot 0.1538)^{1/2} = 0.433$.

For Design 4.3.2, following a similar approach leads to

$$\text{Cov}(\hat{\tau}_1,\hat{\lambda}_1) = \begin{pmatrix} 0.0968 & 0.0242 \\ 0.0242 & 0.1935 \end{pmatrix}$$

with a correlation coefficient of $0.0242/(0.0968 \cdot 0.1935)^{1/2} = 0.177$.

For Design 4.3.3, a similar approach gives

$$\text{Cov}(\hat{\tau}_1,\hat{\lambda}_1) = \begin{pmatrix} 0.1875 & 0.1875 \\ 0.1875 & 0.3750 \end{pmatrix}$$

with a correlation coefficient of $0.1875/(0.1875 \cdot 0.3750)^{1/2} = 0.707$.

We now have the information to evaluate Designs 4.3.1 – 4.3.3 relative to each other. The major considerations are the relative values of the variances of $\hat{\tau}_1$ and $\hat{\lambda}_1$, and their correlation. These are summarized for easy comparison in the table below.

	Design 4.3.1	Design 4.3.2	Design 4.3.3
$\text{Var}(\hat{\tau}_1)$	0.1154	0.0968	0.1875
$\text{Var}(\hat{\lambda}_1)$	0.1538	0.1935	0.3750
$\text{Corr}(\hat{\tau}_1,\hat{\lambda}_1)$	0.433	0.177	0.707

Design 4.3.3 clearly emerges as the worst of the three designs, but the question of which of 4.3.1 or 4.3.2 is best is less obvious. Design 4.3.2 has a smaller variance for the treatment parameter and a much smaller correlation

between the treatment and carryover parameters, but Design 4.3.1 has a smaller variance for the carryover parameter. Since the treatment parameter is presumably of more interest than the carryover parameter, that fact, coupled with the lower correlation coefficient for Design 4.3.2 probably swings the balance in its favor. Another advantage of Design 4.3.2 is that if the experiment has to be terminated for some reason before the third treatment period can be carried out, the design collapses to

		Period 1 2
Sequence	1	A B
	2	B A
	3	A A
	4	B B

which is the Balaam design of Section 4.2. Design 4.3.1, on the other hand, collapses to

		Period 1 2
Sequence	1	A B
	2	B A
	3	A B
	4	B A

which, recognizing that Sequences 1 and 3 are equivalent, and that Sequences 2 and 4 are equivalent, is the same as the 2–treatment, 2–period, 2–sequence (Grizzle/Hills–Armitage) design. Despite the imperfections of the Balaam design, it is nevertheless superior to the Grizzle/Hills–Armitage design, as described in Section 4.2.

The question must be asked about how Design 4.3.2 compares in efficiency with that of Design 4.1.1, which uses only two sequences. The variances of the estimators of τ_1 and λ_1 in both designs can readily be compared using results already obtained, after making a correction for sample size. Assuming that the same total number of subjects N would be available for an experiment, Design 4.1.1, with two sequences, would have N/2 subjects allocated to each sequence.

Design 4.3.2, with four sequences, would have N/4 subjects allocated to each sequence. Since the variances of the estimators are given by the elements of the $(\mathbf{X}'\mathbf{X})^{-1}$ matrix multiplied by $\hat{\sigma}^2/n$, where n is the number of subjects in each sequence, one has only to double the variances and covariances of the estimators in Design 4.3.2 to compare them with those of Design 4.1.1. These results are summarized in the table below:

	Design 4.1.1	Design 4.3.2
$Var(\hat{\tau}_1)$	0.1875	0.1935
$Var(\hat{\lambda}_1)$	0.250	0.3870
$Cov(\hat{\tau}_1,\hat{\lambda}_1)$	0.0	0.0484

These results show that although the variance of $\hat{\tau}_1$ is only slightly less in Design 4.1.1 than in Design 4.3.2, the variance of $\hat{\lambda}_1$ is considerably less in the 2–sequence design, and furthermore, the estimators $\hat{\tau}_1$ and $\hat{\lambda}_1$ are uncorrelated. Thus, there really is no reason to use any of the 4–sequence, 3–period designs considered in this section. Of course, it is necessary to point out that, in this chapter as in previous chapters, we have confined ourselves to carryover effects only of the first order, that is, allowing the residual effect of a treatment in a treatment period to persist only into the next treatment period and not beyond. The justification for this is that it is difficult to find practical data sets for which a second–order carryover effect can be demonstrated. It can also reasonably be argued that if one is in an experimental situation where a second–order or higher order carryover effect is likely to occur, then one probably needs to modify the experimental conditions.

4.4. Other Designs With More Than Two Periods and/or Two Sequences

Sections 4.1–4.3 explored various possibilities of extending the basic 2–treatment, 2–period, 2–sequence design to overcome the problem of over–parameterization that was inherent in that design. In Section 4.1, the use of an extra treatment period was studied, and it was found that this provides an excellent solution to the basic problem, but in that trial the subjects have to report for three treatment periods. It often happens in clinical trials that subjects drop out of the trial before the completion of the experiment. Dropouts are caused by a variety of factors; one factor may relate to a perception on the part of some subjects that if they have received no benefit from the treatments in the initial treatment periods, they may feel that the trial is not worth pursuing. For other subjects, there may be time conflicts which prevent them from attending on more than two occasions. If dropouts result in having to purge all previous records for those subjects at the data analysis stage, as is often the case, much potentially useful information is wasted.

Design 4.2 retained the two periods of the basic 2 x 2 design, using four sequences instead of two. That design did not fare well compared with the extra period design of Section 4.1. The correlation of $\hat{\tau}_1$ with $\hat{\lambda}_1$ was a high 0.707, and the variances of $\hat{\tau}_1$ and $\hat{\lambda}_1$ were also high.

In this section, we examine further modifications to these designs, subject to the condition that the number of treatments remains at two. One such design involves use of six sequences of three periods, made up by joining together Designs 4.1.1 – 4.1.3 as follows:

Design 4.4.1

Seq.	Period 1	2	3
1	A	B	B
2	B	A	A
3	A	B	A
4	B	A	B
5	A	A	B
6	B	B	A

Setting up 18 linear equations in terms up the unknown parameters, following the procedure frequently employed in this chapter, one can obtain from the appropriate elements of the $(X^{'}X)^{-1}$ matrix, the following estimates of the variances and covariances of $\hat{\tau}_1$ and $\hat{\lambda}_1$ (except for a factor $\hat{\sigma}^2/n$),

$$\text{Cov}(\hat{\tau}_1,\hat{\lambda}_1) = \begin{pmatrix} 0.0772 & 0.0441 \\ 0.0441 & 0.1324 \end{pmatrix} .$$

To make a comparison to the variances and covariances for $\hat{\tau}_1$ and $\hat{\lambda}_1$ for Design 4.1.1, the above values need to be multiplied by a factor of 3. This leads to the following table, where results for Design 4.3.2 are also included for comparison.

	Design 4.1.1	Design 4.3.2	Design 4.4.1
$\text{Var}(\hat{\tau}_1)$	0.1875	0.1935	0.2316
$\text{Var}(\hat{\lambda}_1)$	0.250	0.3870	0.3971
$\text{Cov}(\hat{\tau}_1,\hat{\lambda}_1)$	0.0	0.0484	0.1324

From this table, it is clear that Design 4.4.1 is inferior to the much simpler extra–period Design 4.1.1, and is also not as good as Design 4.3.2, the best of the 3–period designs with four sequences. Intuitively, it does not appear that adding extra sequences can improve on the 2–sequence Design 4.1.1.

We now examine the consequences of having of fourth period in the design, and try to deduce some general conclusions from the statistical examination of such designs. Consider the eight possible dual–balanced designs having two sequences of four periods.

Design 4.4.2

Seq. ↓	Period 1	2	3	4
1	A	B	A	B
2	B	A	B	A

Design 4.4.3

Seq. ↓	Period 1	2	3	4
1	A	B	B	A
2	B	A	A	B

Design 4.4.4

Seq. ↓	Period 1	2	3	4
1	A	B	B	B
2	B	A	A	A

Design 4.4.5

Seq. ↓	Period 1	2	3	4
1	A	A	B	B
2	B	B	A	A

(continued on next page)

Design 4.4.6

Period
Seq.↓	1	2	3	4
1	A	B	A	A
2	B	A	B	B

Design 4.4.7

Period
Seq.↓	1	2	3	4
1	A	A	B	A
2	B	B	A	B

Design 4.4.8

Period
Seq.↓	1	2	3	4
1	A	A	A	B
2	B	B	B	A

Design 4.4.9

Period
Seq.↓	1	2	3	4
1	A	A	A	A
2	B	B	B	B

Using Model 4.1, as we have done throughout this chapter, but with the period index j ranging from 1 to 4, and representing the observed cell means by $\overline{Y}_{ij}$, the subscripts identifying the ith sequence and jth period, a data set can be represented as follows:

Period

Sequence	1	2	3	4
1	$\overline{Y}_{11}$	$\overline{Y}_{12}$	$\overline{Y}_{13}$	$\overline{Y}_{14}$
2	$\overline{Y}_{21}$	$\overline{Y}_{22}$	$\overline{Y}_{23}$	$\overline{Y}_{24}$

The eight resulting linear equations in terms of model parameters can be expressed in matrix form, and the elements of the $(\mathbf{X}'\mathbf{X})^{-1}$ matrix obtained, as has been illustrated many times in this chapter. When the parameter vector consists of μ, π_1, π_2, π_3, τ_1, and λ_1, the resulting variances and covariances (except for a multiplying factor $\hat{\sigma}^2/n$) of the estimators of the treatment parameter τ_1 and the carryover parameter λ_1, and their correlation coefficient, may be summarized in the following table.

Design		$\text{Var}(\hat{\tau}_1)$	$\text{Var}(\hat{\lambda}_1)$	$\text{Cov}(\hat{\tau}_1,\hat{\lambda}_1)$	$\text{Corr}(\hat{\tau}_1,\hat{\lambda}_1)$
4.4.2	ABAB/BABA	0.6875	1.0000	0.7500	0.905
4.4.3	ABBA/BAAB	0.1375	0.2000	0.0500	0.302
4.4.4	ABBB/BAAA	0.1719	0.1875	−0.0313	−0.174
4.4.5	AABB/BBAA	0.1375	0.2000	−0.0500	−0.302
4.4.6	ABAA/BABB	0.2292	0.2500	0.1250	0.522
4.4.7	AABA/BBAB	0.2292	0.2500	0.1250	0.522
4.4.8	AAAB/BBBA	0.1875	0.7500	0.1250	0.333
4.4.9	AAAA/BBBB	∞	∞	∞	−

The worst designs are 4.4.2 and 4.4.9, the latter giving rise to a singular matrix as a result of the total confounding of sequence and treatment effects. The best designs are to be found among 4.4.3, 4.4.4 and 4.4.5. Design 4.4.4 has the smallest variance for $\hat{\lambda}_1$ as well as the smallest correlation between $\hat{\tau}_1$ and $\hat{\lambda}_1$, but it has a slightly larger variance for $\hat{\tau}_1$. Note that Designs 4.4.3 and 4.4.4 are both extensions of the extra-period Design 4.1.1 of Section 4.1. Since that design had $\text{Var}(\hat{\tau}_1) = 0.1875$ and $\text{Var}(\hat{\lambda}_1) = 0.250$, any one of the three Designs 4.4.3 − 4.4.5 is slightly more efficient. However, the subjects are required to take part in an extra treatment period, and it is doubtful whether the gain in precision justifies the extra effort required.

Designs 4.4.2, 4.4.3, and 4.4.5 each have both treatments appearing twice in each sequence. Hence, the measure of separability S defined in Section 1.2 may be used. This leads to the following values:

Design		V	Separability, S
4.4.2	ABAB/BABA	0.866	13.4%
4.4.3	ABBA/BAAB	0.289	71.1
4.4.5	AABB/BBAA	0.289	71.1

The interpretation of these results is consistent with previous conclusions. Design 4.4.2 is rather a poor one for separating treatment and carryover effects because of the alternation of Treatments A and B in the sequences ABAB and BABA, whereas Designs 4.4.3 and 4.4.5 are virtually identical in performance and are relatively good ones for separating treatment and carryover. It is not necessarily obvious how to predict intuitively that other design sequences may be ineffective. Design 4.4.8 can be seen to fail because all carryovers from Treatment A occur in its first sequence AAAB, and all carryovers from Treatment B occur in its second sequence BBBA. This results in a high degree of confounding between sequence and carryover. Thus, two general principles emerge, (1) alternations like ABABAB etc. in sequences lead to poor separation of treatment and carryover effects, (2) imbalances in the sequence–by–carryover sub–table such as AAAB/BAAA also lead to poor effects separability.

We now look at designs with four periods that have four or six sequences. Since Designs 4.4.3 – 4.4.5 were the best designs having two sequences, it is logical to combine them in pairs to produce the following 4–sequence designs:

	Design 4.4.10				Design 4.4.11			

	Period					Period			
Seq. ↓	1	2	3	4	Seq. ↓	1	2	3	4
1	A	B	B	A	1	A	B	B	A
2	B	A	A	B	2	B	A	A	B
3	A	B	B	B	3	A	A	B	B
4	B	A	A	A	4	B	B	A	A

Design 4.4.12

	Period			
Seq. ↓	1	2	3	4
1	A	B	B	B
2	B	A	A	A
3	A	A	B	B
4	B	B	A	A

If we use the same methods widely employed previously in this chapter, the sixteen cell means that result from data obtained on any of the Designs 4.4.10 – 4.4.12 give rise to 16 linear equations in terms of the 9 parameters μ, γ_1, γ_2, γ_3, π_1, π_2, π_3, τ_1, and λ_1. The following table (see next page) presents the variances and covariances, except for a multiplying factor $\hat{\sigma}^2/n$, of the $(X'X)^{-1}$ matrix relating to τ_1 and λ_1 for each of these designs.

Design	$\text{Var}(\hat{\tau}_1)$	$\text{Var}(\hat{\lambda}_1)$	$\text{Cov}(\hat{\tau}_1,\hat{\lambda}_1)$	$\text{Corr}(\hat{\tau}_1,\hat{\lambda}_1)$
4.4.10	0.0719	0.0915	0.0065	0.081
4.4.11	0.0625	0.0909	0.0000	0.000
4.4.12	0.0759	0.0966	–0.0207	–0.242

Design 4.4.11 is the best of the 4–sequence designs considered. In addition to having the lowest variances for $\hat{\tau}_1$ and $\hat{\lambda}_1$, the correlation between these two parameters is zero. The estimator of $\hat{\tau}_1$ is an equally–weighted contrast between cell means to which Treatment A was applied and cell means to which Treatment B was applied. Design 4.4.11 is the optimal design having four sequences and four periods (Cheng and Wu, 1980; Laska *et al.*, 1983, Laska and Meisner, 1985, and Matthews, 1987). Since both treatments appear twice within each sequence, we can use the measure S defined in Section 1.2.1 to examine the separability of $\hat{\tau}_1$ and $\hat{\lambda}_1$. The relevant table is

		Carryover		
		0	A	B
Treatment	A	2	3	3
	B	2	3	3

for which $\chi^2 = 0$, $V = 0$, and $S = 100(1 - V) = 100$, consistent with the complete separability of the two effects that was demonstrated above.

Designs 4.4.10 and 4.4.12, as well as various other 4–sequence designs not given here, have non–zero correlations between $\hat{\tau}_1$ and $\hat{\lambda}_1$. In addition, the sequences ABBB and BAAA, which are present in those designs, cause the cell means to contribute unequally to the overall $\hat{\tau}_1$ contrast.

We now examine a four period design with six sequences, obtained by concatenating Designs 4.4.3, 4.4.4 and 4.4.5 to form the following design:

Design 4.4.13

Seq. ↓	Period 1	2	3	4
1	A	B	B	A
2	B	A	A	B
3	A	A	B	B
4	B	B	A	A
5	A	B	B	B
6	B	A	A	A

Other designs having six sequences and four periods are also available, such as the following,

Design 4.4.14

Seq. ↓	Period 1	2	3	4
1	A	B	A	A
2	B	A	B	B
3	A	A	B	B
4	B	B	A	A
5	A	B	B	B
6	B	A	A	A

Jones and Kenward (1989) studied both of the above designs, and concluded that overall the two designs are very similar in terms of their variances and covariances, if carryover up to the first order is considered. If higher order carryover effects are suspected, or if treatment–by–period interactions are of prime interest, Design 4.4.14 may be slightly superior. In this book, we have adopted the attitude that, in general, interest is mainly directed towards treatment effects and carryover effects of the first order.

Use of the methods widely employed in this chapter leads to the following variances and covariances (in multiples of $\hat{\sigma}^2/n$) of $\hat{\tau}_1$ and $\hat{\lambda}_1$.

Design	$\text{Var}(\hat{\tau}_1)$	$\text{Var}(\hat{\lambda}_1)$	$\text{Cov}(\hat{\tau}_1,\hat{\lambda}_1)$	$\text{Corr}(\hat{\tau}_1,\hat{\lambda}_1)$
4.4.13	0.0456	0.0608	−0.0028	−0.052
4.4.14	0.0500	0.0606	0.0000	0.000

Although these variances and covariances appear to be low, they are best assessed in comparison with 4-sequence and 2-sequence designs. For comparability, the above values need to be multiplied by a factor of 3, and values for Design 4.4.11 need to be multiplied by a factor of 2 to make comparisons with Designs 4.4.3 – 4.4.5, the best of the 2-sequence designs.

Design		$\text{Var}(\hat{\tau}_1)$	$\text{Var}(\hat{\lambda}_1)$	$\text{Cov}(\hat{\tau}_1,\hat{\lambda}_1)$	$\text{Corr}(\hat{\tau}_1,\hat{\lambda}_1)$
4.4.13	(6-seq.)	0.1367	0.1823	–0.0083	–0.052
4.4.14	(6-seq.)	0.1500	0.1818	0.0000	0.000
4.4.11	(4-seq.)	0.1250	0.1818	0.0000	0.000
4.4.3	ABBA/BAAB	0.1375	0.2000	0.0500	0.302
4.4.4	ABBB/BAAA	0.1719	0.1875	–0.0313	–0.174
4.4.5	AABB/BBAA	0.1375	0.2000	–0.0500	–0.302

The above results show that the 6-sequence designs 4.4.13 and 4.4.14 are not quite as efficient as Design 4.4.11, the best design of the 4-sequence class. This suggests that there is no advantage to using six sequences. With a 4-sequence design such as 4.4.11, one can obtain a somewhat greater efficiency than with the use of Design 4.1.1, at the cost of greater complexity and requiring the use of four treatment periods. If an investigator is interested only in establishing whether or not there is evidence of a direct treatment effect and of a first order carryover effect, the 2-sequence extra-period design 4.1.1, having but three treatment periods, is recommended. Its properties were studied in detail in Section 4.1, including the case where baseline measurements are also available, although the inclusion of baseline measurements scarcely improves the design's efficiency.

Designs such as 4.4.11 (and indeed also Designs 4.4.13 and 4.4.14) come into consideration when the investigator wants more than just an indication of treatment and first order carryover. Designs such as 4.4.10 – 4.4.13 enable the investigator to examine such questions as whether or not there is second order carryover (that is, a residual effect of a treatment two periods later) or an interaction between treatment and period, or an interaction between treatment and carryover effect. Jones and Kenward (1989, Chapter 4) study these situations in detail and present considerable information to enable the user to decide which design to use. It is important to realize that a design which is the most efficient for studying treatment effects and first order carryover effects may not be the best for estimating second order carryover effects or a treatment–by–period interaction. In general, those designs that are best for first order carryover generally are not very efficient for estimating second order carryover.

Some designs are restricted to the number of parameters that can be estimated by the number of cell means available. Design 4.1.1 has but six cell means, so it can estimate a maximum of six parameters. Since the overall mean, a differential sequence effect, and two period effects account for four of these parameters, only a differential treatment effect and first order carryover effect may be estimated. Moving up to a 2–sequence, 4–period design brings in two extra cell means, but since the design adds an extra period effect, only one extra parameter may be estimated, such as a second order carryover effect, a treatment–by–period interaction, a treatment–by–first order carryover interaction, but not two or all of these.

As Jones and Kenward (1989, pp. 178–180) show, a design such as 4.4.5 which does very well for effects up to first order carryover, is incapable of estimating second order carryover because that effect is confounded with sequence effects. The best design for estimating second order carryover, among the 2–sequence, 4–period designs, is Design 4.4.4. Since that design is also an efficient one when second order carryover is excluded from the model, it might be sensible to choose that design if the user feels that there is a possibility of second order carryover. For 4–sequence designs of four periods, the optimal design is Design 4.4.11 for effects up to first order carryover, but is not the best design of its class when second order carryover needs to be estimated. Instead, Design 4.4.10, which is only slightly less efficient than Design 4.4.11 for effects up to first order carryover, should be employed. Difficulties are created by the fact that those considerations, involving treatment sequences that have Treatment A followed by itself equally as often as it is followed by Treatment B, which promote the efficiency of estimating first order carryover, may work in the opposite direction when second order carryover is concerned. Therefore, it is difficult to find designs that are optimal both for first and second order carryover estimation, not to mention various interactions involving treatment, carryover and periods. See Jones and Kenward (1989) for additional information on these designs.

In this book, we have adopted the position that second order carryover is much less likely to occur in practical research situations than first order carryover. Indeed, it is a difficult matter to find practical cases of significant second order carryover effects. If second order carryover, or more generally, prolonged residual effects of treatment, is indeed an object of concern in an investigation, then cross-over trials are probably not the most efficient trials to use to study that phenomenon.

4.5. Recommendations About the Use of the Designs in This Chapter

In Chapter 3, we examined the basic 2 x 2 (Grizzle/Hills-Armitage) design in great detail, and concluded that the design was only viable in the presence of carryover effects if baseline and two wash-out periods are included in the design. The properties of the design are summarized in Section 3.6. In this chapter, we have looked at other modifications of the basic design. Amongst the designs with two treatments and three periods, Design 4.1.1,

```
                      Period
          Seq.        1  2  3
           ↓
           1        | A  B  B |
           2        | B  A  A |
```
                                    ,

was the best, being a fully-efficient "extra treatment period" design proposed over three decades ago by Patterson and Lucas (1959). The correlation between the "direct" and "residual" treatment parameter estimators, $\hat{\tau}_1$ and $\hat{\lambda}_1$, respectively, is zero, meaning that the effects of treatment and carryover are orthogonal and therefore fully separable. The only way to improve on such estimators is to reduce their variances while retaining their orthogonality. In Section 4.2, we examined whether the use of four sequences with just two periods (Design 4.2) would result in any improvement over the extra-period design, but it was found that this design were markedly inferior to Design 4.1.1. Addition of baseline measurements to Design 4.2 improved its performance, but it was still inferior to Design 4.1.1. Therefore, one should not consider using Design 4.2 unless it is difficult or impossible to have the experimental subjects return for a third treatment period, making it imperative to restrict the trial to two treatment periods.

Section 4.3 examined designs with four periods and three sequences, and it was found that none of those designs was as good as Design 4.1.1, provided that consideration is restricted to carryover effects of first order only. In Section 4.4, other designs with more than two periods and/or two sequences were considered. Such designs may be better than Design 4.1.1 only if one ignores the facts that there is an extra period and extra complexity involved in carrying out trials with those designs. It seems that Design 4.1.1 is the best one to use to gain some insight into whether carryover effects are present or not. Its main restriction is that there is only one parameter to measure carryover, so that tests of other potentially significant effects, such as a treatment–by–period interaction, or a treatment–by–carryover interaction, cannot be made.

Exercises

4.1. The following data on milk protein production is taken from Gill (1978), Table 8.12 (reproduced with the permission of The Iowa State University Press).

Sequence ABA				Sequence BAB			
	Period				Period		
Cow↓	1	2	3	Cow↓	1	2	3
1	6.46	4.96	5.87	6	8.38	9.34	8.75
2	8.62	7.85	8.38	7	10.02	10.90	9.72
3	10.31	9.20	9.98	8	7.13	7.64	6.10
4	8.89	7.30	8.18	9	6.12	7.02	5.67
5	8.22	7.19	8.19	10	7.85	8.60	7.63

(a) As this data set corresponds to Design 4.1.2, one may expect a poor degree of separability between treatment and carryover effects. How might this effect the interpretation of the results of the analysis of variance of the above data set?

(b) Calculate the variance inflation factor VIF, in going from a model with no carryover parameters to one having carryover parameters in the model.

(c) Carry out the analysis of variance of the above data set. In view of your answer to the question in part (a), is there evidence that a serious multicollinearity situation prevails here?

4.2. In Table 4.1, the data of Ebbutt (1984) for the sequences ABB and BAA of the four–sequence design, Design 4.3.1,

Seq. ↓	Period 1 2 3
1	A B B
2	B A A
3	A B A
4	B A B

were presented. Below, we present the remainder of Ebbutt's data, for the sequences ABA and BAB, representing systolic blood pressures.

[Data of Ebbutt (1984), reproduced with permission.]

Sequence ABA

Subject	Period 1	2	3
1	184	154	145
2	210	160	140
3	250	210	190
4	180	110	112
5	165	130	140
6	210	180	190
7	175	155	120
8	186	170	164
9	178	170	140
10	150	155	130
11	130	115	110
12	155	180	136
13	140	130	120
14	180	135	140
15	162	148	148
16	185	180	180
17	220	190	155
18	170	178	152
19	220	172	178
20	172	164	150
21	200	170	140
22	154	168	176
23	150	130	120

Sequence BAB

Subject	Period 1	2	3
1	140	160	145
2	156	156	152
3	215	195	195
4	150	130	126
5	170	130	136
6	170	140	140
7	198	160	160
8	210	140	180
9	170	140	135
10	160	100	129
11	168	148	164
12	200	150	170
13	240	205	240
14	155	140	140
15	180	154	180
16	160	150	130
17	150	140	130

(a) From the information provided on this design and other 2–treatment, 3–period, 4–sequence designs presented in Section 4.3, discuss the advantages and disadvantages of this design (Design 4.3.1) compared to the others having the same number of treatments, periods and sequences.

(b) Carry out the analysis of variance of the complete data set of four sequences, testing whether or not there are significant treatment and carryover effects.

4.3. In this exercise, we consider a comment due to Fleiss (1989), who drew attention to the possibility that a carryover effect of a treatment onto itself, say

Treatment A followed in the next period by Treatment A, may be different from the effect of a treatment onto another treatment, say Treatment A followed in the next period by Treatment B. He cautioned against the possibility of misinterpretation of the results of the experiment if such an effect was in force.

(a) Show that the method of coding of Section 1.1.1 easily handles the above situation, requiring only the introduction of a different "code letter" for carryover for the combination AA than for the combination AB, and so forth.

(b) Apply that coding method to Design 4.3.1 (see Exercise 4.2), allowing for different carryover parameters for A to A, A to B, B to B and B to A. How many degrees of freedom should one expect the carryover sum of squares term to have?

(c) Analyze the data set of Exercise 4.2 using the above coding.

5

Cross-over Designs with Variance Balance

In the present chapter, we catalog a large number of designs that possess a certain property of balance known as "variance balance". A variance–balanced design is one in which *all* treatment contrasts are equally precise. Thus, in a design with three treatments A, B and C, say, it follows that

$$\mathrm{Var}(\tau_A - \tau_B) = \mathrm{Var}(\tau_A - \tau_C) = \mathrm{Var}(\tau_B - \tau_C).$$

In addition, the contrasts of the carryover effects are also of equal precision, and the treatment and carryover effects of a given treatment are negatively correlated. Designs for which the variance of treatment contrasts are not all equal, which include designs based upon partially balanced incomplete block designs and upon cyclic block designs, will be presented in Chapter 6.

5.1. The Variance–Balanced Cross–over Designs of Patterson and Lucas

Many of the designs that possess variance balance were catalogued by Patterson and Lucas (1962). That monumental study also included many designs not possessing variance balance, the treatment of which will be deferred until Chapter 6. In a similar fashion to the way in which the designs were listed in that paper, we present an index in Appendix 5.A of all of the Patterson–Lucas designs in this chapter. In further appendices, we list examples of each of these designs, using letters of the alphabet to indicate the treatments. In some cases, there are other designs that are of the same type, that cannot be constructed merely by a randomization of the letters. In these cases, that information will be given.

For all the designs in this chapter, the treatments are to be assigned randomly to the letters of the alphabet. In addition, some other forms of randomization may be required. For example, in some designs, the rows may be permuted, provided that the permutation is done for the same rows in each of the blocks. This will be indicated wherever such a randomization is permissible.

Appendix 5.A lists a set of efficiencies, expressed as a percentage, for each design in this chapter. The efficiency E_t is that, compared to a complete Latin square design, when carryover effects are not in the model. Designs PL2, PL5, PL6, PL10, PL11, PL14 and PL15 have 100% efficiency since they are either a single Latin square, a set of digram-balanced Latin squares (see Chapter 2) or a set of mutually orthogonal Latin squares. Thus, all forms of Latin square are equally 100% efficient and become the basis for comparing all other designs.

The efficiencies E_d for "direct" treatment effects and E_r for "residual" treatment effects (carryover) apply when both treatment and carryover parameters are in the model, and are less than 100% even for Latin squares.

However, the above efficiencies are not the sole measures of a cross-over design's performance. We have seen in Section 1.3 that "separability" of treatment and carryover effects is vitally important for a design to be considered to be a good one. In very extreme cases, poor separability may lead to erroneous conclusions about the significance of treatment effects. Hence, in Appendix 5.B we present a variance inflation factor VIF for each design. This quantity is the factor by which the variance of the treatment parameters inflate as a result of adding carryover parameters to a model which lacked these parameters. There is, in fact, a close connection between separability, VIF's, and the efficiencies E_d and E_r, and the power of a design to detect treatment differences. The designs with low values of E_d and E_r are also the ones that have high VIF's and low power. The virtue of the VIF is that is a direct measure of how much the variance increases if one includes carryover parameters in the model.

Appendix 5.B lists the balanced designs studied by Patterson and Lucas (1962) for which the number of units per block k equals the number of treatments t, and which do not possess an extra period. These 29 designs are designated PL1–29, and a typical representative of each design is given. In some cases, other designs could have been formed by different random selections from a design having more periods. For example, PL16 is constructed by deleting any four periods of Design PL23. Since the latter has six periods, the selection can be made in any of 15 ways.

The concept of a block occurs throughout the whole of experimental design, and is important for minimizing the errors in experimentation. A block generally consists of units which are as homogeneous as possible within the block. In field plot experimentation, blocks are chosen with a view to having a small degree of variation of such factors as soil fertility, moisture, aspect, drainage, etc., *within* the units of the block. The blocks themselves may differ greatly with respect to these factors, as long as the k units within a block are relatively homogeneous. All the designs of Appendix 5.B have $k = t$, so that each treatment is represented once and once only in each block.

An example of an experiment containing blocks in a cross–over trial is that of Clarke and Ratkowsky (1990). The purpose of the experiment was to compare the effect of two agents for the relief of asthma, sodium cromoglycate (SCG) and fenoterol hydrobromide. It was of interest to compare the two agents alone as well as in combination, and the use of a placebo was considered desirable, as in other clinical trials, to act as a control. Since sodium cromoglycate was administered as a dry powder, and fenoterol as a pressurized aerosol, two placebos had to be used, and these were designed to be indistinguishable in taste and appearance from the active medications. The following set of four treatments was employed,

A. The two placebos,
B. 20 mg SCG plus its placebo,
C. 100 μg fenoterol plus its placebo,
D. 100 μg fenoterol plus 20 mg SCG.

A set of five 4x4 Latin squares were chosen to accommodate a total of 20 patients in the trial. These were the first 20 people to volunteer to take part in the trial, provided that each of them had experienced a fall of at least 15% in their forced expiratory volumes in one second (FEV_1) after exercise. Since these 20 people became available over a long period of time, it was necessary to incorporate each patient into the study as that patient became available. The basic 4 x 4 Latin square used was a variant of Design PL5 in Appendix 5.B:

```
                    Block
                      1
    Subject  →    1   2   3   4
              ┌─────────────────┐
          1   │  A   B   C   D  │
    Period 2  │  B   C   D   A  │
          3   │  D   A   B   C  │
          4   │  C   D   A   B  │
              └─────────────────┘
```

The first patient to enter the trial was placed on the sequence for Subject 1 in the above design. The second patient was given the sequence for Subject 2, and so on. Keeping the patients in $b = 5$ blocks of $k (= t) = 4$ units per block helps guard against the possibility of unforeseen changes occurring, either in the techniques used for administering the treatments, or in the assessment of the effects of the treatments, from the start of the experiment to the end of the experiment, which was conducted over a period of two years. That is, any differences between the five squares can be taken into account in the analysis stage, by removing an effect for differences between squares. In that experiment, five identical squares were used to make it easier for the Registered Nurse who carried out the measurements to administer the doses with less chance of making a mistake. Although different squares could have consisted of different sequences, the chances of an error in the order of dispensation of the medications would doubtlessly increase.

For the extra–period designs PL30–54 in Appendix 5.C, where the treatments in the extra period are the same as in the penultimate period, the variance inflation factor VIF is unity in all cases and is therefore not reported. This is because the treatment and carryover parameters are orthogonal to each other. In this very desirable condition, dropping the carryover parameters from the model leaves the estimates of the treatment parameters and their variances unchanged. Included in Appendix 5.C for each design is a pair of "improvement" factors IF_d and IF_r, for the treatment and carryover effects, respectively, representing the multiplying factor by which the variance decreases as a result of the extra period being added. Improvement factors are very large when a third period is added to a two–period design. Small improvement factors occur when the original design already has a large number of periods.

Balanced designs (excluding extra period designs) with the number of treatments t exceeding the block size k are listed in Appendix 5.D. These designs, labelled PL55–75, like those of Appendix 5.B, do not have the carryover effects orthogonal to the treatment effects, and there is a variance inflation factor VIF to compare the design when carryover parameters are in the model to the design when they are absent. As was the case when $t = k$, designs with two periods experience a greater inflation of variance than those with more periods.

Appendix 5.E contains designs PL76–93, and represent designs for $t > k$ which have an extra period, the treatments in the extra period being identical to those in the penultimate period. As these designs have treatment effects orthogonal to the carryover effects, VIF is unity in each case and is therefore not reported. Instead, as was the case for the designs in Appendix 5.E, "improvement factors" IF_d and IF_r (for treatment and carryover, respectively) are given, representing the multiplying factor by which the variance decreases as a result of the extra period being added. These improvement factors are very large when the original design has only two periods, and very small when the basic design has a large number of periods.

Appendix 5.F contains designs PL94–98, which is a small set of balanced extra–period designs for block size $k = 2$. In these designs, the "parent" design is not to be found in Appendix 5.D because carryover and treatment cannot simultaneously be estimated. Thus, no improvement factors can be listed for designs PL94–98.

5.2. Analysis of Balanced Designs of Patterson and Lucas

The analyses of all the designs catalogued in Appendix 5.A, examples of which are given in Appendices 5.B – 5.F, are carried out in an identical fashion in a manner similar to that described in Section 1.1 of Chapter 1. To illustrate, consider the data set that appeared in Table 3.1 of Patterson and Lucas (1962), which is a version of design PL56 ($t = 4$, $p = 3$, $k = 3$, $b = 4$, $n = 12$), reproduced here as Table 5.1. The 12 experimental units were divided up at random into four blocks of three cows each and the three sequences within each square were separately allocated to cows at random. Periods were five weeks long and the response variable is the average daily production of fat–corrected milk (FCM).

Table 5.1. Data from Patterson and Lucas (1962), illustrating analysis of data from a balanced cross–over trial. (Reproduced here with the permission of the North Carolina State University.) Treatments given in parentheses.

		Block 1				Block 2		
		Cow 1	Cow 2	Cow 3	Cow 1	Cow 2	Cow 3
	1	38.7(A)	48.9(D)	35.2(C)	34.6(A)	32.9(C)	30.4(B)
Period	2	37.4(D)	46.9(C)	33.5(A)	32.3(C)	33.1(B)	29.5(A)
	3	34.3(C)	42.0(A)	28.4(D)	28.5(B)	27.5(A)	26.7(C)

		Block 3			Block 4		
		Cow 1	Cow 2	Cow 3	Cow 1	Cow 2	Cow 3
	1	25.7(C)	30.8(D)	25.4(B)	21.8(A)	21.4(B)	22.8(D)
Period	2	26.1(D)	29.3(B)	26.0(C)	23.9(B)	22.0(D)	21.0(A)
	3	23.4(B)	26.4(C)	23.9(D)	21.7(D)	19.4(A)	18.6(B)

The analysis using SAS© involves first the creation of a "data" step:

```
DATA FCM;
 INPUT BLOCK UNIT PERIOD TREAT$ CARRY$ FCM @@;
IF CARRY = '0' THEN CARRY = 'D';
CARDS;
1 1 1 A 0 38.7      1 1 2 D A 37.4      1 1 3 C D 34.3
1 2 1 D 0 48.9      1 2 2 C D 46.9      1 2 3 A C 42.0
1 3 1 C 0 35.2      1 3 2 A C 33.5      1 3 3 D A 28.4
2 1 1 A 0 34.6      2 1 2 C A 32.3      2 1 3 B C 28.5
2 2 1 C 0 32.9      2 2 2 B C 33.1      2 2 3 A B 27.5
2 3 1 B 0 30.4      2 3 2 A B 29.5      2 3 3 C A 26.7
3 1 1 C 0 25.7      3 1 2 D C 26.1      3 1 3 B D 23.4
3 2 1 D 0 30.8      3 2 2 B D 29.3      3 2 3 C B 26.4
3 3 1 B 0 25.4      3 3 2 C B 26.0      3 3 3 D C 23.9
4 1 1 A 0 21.8      4 1 2 B A 23.9      4 1 3 D B 21.7
4 2 1 B 0 21.4      4 2 2 D B 22.0      4 2 3 A D 19.4
4 3 1 D 0 22.8      4 3 2 A D 21.0      4 3 3 B A 18.6
RUN;
```

The IF statement is needed to ensure estimability of the treatment and carryover effects using the LSMEANS statement in the following procedure. The period effect can be assumed to be the same in each block, or, following Patterson and Lucas (1962), can be assumed to be different in each block. Those authors fitted a "period" effect plus a "period–by–block" interaction; we choose to use a nested term, periods within blocks, which results in a Type I sum of squares which is equal to that of "periods" plus the "period–by–block" interaction. These are two distinct ways of expressing the view that the time frame either may not be the same in the four blocks, or that there may be a difference in the effect of time in the four blocks.

```
PROC GLM;
  CLASS BLOCK UNIT PERIOD TREAT CARRY;
  MODEL FCM = BLOCK UNIT(BLOCK) PERIOD(BLOCK) TREAT CARRY/
      SOLUTION SS1 SS2 E1 E2;
  RANDOM UNIT(BLOCK);
  LSMEANS TREAT CARRY / STDERR PDIFF;
RUN;
```

This code results in the following edited output:

Dependent Variable: FCM

Source	DF	Sum of Squares	Mean Square	F Value	Pr > F
Model	25	1861.3442	74.4538	88.87	0.0001
Error	10	8.3780	0.8378		
Corrected Total	35	1869.7222			

Source	DF	Type I SS	Mean Square	F Value	Pr > F
BLOCK	3	1395.673333	465.224444	555.29	0.0001
UNIT(BLOCK)	8	334.695556	41.836944	49.94	0.0001
PERIOD(BLOCK)	8	125.802222	15.725278	18.77	0.0001
TREAT	3	3.099167	1.033056	1.23	0.3484
CARRY	3	2.073958	0.691319	0.83	0.5095

Source	DF	Type II SS	Mean Square	F Value	Pr > F
TREAT	3	4.648625	1.549542	1.85	0.2020
CARRY	3	2.073958	0.691319	0.83	0.5095

Neither the carryover effect nor the treatment effect is significant. Since the design used (PL56) has the relatively small variance inflation factor VIF = 1.25, there is no reason to suspect that when carryover is dropped from the model the treatment effect might become significant. This was confirmed by a further run with carryover not in the model, which showed that the P–value for treatment was P = 0.3209. A further run using "period" instead of "period within blocks" reduced the explained sum of squares for periods from 125.802 to 109.962, a non–significant decrease. Hence, it is permissible to consider that there is no interaction between blocks and periods, that is, the same time frame can be considered to apply to each block.

5.3. The Designs of Quenouille, Berenblut and Patterson

These designs generally contain a very large number of periods and would not likely be used in animal feeding trials and clinical trials with humans. However, such designs may find use in psychological experiments, as designs with up to 128 periods are known (Shoben *et al.*, 1989).

Quenouille (1953) presented some designs for $t = 2$, 3 and 4 treatments which utilized four periods and four subjects for $t = 2$, six periods and 18 subjects for $t = 3$, and eight periods and 16 subjects for $t = 4$. All these designs have treatment and carryover effects orthogonal to each other. Berenblut (1964) showed that treatment and carryover effects can be made to be orthogonal in a three treatment design with only half as many subjects as in Quenouille's design for $t = 3$. In general, he showed that with t treatments, designs were possible with $2t$ periods and t^2 subjects, and presented a general format for generating such designs. Patterson (1973) deduced Quenouille's method of construction and generalized it to apply to any number of treatments. For an odd number of treatments, he found a simple modification of Quenouille's method that reduced the number of subjects to t^2. Thus, for $2 < t < 8$, the following table summarizes the characteristics of these designs.

t, treatments	p, periods	n, subjects
2	4	4
3	6	9
4	8	16
5	10	25
6	12	36
7	14	49
8	16	64 .

The design for $t = 2$ was presented by Quenouille (1953, Table 9.6a), and was discussed in Chapter 4 as Design 4.4.11:

Design 4.4.11

```
            Period
Seq.     1  2  3  4
 ↓
 1      ┌──────────┐
        │ A  B  B  A │
 2      │ B  A  A  B │
 3      │ A  A  B  B │
 4      │ B  B  A  A │
        └──────────┘
```

Examples of this class of design for $t = 3$, 4, 5 and 6 are given in Appendix 5.G. The analysis of data from such designs is carried out in an identical fashion to the analysis of any other cross–over design. No special procedures are necessary.

5.4. The Designs of Balaam

We have already encountered the designs of Balaam (1968) in Chapter 4, Section 4.2, in the form of Design 4.2, reproduced below.

Design 4.2

```
                 Period
                 1   2
             1 ┌───────┐
             1 │ A   B │
Sequence     2 │ B   A │
             3 │ A   A │
             4 │ B   B │
               └───────┘
```

In general, Balaam designs have t treatments, t^2 sequences, and only two periods in which the treatments appear in all combinations, including with itself. Experience with the two period designs of Patterson and Lucas (1962) catalogued in Appendix 5.A shows that they are always of low efficiency, and therefore one might expect Balaam designs to be inefficient. This is confirmed in practice. Modifications to Balaam designs to improve their efficiency involve adding an extra period, or using baseline measurements.

For three treatments, the Balaam design will have the following form:

Design 5.1

Period
1   2

		1	2
	1	A	A
	2	A	B
	3	A	C
	4	B	A
Sequence	5	B	B
	6	B	C
	7	C	A
	8	C	B
	9	C	C

By rearranging the order of the sequences, one can see that this design is really the first two periods of the Quenouille–Berenblut–Patterson design QBP1 in Appendix 5.G. Also, Design 4.2 is simply the first two periods of Quenouille's Design 4.4.11, and, in general, Balaam designs consist of the first two periods of a QBP design.

Balaam designs always have the property of variance balance, as can be seen from the incidence matrices for treatment versus carryover. Each treatment is preceded equally often by each other treatment. To examine the variances and covariances of the estimators of treatment and carryover parameters, we write down a model for this design,

$$Y_{ijk} = \mu + \gamma_i + \pi_j + \tau_t + \xi_{i(k)} + \delta_j \lambda_r + \epsilon_{ijk} \tag{5.1}$$

where $i=1,2,...,9$, $j=1,2$, $t,r=1,2,3$ and $k=1,2,...,n_j$. Denoting the observed cell means by $\overline{Y}_{ij}$, the subscripts identifying the ith sequence and the jth period, one has, from a set of data, assuming the n's are equal,

Period
1   2

		Period 1	2
	1	$\overline{Y}_{11}$	$\overline{Y}_{12}$
	2	$\overline{Y}_{21}$	$\overline{Y}_{22}$
Sequence	3	$\overline{Y}_{31}$	$\overline{Y}_{32}$
	4	$\overline{Y}_{41}$	$\overline{Y}_{42}$
	5	$\overline{Y}_{51}$	$\overline{Y}_{52}$

Period
1   2

	Period 1	2
6	$\overline{Y}_{11}$	$\overline{Y}_{12}$
7	$\overline{Y}_{21}$	$\overline{Y}_{22}$
8	$\overline{Y}_{31}$	$\overline{Y}_{32}$
9	$\overline{Y}_{41}$	$\overline{Y}_{42}$

The matrix equations $X\beta = \mu$ can be written as follows:

$$
\begin{pmatrix}
1 & 1 & 0 & 0 & 0 & 0 & 0 & 0 & 0 & 1 & 1 & 0 & 0 & 0 \\
1 & 1 & 0 & 0 & 0 & 0 & 0 & 0 & 0 & -1 & 1 & 0 & 1 & 0 \\
1 & 0 & 1 & 0 & 0 & 0 & 0 & 0 & 0 & 1 & 1 & 0 & 0 & 0 \\
1 & 0 & 1 & 0 & 0 & 0 & 0 & 0 & 0 & -1 & 0 & 1 & 1 & 0 \\
1 & 0 & 0 & 1 & 0 & 0 & 0 & 0 & 0 & 1 & 1 & 0 & 0 & 0 \\
1 & 0 & 0 & 1 & 0 & 0 & 0 & 0 & 0 & -1 & -1 & -1 & 1 & 0 \\
1 & 0 & 0 & 0 & 1 & 0 & 0 & 0 & 0 & 1 & 0 & 1 & 0 & 0 \\
1 & 0 & 0 & 0 & 1 & 0 & 0 & 0 & 0 & -1 & 1 & 0 & 0 & 1 \\
1 & 0 & 0 & 0 & 0 & 1 & 0 & 0 & 0 & 1 & 0 & 1 & 0 & 0 \\
1 & 0 & 0 & 0 & 0 & 1 & 0 & 0 & 0 & -1 & 0 & 1 & 0 & 1 \\
1 & 0 & 0 & 0 & 0 & 0 & 1 & 0 & 0 & 1 & 0 & 1 & 0 & 0 \\
1 & 0 & 0 & 0 & 0 & 0 & 1 & 0 & 0 & -1 & -1 & -1 & 0 & 1 \\
1 & 0 & 0 & 0 & 0 & 0 & 0 & 1 & 0 & 1 & -1 & -1 & 0 & 0 \\
1 & 0 & 0 & 0 & 0 & 0 & 0 & 1 & 0 & -1 & 1 & 0 & -1 & -1 \\
1 & 0 & 0 & 0 & 0 & 0 & 0 & 0 & 1 & 1 & -1 & -1 & 0 & 0 \\
1 & 0 & 0 & 0 & 0 & 0 & 0 & 0 & 1 & -1 & 0 & 1 & -1 & -1 \\
1 & -1 & -1 & -1 & -1 & -1 & -1 & -1 & -1 & 1 & -1 & -1 & 0 & 0 \\
1 & -1 & -1 & -1 & -1 & -1 & -1 & -1 & -1 & -1 & -1 & -1 & -1 & -1
\end{pmatrix}
\begin{pmatrix}
\mu \\ \gamma_1 \\ \gamma_2 \\ \gamma_3 \\ \gamma_4 \\ \gamma_5 \\ \gamma_6 \\ \gamma_7 \\ \gamma_8 \\ \pi_1 \\ \tau_1 \\ \tau_2 \\ \lambda_1 \\ \lambda_2
\end{pmatrix}
=
\begin{pmatrix}
\mu_{11} \\ \mu_{12} \\ \mu_{21} \\ \mu_{22} \\ \mu_{31} \\ \mu_{32} \\ \mu_{41} \\ \mu_{42} \\ \mu_{51} \\ \mu_{52} \\ \mu_{61} \\ \mu_{62} \\ \mu_{71} \\ \mu_{72} \\ \mu_{81} \\ \mu_{82} \\ \mu_{91} \\ \mu_{92}
\end{pmatrix} .
$$

The portion of the inverse of the $X'X$ matrix relating to the treatment and carryover parameters is

	τ_1	τ_2	λ_1	λ_2
τ_1	0.4444	−0.2222	0.4444	−0.2222
τ_2	−0.2222	0.4444	−0.2222	0.4444
λ_1	0.4444	−0.2222	0.8889	−0.4444
λ_2	−0.2222	0.4444	−0.4444	0.8889

,

which tells potential users of this design that the variance of the carryover parameters is twice that of the variance of the treatment parameters, and that the correlation between treatment and carryover parameters is

$$
\text{Corr}(\tau_2, \lambda_1) = -0.2222/(0.4444 \cdot 0.8889)^{1/2} = -0.3536,
$$

which is not a particularly high value.

Repeating the above exercise with the same X matrix except for omission of the carryover parameters leads to the following portion of the inverse of the $X'X$ matrix relating to the treatment parameters only:

$$\begin{array}{cc} & \tau_1 \qquad\qquad \tau_2 \\ \begin{array}{c} \tau_1 \\ \tau_2 \end{array} & \left(\begin{array}{cc} 0.2222 & -0.1111 \\ -0.1111 & 0.2222 \end{array} \right) . \end{array}$$

Since the variances are halved, this means that with the carryover parameters in the model, there is a variance inflation factor VIF of 2.0. Compared to the Quenouille–Berenblut–Patterson designs, where there is no inflation of variance, this is considerable. The Balaam design can be improved by adding an extra period or by having a baseline observation of the same response variable on each individual. We look at each modification in turn.

Consider first the design with a third period which is identical to the second period.

Design 5.2

		Period	
	1	2	3
1	A	A	A
2	A	B	B
3	A	C	C
4	B	A	A
Sequence 5	B	B	B
6	B	C	C
7	C	A	A
8	C	B	B
9	C	C	C

Using a similar procedure to that used previously, the elements of the inverse of the $X'X$ matrix relating to the treatment and carryover parameters is

$$\begin{array}{c@{\quad}c@{\quad}c@{\quad}c@{\quad}c}
 & \tau_1 & \tau_2 & \lambda_1 & \lambda_2 \\
\begin{array}{c} \tau_1 \\ \tau_2 \\ \lambda_1 \\ \lambda_2 \end{array}
\left(\begin{array}{cccc}
0.1667 & -0.0833 & 0 & 0 \\
-0.0833 & 0.1667 & 0 & 0 \\
0 & 0 & 0.1667 & -0.0833 \\
0 & 0 & -0.0833 & 0.1667
\end{array}\right)
\end{array}.$$

As expected, the treatment and carryover parameters are orthogonal, and the variances of the treatment parameters are the same as that of the carryover parameters. Jones and Kenward (1989, p. 204) were apparently the first to note that the efficiency of estimating carryover effects is identical to that of estimating treatment effects when one adds an extra period to any Balaam design. Compared with the basic design before addition of the extra period, one can deduce that the improvement factors for treatment and carryover are, respectively,

$$IF_d = 0.4444/0.1667 = 2.667,$$

and

$$IF_r = 0.8889/0.1667 = 5.333 \quad .$$

Thus, big improvements in efficiency are obtained by adding the extra period.

Now let us look at the modification to Design 5.1 by taking a baseline reading prior to the first treatment period. The design has the following form:

Design 5.3

		Period		
		1	2	3
	1	–	A	A
	2	–	A	B
	3	–	A	C
	4	–	B	A
Sequence	5	–	B	B
	6	–	B	C
	7	–	C	A
	8	–	C	B
	9	–	C	C

The elements of the inverse of the $\mathbf{X'X}$ matrix relating to the treatment and carryover parameters for this design is

$$
\begin{array}{cccc}
\tau_1 & \tau_2 & \lambda_1 & \lambda_2
\end{array}
$$

$$
\begin{array}{c}
\tau_1 \\ \tau_2 \\ \lambda_1 \\ \lambda_2
\end{array}
\left(
\begin{array}{cccc}
0.1905 & -0.0952 & 0.0952 & -0.0476 \\
-0.0952 & 0.1905 & -0.0476 & 0.0952 \\
0.0952 & -0.0476 & 0.3810 & -0.1905 \\
-0.0476 & 0.0952 & -0.1905 & 0.3810
\end{array}
\right) .
$$

Removing the carryover parameters results in

$$
\begin{array}{cc}
\tau_1 & \tau_2
\end{array}
$$

$$
\begin{array}{c}
\tau_1 \\ \tau_2
\end{array}
\left(
\begin{array}{cc}
0.1667 & -0.0833 \\
-0.0833 & 0.1667
\end{array}
\right) .
$$

The variances and covariances are exactly the same as those for the extra-period Design 5.2. The difference is that these low variances also apply when carryover parameters are in the model, due to the orthogonality of those two effects, whereas in Design 5.3, the low variance prevails only when carryover is negligible. Note, however, that even with carryover in the model, the variances of the treatment parameters are only 0.1905, compared with 0.4444 in Design 5.1. Thus, the use of a baseline reading on the response variable results in an improvement factor of 2.333 for treatments, which is only somewhat less than the improvement factor of 2.667 when an extra period is added. It is in the efficiency of measuring carryover effects that the two modifications of the Balaam design differ the most. For the extra–period design, the improvement factor IF_r for carryover effects is 5.333, whereas for the design with baseline, it is only 2.333.

It will be seen in the next section that this is a typical result when a baseline observation precedes a two–period design. Improvement in efficiencies will result, but the improvement will be greater for treatment effects than for carryover effects. Generally speaking, it is better from the point of view of efficiency to add an extra period than to use baselines. However, the price that the experimenter has to pay is that, with the extra period design, three active treatment periods are required, with the final measurement to be made at a fourth visit. For the design involving a baseline reading, a measurement of the response variable is taken before commencement of active treatments, so no third treatment period and fourth visit are required.

5.5. Improving The Efficiencies of Two–Period Designs

Appendix 5.A lists nine balanced designs of Patterson and Lucas (1962) that contain only two periods. The efficiencies of these designs are the lowest ones in the whole table. This is not surprising, as the efficiencies are relative to Latin square designs with $t = p = k$. There are generally two methods for improving the efficiency of a two–period design. One method is to add an extra period to each experimental subject, the treatments of which are identical to those used in the last period. For a balanced design, this results in orthogonality of the treatment effects and carryover effects in the extra–period design. The second method is to take a "baseline" reading from each subject on the same response variable that is going to be measured after applying the treatments. This has the advantage over the first method of not requiring the subject to undergo another treatment period, nor of returning after that period to have another reading taken. The disadvantage is that the improvement in efficiency is not as great as when a third treatment period is used. Results supporting these conclusions were given in Section 5.4 for Balaam designs, which also have two periods.

We consider now an example to study the effect of adding an extra period to a design or of using baselines. Consider Design PL1 in Appendix 5.B, which is reproduced below.

Design PL1. $t = k = 3$, $p = 2$, $b = 2$, $n = 6$, VIF $= 4.0$

An example is

		Block				
		1			2	
Subject →	1	2	3	4	5	6
Period 1	A	B	C	A	B	C
2	B	C	A	C	A	B

Adding an extra period results in design PL30

		Block				
		1			2	
Subject →	1	2	3	4	5	6
Period 1	A	B	C	A	B	C
2	B	C	A	C	A	B
3	B	C	A	C	A	B

whereas preceding the first period with a baseline results in

$$
\begin{array}{c}
\text{Block} \\
\end{array}
$$

		1			2	
Subject →	1	2	3	4	5	6
Period 1	−	−	−	−	−	−
2	A	B	C	A	B	C
3	B	C	A	C	A	B

Setting up matrix equations for the original design PL1 in a manner similar to that used in Chapter 4 and in Section 5.4, the following is obtained,

$$
\begin{pmatrix}
1 & 1 & 1 & 0 & 0 & 0 & 1 & 1 & 1 & 0 & 0 & 0 \\
1 & 1 & 1 & 0 & 0 & 0 & -1 & -1 & 0 & 1 & 1 & 0 \\
1 & 1 & 0 & 1 & 0 & 0 & 1 & 1 & 0 & 1 & 0 & 0 \\
1 & 1 & 0 & 1 & 0 & 0 & -1 & -1 & -1 & -1 & 0 & 1 \\
1 & 1 & -1 & -1 & 0 & 0 & 1 & 1 & -1 & -1 & 0 & 0 \\
1 & 1 & -1 & -1 & 0 & 0 & -1 & -1 & 1 & 0 & -1 & -1 \\
1 & -1 & 0 & 0 & 1 & 0 & 1 & -1 & 1 & 0 & 0 & 0 \\
1 & -1 & 0 & 0 & 1 & 0 & -1 & 1 & -1 & -1 & 1 & 0 \\
1 & -1 & 0 & 0 & 0 & 1 & 1 & -1 & 0 & 1 & 0 & 0 \\
1 & -1 & 0 & 0 & 0 & 1 & -1 & 1 & 1 & 0 & 0 & 1 \\
1 & -1 & 0 & 0 & -1 & -1 & 1 & -1 & -1 & -1 & 0 & 0 \\
1 & -1 & 0 & 0 & -1 & -1 & -1 & 1 & 0 & 1 & -1 & -1
\end{pmatrix}
\begin{pmatrix}
\mu \\
\gamma_1 \\
\zeta_{11} \\
\zeta_{12} \\
\zeta_{21} \\
\zeta_{22} \\
\pi_1 \\
\gamma_1 \pi_1 \\
\tau_1 \\
\tau_2 \\
\lambda_1 \\
\lambda_2
\end{pmatrix}
=
\begin{pmatrix}
\mu_{11} \\
\mu_{12} \\
\mu_{21} \\
\mu_{22} \\
\mu_{31} \\
\mu_{32} \\
\mu_{41} \\
\mu_{42} \\
\mu_{51} \\
\mu_{52} \\
\mu_{61} \\
\mu_{62}
\end{pmatrix} .
$$

The model, following Patterson and Lucas (1962), corresponding to the above set of equations is a "saturated" model, that is, the number of parameters to be estimated equals the number of cell means, which in this case is 12. Although Design PL1 specifies subjects-by-periods, the entire two blocks of the design may be replicated many times, and the relevant information for estimation of the parameters is contained in the cell means of each of the 12 sequence-by-period combinations. The same is true for any of the designs listed in Appendix 5.A.

To be more precise, the model for the design is written as

$$
Y_{ijk} = \mu + \gamma_i + \zeta_{il(k)} + \pi_j + \gamma_i \pi_j + \tau_t + \delta_j \lambda_r + \epsilon_{ijk} \tag{5.1}
$$

where $i=1,2$, $j=1,2$, $t,r=1,2,3$, $l=1,2,,3$ and $k=1,2,...,n_i$. The index i denotes the two blocks, each of which has three sequences containing $k \geq 1$ subjects each. The ζ_{il} parameters denote the three sequences within each of the two blocks. After imposition of the sum–to–zero conditions, the number of parameters are reduced to the number of degrees of freedom, so that we end up with a grand mean, four (nested) sequence parameters, a period parameter, a block–by–period interaction parameter as used by Patterson and Lucas (1962), two treatment parameters and two carryover parameters.

The portion of the inverse of the $\mathbf{X'X}$ matrix relating to the treatment and carryover parameters is given by

$$
\begin{array}{c}
\tau_1 \\ \tau_2 \\ \lambda_1 \\ \lambda_2
\end{array}
\begin{pmatrix}
0.8889 & -0.4444 & 1.3333 & -0.6667 \\
-0.4444 & 0.8889 & -0.6667 & 1.3333 \\
1.3333 & -0.6667 & 2.6667 & -1.3333 \\
-0.6667 & 1.3333 & -1.3333 & 2.6667
\end{pmatrix}.
$$

with column headings $\tau_1 \quad \tau_2 \quad \lambda_1 \quad \lambda_2$.

Eliminating the carryover parameters results in

$$
\begin{array}{c}
\tau_1 \\ \tau_2
\end{array}
\begin{pmatrix}
0.2222 & -0.1111 \\
-0.1111 & 0.2222
\end{pmatrix}
$$

with column headings $\tau_1 \quad \tau_2$.

which is a reduction in the variance of the treatment parameters by a factor of 4.0. Hence, the variance inflation factor VIF is 4, as appears in the heading for design PL 1. Had we not used a block–by–period interaction in the above design, we could have obtained an equivalent result to the above by considering the two time frames to differ in the two blocks. That is, instead of having a "period" effect and a "block–by–period" interaction, each with one degree of freedom, we could have had two degrees of freedom for period, one for each block. Or, we could have considered the time frame to be the same in each block and used only a single period parameter, for one degree of freedom, with the same results for the treatment and carryover parameters.

Repeating the above exercise for design PL1 with an extra treatment period (that is, design PL30) results in the following portion of the inverse of the $X'X$ matrix relating to the treatment and carryover parameters

$$
\begin{array}{c c}
 & \begin{array}{cccc} \tau_1 & \tau_2 & \lambda_1 & \lambda_2 \end{array} \\
\begin{array}{c} \tau_1 \\ \tau_2 \\ \lambda_1 \\ \lambda_2 \end{array} &
\left(\begin{array}{cccc}
0.1667 & -0.0833 & 0 & 0 \\
-0.0833 & 0.1667 & 0 & 0 \\
0 & 0 & 0.2000 & -0.1000 \\
0 & 0 & -0.1000 & 0.2000
\end{array}\right)
\end{array} .
$$

Not only are the treatment parameters orthogonal to the carryover parameters, but there has been a tremendous reduction in both the treatment and carryover parameter variances. The improvement factors are, respectively,

$$IF_d = 0.8889/0.1667 = 5.333,$$

and

$$IF_r = 2.6667/0.2 = 13.333.$$

Thus, the use of an extra period increases the efficiencies greatly.

Repeating the exercise once again, this time with a baseline added to the basic design (but without an extra treatment period), the portion of the inverse of the $X'X$ matrix relating to the treatment and carryover parameters is

$$
\begin{array}{c c}
 & \begin{array}{cccc} \tau_1 & \tau_2 & \lambda_1 & \lambda_2 \end{array} \\
\begin{array}{c} \tau_1 \\ \tau_2 \\ \lambda_1 \\ \lambda_2 \end{array} &
\left(\begin{array}{cccc}
0.3333 & -0.1667 & 0.3333 & -0.1667 \\
-0.1667 & 0.3333 & -0.1667 & 0.3333 \\
0.3333 & -0.1667 & 0.8333 & -0.4167 \\
-0.1667 & 0.3333 & -0.4167 & 0.8333
\end{array}\right)
\end{array} .
$$

The improvement factors are, respectively,

$$IF_d = 0.8889/0.3333 = 2.667,$$

and

$$IF_r = 2.6667/0.8333 = 3.200,$$

which are not nearly so dramatic as when an extra treatment period is used. Nevertheless, a baseline measurement is relatively easy to obtain in many experimental situations involving cross–over trials. The above results show that the improvement in efficiency is of the order of 3 for both treatment and carryover effects. Comparing the relative efficiencies of the two forms of modification to the basic design PL1, the addition of an extra treatment period is twice more efficient for treatment effects, and 4.167 times as efficient for carryover effects.

5.6. The Efficiency of an Extra–Period Design versus a Complete Latin Square

In Section 5.5, we studied the effect of adding an extra period to a two–period balanced design and found that there were dramatic improvements in the efficiencies of the treatment and carryover parameters. We now wish to compare the extra–period design (Design PL30) which results by adding an extra period to Design PL1 with the pair of 3 x 3 Latin squares, Design PL2. That is, we wish to compare Design PL30,

		Block					
			1			2	
Subject	→	1	2	3	4	5	6
Period	1	A	B	C	A	B	C
	2	B	C	A	C	A	B
	3	B	C	A	C	A	B

to Design PL2,

		Block					
			1			2	
Subject	→	1	2	3	4	5	6
	1	A	B	C	A	B	C
Period	2	B	C	A	C	A	B
	3	C	A	B	B	C	A

In Section 5.5, it was found that the portion of the inverse of the $X'X$ matrix relating to the treatment and carryover parameters for Design PL30 was as follows:

	τ_1	τ_2	λ_1	λ_2
τ_1	0.1667	−0.0833	0	0
τ_2	−0.0833	0.1667	0	0
λ_1	0	0	0.2000	−0.1000
λ_2	0	0	−0.1000	0.2000

If the same exercise is carried out for Design PL2, the following is obtained:

	τ_1	τ_2	λ_1	λ_2
τ_1	0.1389	−0.0694	0.0833	−0.0417
τ_2	−0.0694	0.1389	−0.0417	0.0833
λ_1	0.0833	−0.0417	0.2500	−0.1250
λ_2	−0.0417	0.0833	−0.1250	0.2500

Comparing the two matrices, certain conclusions emerge. In Design PL30, the treatment and carryover parameters are orthogonal, but the variance of the treatment parameters, 0.1667, is higher than that in Design PL2, 0.1389. However, the variance of the carryover parameters, 0.20, in Design PL30 is greater than that, 0.25, in Design PL2. Thus, the extra–period Design PL30 might be preferred if the investigators are particularly interested in testing for carryover effects. If, however, as is the more usual case, the investigators wish to focus on treatment effects, the complete Latin square PL2 is to be preferred. In Appendix 5.A, the efficiencies are given, and it is easy to check that the ratio of 80 to 67 is, except for rounding error, the same as the ratio of 0.1667 to 0.1389. Similarly, the ratio of 56 to 44 is the same, except for rounding error, as the ratio of 0.25 to 0.20.

Exercises

5.1. Consider a design due to Koch *et al.* (1989) involving three treatments and two periods. Those authors were considering a situation involving a chronic health disorder, in which the first two treatments were active agents (labelled A and B), with the third "treatment" being a placebo, labelled P. With six sequences of subjects and equal allocation of subjects to sequences, the following treatment arrangements apply:

Period
1   2

		1	2
	1	A	B
	2	B	A
Sequence	3	A	P
	4	P	A
	5	B	P
	6	P	B

Compare this design with the balanced designs of Patterson and Lucas (1962) given in Appendix 5.B and find one to which the above design is similar. In what way do these two designs differ?

5.2. Koch *et al.* (1989) considered that the sequences in the design in Exercise 5.1 should be allocated in the proportions 3:3:1:1:1:1, rather than equally, so that the sequences involving a placebo as one of the treatments appear one-third as frequently as the sequences containing active treatments in both of its periods. It is convenient to give Sequences 1 and 2 three times the weight of the other sequences by repeating these two sequences three times each, resulting in a design with ten sequences as follows:

Period
1   2

		1	2
	1	A	B
	2	A	B
	3	A	B
	4	B	A
Sequence	5	B	A
	6	B	A
	7	A	P
	8	P	A
	9	B	P
	10	P	B

Using the approach employed in Section 5.4 and in other chapters, one can write a model for this design as

$$Y_{ijk} = \mu + \gamma_i + \pi_j + \tau_t + \xi_{i(k)} + \delta_j \lambda_r + \epsilon_{ijk}$$

where $i=1,2,...10$, $j=1,2$, $t,r= 1,2,3$ and $k=1,2,...,n_i$, with $n_i=n$, for all i.

(a) Write down 20 equations in terms of model parameters and population cell means in matrix form, and adopt the "sum–to–zero" convention to obtain the design matrix **X**.

(b) Find the portion of the inverse of the $\mathbf{X'X}$ matrix relating to the treatment and carryover parameters for this design.

(c) From the results of part (b), calculate the correlations between the treatment parameters and the carryover parameters and comment on their magnitudes. Also, comment on the variances of the treatment and carryover parameters.

5.3. Consider the following two–period cross–over experiment used to study methionine requirements in dairy calves. Suppose 5 doses are studied (A = 70%, B = 85%, C = 100%, D = 115%, E = 130% of methionine content of milk). Gill (1978), Exercise 8.11, pp. 248–9, considered putting 20 calves in ten 2 x 2 Latin squares to permit a balanced design. He presented the following data (reproduced here with the permission of the Iowa State University Press):

		Period				Period	
		1	2			1	2
Calf No.	1	0.49(A)	0.80(B)		11	0.80(B)	0.97(D)
	2	0.42(B)	0.45(A)		12	0.65(D)	0.50(B)
	3	0.45(A)	0.92(C)		13	0.48(B)	0.74(E)
	4	0.73(C)	0.48(A)		14	0.70(E)	0.77(B)
	5	0.20(A)	0.67(D)		15	0.86(C)	1.02(D)
	6	0.66(D)	0.40(A)		16	1.18(D)	1.22(C)
	7	0.54(A)	0.96(E)		17	0.87(C)	0.93(E)
	8	0.81(E)	0.69(A)		18	0.63(E)	0.67(C)
	9	0.71(B)	0.80(C)		19	1.20(D)	1.26(E)
	10	0.76(C)	0.69(B)		20	1.19(E)	1.38(D)

(a) The above is not the only way of blocking five treatments into two periods using 20 subjects. Examine the index to balanced cross–over designs in Appendix 5.A and identify two other ways that 20 subjects cannot be blocked with five treatments and two periods.

(b) Find plans for these designs in subsequent appendices and note the variance inflation factors (VIF's) for these designs. Does the VIF increase or decrease with k, the number of units per block? What does this suggest about the advisability of using 2 x 2 Latin squares, as in the above layout, if large block sizes are available?

(c) Analyze the above design and determine whether there are significant treatment and carryover effects.

5.4. Consider the following data on milk production in kg/day for an incomplete block design with four treatments, 12 subjects, and two periods (reproduced here with the permission of the Iowa State University Press). There is no indication that the animals were grouped in blocks.

		Period				Period	
		1	2			1	2
Cow No.	1	43.55(A)	32.17(B)	7		36.04(A)	29.50(D)
	2	38.03(B)	32.07(A)	8		29.62(D)	26.13(A)
	3	35.33(A)	22.72(C)	9		25.14(B)	18.49(D)
	4	26.17(C)	28.78(A)	10		36.80(D)	32.70(B)
	5	34.87(B)	23.98(C)	11		26.44(C)	22.88(D)
	6	24.77(C)	25.04(B)	12		36.86(D)	29.14(C)

(a) Compare this design with Design PL55 of Appendix 5.D. What similarities and what differences can be noted? From the information provided in this appendix and Appendix 5.A, is this likely to be a good design for the separation of treatment and carryover effects?

(b) Do an analysis of variance of the above data set and test whether or not there are significant treatment and carryover differences.

Appendix 5.A. Index to Balanced Cross-over Designs

(t = no. of treatments, p = no. of periods, k = no. of units per block, b = no. of blocks, n = total number of subjects. Efficiencies of designs are compared to a complete Latin square: E_t = efficiency when carryover parameters are not in the model, E_d = efficiency of treatment effects in the presence of carryover, E_r = efficiency of carryover effects in the presence of treatment).

t	p	k	b	n	E_t	E_d	E_r	Design No.	Appendix No.
3	2	3	2	6	75	19	6	PL1	5.B
3	3	2	3	6	67	67	50	PL94	5.F
3	3	3	2	6	100	80	44	PL2	5.B
3	3	3	2	6	67	67	56	PL30	5.C
3	4	3	2	6	94	94	75	PL31	5.C
4	2	3	4	12	67	17	6	PL55	5.D
4	2	4	3	12	67	22	8	PL3	5.B
4	3	2	6	12	59	59	44	PL95	5.F
4	3	3	4	12	89	71	40	PL56	5.D
4	3	3	4	12	59	59	49	PL76	5.E
4	3	4	3	12	89	72	42	PL4	5.B
4	3	4	3	12	59	59	52	PL32	5.C
4	4	3	4	12	83	83	67	PL77	5.E
4	4	4	3	12	83	83	69	PL33	5.C
4	4	4	1	4	100	91	62	PL5	5.B
4	4	4	3	12	100	91	62	PL6	5.B
4	5	4	1	4	96	96	80	PL34	5.C
4	5	4	3	12	96	96	80	PL35	5.C
5	2	4	5	20	62	21	8	PL57	5.D
5	2	5	4	20	62	23	9	PL7	5.B
5	3	2	10	20	56	56	42	PL96	5.F
5	3	4	5	20	83	67	39	PL58	5.D
5	3	4	5	20	56	56	49	PL78	5.E
5	3	5	4	20	83	68	41	PL8	5.B
5	3	5	4	20	56	56	50	PL36	5.C
5	4	4	5	20	94	85	59	PL59	5.D
5	4	4	5	20	78	78	64	PL79	5.E
5	4	5	4	20	94	85	60	PL9	5.B
5	4	5	4	20	78	78	66	PL37	5.C
5	5	4	5	20	90	90	75	PL80	5.E
5	5	5	2	10	100	95	72	PL10	5.B
5	5	5	4	20	100	95	72	PL11	5.B
5	5	5	4	20	90	90	76	PL38	5.C
5	6	5	2	10	97	97	83	PL39	5.C
5	6	5	4	20	97	97	83	PL40	5.C
6	2	6	5	30	60	24	10	PL12	5.B
6	3	2	15	30	53	53	40	PL97	5.F
6	3	6	5	30	53	53	49	PL41	5.C
6	5	5	6	30	96	91	69	PL60	5.D
6	5	6	5	30	96	91	70	PL13	5.B
6	6	5	6	30	93	93	80	PL81	5.E
6	6	6	1	6	100	97	78	PL14	5.B
6	6	6	5	30	100	97	78	PL15	5.B
6	6	6	5	30	93	93	81	PL42	5.C
7	2	3	14	42	58	15	5	PL61	5.D
7	2	6	7	42	58	23	10	PL62	5.D
7	2	7	6	42	58	24	10	PL16	5.B
7	3	2	21	42	52	52	39	PL98	5.F
7	3	3	14	42	78	62	35	PL63	5.D
7	3	3	14	42	52	52	43	PL82	5.E
7	3	6	7	42	78	64	39	PL64	5.D

(continued on following page)

t	p	k	b	n	E_t	E_d	E_r	Design No.	Appendix No.
7	3	6	7	42	52	52	48	PL83	5.E
7	3	7	3	21	78	64	40	PL17	5.B
7	3	7	6	42	78	64	40	PL18	5.B
7	3	7	6	42	52	52	48	PL43	5.C
7	4	3	14	42	73	73	58	PL84	5.E
7	4	4	7	28	88	80	55	PL65	5.D
7	4	6	7	42	88	80	57	PL66	5.D
7	4	6	7	42	73	73	62	PL85	5.E
7	4	7	2	14	88	80	57	PL19	5.B
7	4	7	3	21	73	73	62	PL44	5.C
7	4	7	6	42	88	80	57	PL20	5.B
7	4	7	6	42	73	73	62	PL45	5.C
7	5	4	7	28	84	84	70	PL86	5.E
7	5	6	7	42	93	88	68	PL67	5.D
7	5	6	7	42	84	84	72	PL87	5.E
7	5	7	2	14	84	84	72	PL46	5.C
7	5	7	3	21	93	88	68	PL21	5.B
7	5	7	6	42	93	88	68	PL22	5.B
7	5	7	6	42	84	84	72	PL47	5.C
7	6	6	7	42	97	94	76	PL68	5.D
7	6	6	7	42	91	91	78	PL88	5.E
7	6	7	3	21	91	91	79	PL48	5.C
7	6	7	6	42	97	94	76	PL23	5.B
7	6	7	6	42	91	91	79	PL49	5.C
10	4	4	15	60	83	76	52	PL69	5.D
10	5	4	15	60	80	80	67	PL89	5.E
11	3	5	11	55	73	60	36	PL70	5.D
11	3	11	5	55	73	61	39	PL24	5.B
11	4	5	11	55	69	69	58	PL90	5.E
11	4	11	5	55	69	69	60	PL50	5.C
11	5	5	11	55	88	83	63	PL71	5.D
11	5	11	5	55	88	84	65	PL25	5.B
11	6	5	11	55	86	86	73	PL91	5.E
11	6	6	11	66	92	89	71	PL72	5.D
11	6	11	2	22	92	89	72	PL26	5.B
11	6	11	5	55	86	86	75	PL51	5.C
13	4	4	13	52	81	74	51	PL73	5.D
13	4	13	4	52	81	74	54	PL27	5.B
13	5	4	13	52	78	78	65	PL92	5.E
13	5	13	3	39	87	82	65	PL28	5.B
13	5	13	4	52	78	78	68	PL52	5.C
13	6	13	3	39	84	84	74	PL53	5.C
16	4	4	20	80	80	73	50	PL74	5.D
16	4	16	5	80	80	73	54	PL29	5.B
16	5	4	20	80	77	77	64	PL93	5.E
16	5	16	5	80	77	77	67	PL54	5.C
16	6	6	16	96	89	86	69	PL75	5.D

Appendix 5.B. Designs PL 1–29. Balanced designs with $t = k$

Design PL1. $t = k = 3$, $p = 2$, $b = 2$, $n = 6$, VIF = 4.000
    Constructed by deleting any one period of Design PL2. An example is

		Block					
			1			2	
Subject	→	1	2	3	4	5	6
Period	1	A	B	C	A	B	C
	2	B	C	A	C	A	B

Further examples involve taking the second and third rows of Design PL2, or the first and third rows of that design.

Design PL2. $t = k = 3$, $p = 3$, $b = 2$, $n = 6$, VIF = 1.200
    A pair of 3 x 3 digram–balanced Latin squares. An example is

		Block					
			1			2	
Subject	→	1	2	3	4	5	6
	1	A	B	C	A	B	C
Period	2	B	C	A	C	A	B
	3	C	A	B	B	C	A

Design PL3. $t = k = 4$, $p = 2$, $b = 3$, $n = 12$, VIF = 3.000
    Constructed by deleting any two periods of Design PL6. An example is

		Block		
		1	2	3
Subject	→	1 2 3 4	1 2 3 4	1 2 3 4
Period	1	D C B A	B A D C	C D A B
	2	C D A B	D C B A	B A D C

Design PL4. $t = k = 4$, $p = 3$, $b = 3$, $n = 12$, VIF = 1.235
    Constructed by deleting any one period of Design PL6. An example is

		Block		
		1	2	3
Subject	→	1 2 3 4	1 2 3 4	1 2 3 4
	1	A B C D	A B C D	A B C D
Period	2	D C B A	B A D C	C D A B
	3	C D A B	D C B A	B A D C

Design PL5. $t = k = 4$, $p = 4$, $b = 1$, $n = 4$, VIF = 1.100
   A single 4 x 4 Latin square. An example is

```
                      Block
                        1
        Subject →    1   2   3   4

               1 ┌ A   B   C   D ┐
        Period 2 │ B   D   A   C │
               3 │ C   A   D   B │
               4 └ D   C   B   A ┘
```

Design PL6. $t = k = 4$, $p = 4$, $b = 3$, $n = 12$, VIF = 1.100
   A set of t-1 mutually orthogonal Latin squares. An example is

```
                          Block
                  1          2          3
        Subject → 1 2 3 4    1 2 3 4    1 2 3 4

               1 ┌ A B C D │ A B C D │ A B C D ┐
        Period 2 │ B A D C │ C D A B │ D C B A │
               3 │ C D A B │ D C B A │ B A D C │
               4 └ D C B A │ B A D C │ C D A B ┘
```

Design PL7. $t = k = 5$, $p = 2$, $b = 4$, $n = 20$, VIF = 2.667
   Constructed by deleting any three periods of Design PL11. An example is

```
                              Block
                  1           2           3           4
        Subject → 1 2 3 4 5   1 2 3 4 5   1 2 3 4 5   1 2 3 4 5

Period 1 ┌ C D E A B │ E A B C D │ B C D E A │ D E A B C ┐
       2 └ D E A B C │ B C D E A │ E A B C D │ C D E A B ┘
```

Design PL8. $t = k = 5$, $p = 3$, $b = 4$, $n = 20$, VIF = 1.227.
   Constructed by deleting any two periods of Design PL11. The remaining periods may then be randomized, provided the same randomization is used for all blocks. An example is

```
                              Block
                  1           2           3           4
        Subject → 1 2 3 4 5   1 2 3 4 5   1 2 3 4 5   1 2 3 4 5

               1 ┌ D E A B C │ B C D E A │ E A B C D │ C D E A B ┐
        Period 2 │ E A B C D │ D E A B C │ C D E A B │ B C D E A │
               3 └ C D E A B │ E A B C D │ B C D E A │ D E A B C ┘
```

Design PL9. $t = k = 5$, $p = 4$, $b = 4$, $n = 20$, VIF $= 1.098$

Constructed by deleting any one period of Design PL11. The periods in the design may then be randomized, provided the same randomization is used for all blocks. An example is

		Block		
	1	2	3	4
Subject →	1 2 3 4 5	1 2 3 4 5	1 2 3 4 5	1 2 3 4 5
Period 1	A B C D E	A B C D E	A B C D E	A B C D E
2	D E A B C	B C D E A	E A B C D	C D E A B
3	E A B C D	D E A B C	C D E A B	B C D E A
4	C D E A B	E A B C D	B C D E A	D E A B C

Design PL10. $t = k = 5$, $p = 5$, $b = 2$, $n = 10$, VIF $= 1.056$

A pair of digram–balanced Latin squares. An example is

		Block	
		1	2
Subject →		1 2 3 4 5	1 2 3 4 5
	1	A B C D E	A B C D E
	2	B C D E A	E A B C D
Period	3	E A B C D	B C D E A
	4	C D E A B	D E A B C
	5	D E A B C	C D E A B

Design PL11. $t = k = 5$, $p = 5$, $b = 4$, $n = 20$, VIF $= 2.500$

A set of t-1 mutually orthogonal Latin squares. An example is

		Block		
	1	2	3	4
Subject →	1 2 3 4 5	1 2 3 4 5	1 2 3 4 5	1 2 3 4 5
1	A B C D E	A B C D E	A B C D E	A B C D E
2	B C D E A	C D E A B	D E A B C	E A B C D
Period 3	C D E A B	E A B C D	B C D E A	D E A B C
4	D E A B C	B C D E A	E A B C D	C D E A B
5	E A B C D	D E A B C	C D E A B	B C D E A

Design PL12. $t = k = 6$, $p = 2$, $b = 5$, $n = 30$, VIF $= 1.056$

Constructed from the first two periods of Design PL15. An example is

			Block		
		1	2	3	4	5
				Subjects		
Per.	1	ABCDEF	ABCDEF	ABCDEF	ABCDEF	ABCDEF
	2	BCDEFA	CDEFAB	DEFABC	EFABCD	FABCDE

Design PL13. $t = k = 6$, $p = 5$, $b = 5$, $n = 30$, VIF $= 1.055$
Constructed by deleting the sixth period of Design PL15. An example is

		Block				
		1	2	3	4	5
				Subjects		
	1	ABCDEF	ABCDEF	ABCDEF	ABCDEF	ABCDEF
	2	BCDEFA	CDEFAB	DEFABC	EFABCD	FABCDE
Per.	3	CDEFAB	DEFABC	EFABCD	CDEFAB	BCDEFA
	4	FABCDE	FABCDE	CDEFAB	FABCDE	EFABCD
	5	DEFABC	EFABCD	BCDEFA	BCDEFA	DEFABC

Design PL14. $t = k = 6$, $p = 6$, $b = 1$, $n = 6$, VIF $= 1.036$
A single 6 x 6 Latin square. An example is

		Block					
				1			
Subject →		1	2	3	4	5	6
	1	A	B	C	D	E	F
	2	B	C	D	E	F	A
Period	3	F	A	B	C	D	E
	4	C	D	E	F	A	B
	5	E	F	A	B	C	D
	6	D	E	F	A	B	C

Design PL15. $t = k = 6$, $p = 6$, $b = 5$, $n = 30$, VIF $= 1.036$
An example is

		Block				
		1	2	3	4	5
				Subjects		
	1	ABCDEF	ABCDEF	ABCDEF	ABCDEF	ABCDEF
	2	BCDEFA	CDEFAB	DEFABC	EFABCD	FABCDE
Per.	3	CDEFAB	DEFABC	EFABCD	CDEFAB	BCDEFA
	4	FABCDE	FABCDE	CDEFAB	FABCDE	EFABCD
	5	DEFABC	EFABCD	BCDEFA	BCDEFA	DEFABC
	6	EFABCD	BCDEFA	FABCDE	DEFABC	CDEFAB

Design PL16. $t = k = 7$, $p = 2$, $b = 6$, $n = 42$, VIF $= 2.400$
Constructed by deleting any four periods of Design PL23. An example is

	Block					
	1	2	3	4	5	6
			Subjects			
1	CDEFGAB	FGABCDE	GABCDEF	BCDEFGA	DEFGABC	FGABCDE
2	DEFGABC	GABCDEF	CDEFGAB	FGABCDE	BCDEFGA	EFGABCD

Design PL17. $t = k = 7$, $p = 3$, $b = 3$, $n = 21$, VIF $= 1.219$
    Constructed from the first three periods of Design PL21. An example is

Block

| | | Subject → | | | 1 | 2 3 4 5 6 7 | 2 | 1 2 3 4 5 6 7 | 3 | 1 2 3 4 5 6 7 |

		1 2 3 4 5 6 7	1 2 3 4 5 6 7	1 2 3 4 5 6 7
	1	A B C D E F G	A B C D E F G	A B C D E F G
Period	2	B C D E F G A	C D E F G A B	E F G A B C D
	3	E F G A B C D	B C D E F G A	C D E F G A B

Design PL18. $t = k = 7$, $p = 3$, $b = 6$, $n = 42$, VIF $= 1.219$
    Constructed by deleting any three periods of Design PL23. The periods in the design may then be randomized, provided the same randomization is used for all blocks. An example is (the rows representing periods)

Block

Subjects

	1	2	3	4	5	6
1	ABCDEFG	ABCDEFG	ABCDEFG	ABCDEFG	ABCDEFG	ABCDEFG
2	EFGABCD	BCDEFGA	FGABCDE	CDEFGAB	GABCDEF	DEFGABC
3	FGABCDE	DEFGABC	BCDEFGA	GABCDEF	EFGABCD	CDEFGAB

Design PL19. $t = k = 7$, $p = 4$, $b = 2$, $n = 14$, VIF $= 1.096$
    An example is

Block

		1	2
	Subject →	1 2 3 4 5 6 7	1 2 3 4 5 6 7
	1	A B C D E F G	A B C D E F G
Period	2	B C D E F G A	G A B C D E F
	3	D E F G A B C	E F G A B C D
	4	G A B C D E F	B C D E F G A

Design PL20. $t = k = 7$, $p = 4$, $b = 6$, $n = 42$, VIF $= 1.096$
    Constructed by deleting any two periods of Design PL23. The periods in the design may then be randomized, provided the same randomization is used for all blocks. An example is (the rows representing periods)

Block

Subjects

	1	2	3	4	5	6
1	ABCDEFG	ABCDEFG	ABCDEFG	ABCDEFG	ABCDEFG	ABCDEFG
2	CDEFGAB	EFGABCD	GABCDEF	BCDEFGA	DEFGABC	FGABCDE
3	DEFGABC	GABCDEF	CDEFGAB	FGABCDE	BCDEFGA	EFGABCD
4	EFGABCD	BCDEFGA	FGABCDE	CDEFGAB	GABCDEF	DEFGABC

Design PL21. $t = k = 7$, $p = 5$, $b = 3$, $n = 21$, VIF = 1.055
　An example is

		Block		
		1	2	3
Subject →		1 2 3 4 5 6 7	1 2 3 4 5 6 7	1 2 3 4 5 6 7
	1	A B C D E F G	A B C D E F G	A B C D E F G
	2	B C D E F G A	C D E F G A B	E F G A B C D
Period	3	E F G A B C D	B C D E F G A	C D E F G A B
	4	G A B C D E F	F G A B C D E	D E F G A B C
	5	F G A B C D E	D E F G A B C	G A B C D E F

Design PL22. $t = k = 7$, $p = 5$, $b = 6$, $n = 42$, VIF = 1.055
　Constructed by deleting any one period of Design PL23. The periods in the design may then be randomized, provided the same randomization is used for all blocks. An example is (the rows representing periods)

			Block		
1	2	3	4	5	6
			Subjects		
ABCDEFG	ABCDEFG	ABCDEFG	ABCDEFG	ABCDEFG	ABCDEFG
BCDEFGA	CDEFGAB	DEFGABC	EFGABCD	FGABCDE	GABCDEF
CDEFGAB	EFGABCD	GABCDEF	BCDEFGA	DEFGABC	FGABCDE
EFGABCD	BCDEFGA	FGABCDE	CDEFGAB	GABCDEF	DEFGABC
FGABCDE	DEFGABC	BCDEFGA	GABCDEF	EFGABCD	CDEFGAB

(rows numbered 1 2 3 4 5)

Design PL23. $t = k = 7$, $p = 6$, $b = 6$, $n = 42$, VIF = 1.036
　An example is (the rows representing periods)

			Block		
1	2	3	4	5	6
			Subjects		
ABCDEFG	ABCDEFG	ABCDEFG	ABCDEFG	ABCDEFG	ABCDEFG
BCDEFGA	CDEFGAB	DEFGABC	EFGABCD	FGABCDE	GABCDEF
CDEFGAB	EFGABCD	GABCDEF	BCDEFGA	DEFGABC	FGABCDE
DEFGABC	GABCDEF	CDEFGAB	FGABCDE	BCDEFGA	EFGABCD
EFGABCD	BCDEFGA	FGABCDE	CDEFGAB	GABCDEF	DEFGABC
FGABCDE	DEFGABC	BCDEFGA	GABCDEF	EFGABCD	CDEFGAB

(rows numbered 1 2 3 4 5 6)

Design PL24. $t = k = 11$, $p = 3$, $b = 5$, $n = 55$, VIF = 1.212
    Constructed by using the first three periods of Design PL25. An example is

| | | **Block** | | |
| | | 1 | 2 | 3 |
		Subjects		
	1	ABCDEFGHIJK	ABCDEFGHIJK	ABCDEFGHIJK
Period	2	BCDEFGHIJKA	DEFGHIJKABC	EFGHIJKABCD
	3	JKABCDEFGHI	FGHIJKABCDE	DEFGHIJKABC

(continued)

| | | **Block** | |
| | | 4 | 5 |
		Subjects	
	1	ABCDEFGHIJK	ABCDEFGHIJK
Period	2	FGHIJKABCDE	JKABCDEFGHI
	3	BCDEFGHIJKA	EFGHIJKABCD

Design PL25. $t = k = 11$, $p = 5$, $b = 5$, $n = 55$, VIF = 1.054
    An example is

| | | **Block** | | |
| | | 1 | 2 | 3 |
		Subjects		
	1	ABCDEFGHIJK	ABCDEFGHIJK	ABCDEFGHIJK
	2	BCDEFGHIJKA	DEFGHIJKABC	EFGHIJKABCD
Period	3	JKABCDEFGHI	FGHIJKABCDE	DEFGHIJKABC
	4	DEFGHIJKABC	JKABCDEFGHI	BCDEFGHIJKA
	5	KABCDEFGHIJ	IJKABCDEFGH	HIJKABCDEFG

(continued)

| | | **Block** | |
| | | 4 | 5 |
		Subjects	
	1	ABCDEFGHIJK	ABCDEFGHIJK
	2	FGHIJKABCDE	JKABCDEFGHI
Period	3	BCDEFGHIJKA	EFGHIJKABCD
	4	EFGHIJKABCD	FGHIJKABCDE
	5	GHIJKABCDEF	CDEFGHIJKAB

Design PL26. $t = k = 11$, $p = 6$, $b = 2$, $n = 22$, VIF = 1.035
  An example is

		Block 1	Block 2
		Subjects	Subjects
	1	ABCDEFGHIJK	ABCDEFGHIJK
	2	BCDEFGHIJKA	KABCDEFGHIJ
Period	3	DEFGHIJKABC	IJKABCDEFGH
	4	GHIJKABCDEF	FGHIJKABCDE
	5	KABCDEFGHIJ	BCDEFGHIJKA
	6	EFGHIJKABCD	HIJKABCDEFG

Design PL27. $t = k = 13$, $p = 4$, $b = 4$, $n = 52$, VIF = 1.094
  An example is (the rows representing periods)

	Block 1	Block 2	Block 3	Block 4
	Subjects			
1	ABCDEFGHIJKLM	ABCDEFGHIJKLM	ABCDEFGHIJKLM	ABCDEFGHIJKLM
2	BCDEFGHIJKLMA	IJKLMABCDEFGH	KLMABCDEFGHIJ	HIJKLMABCDEFG
3	DEFGHIJKLMABC	MABCDEFGHIJKL	GHIJKLMABCDEF	FGHIJKLMABCDE
4	JKLMABCDEFGHI	CDEFGHIJKLMAB	LMABCDEFGHIJK	EFGHIJKLMABCD

Design PL28. $t = k = 13$, $p = 5$, $b = 3$, $n = 39$, VIF = 1.054
  An example is

		Block 1	Block 2	Block 3
		Subjects		
	1	ABCDEFGHIJKLM	ABCDEFGHIJKLM	ABCDEFGHIJKLM
	2	BCDEFGHIJKLMA	DEFGHIJKLMABC	JKLMABCDEFGHI
Period	3	DEFGHIJKLMABC	JKLMABCDEFGHI	BCDEFGHIJKLMA
	4	HIJKLMABCDEFG	IJKLMABCDEFGH	LMABCDEFGHIJK
	5	CDEFGHIJKLMAB	GHIJKLMABCDEF	FGHIJKLMABCDE

Design PL29. $t = k = 16$, $p = 4$, $b = 5$, $n = 80$, VIF $= 1.093$

In this design, there are five blocks of four 4 x 4 Latin squares. An example is (the rows represent periods and the columns subjects within blocks)

Block 1

A B C D	E F G H	I J K L	M N O P
B D A C	F H E G	J L I K	N P M O
C A D B	G E H F	K I L J	O M P N
D C B A	H G F E	L K J I	P O N M

Block 2

A E I M	B F J N	C G K O	D H L P
E M A I	F N B J	G O C K	H P D L
I A M E	J B N F	K C O G	L D P H
M I E A	N J F B	O K G C	P L H D

Block 3

A F K P	E B O L	I N C H	M J G D
F P A K	B L E O	N H I C	J D M G
K A P F	O E L B	C I H N	G M D J
P K F A	L O B E	H C N I	D G J M

Block 4

A N G L	M B K H	E J C P	I F O D
N L A G	B H M K	J P E C	F D I O
G A L N	K M H B	C E P J	O I D F
L G N A	H K B M	P C J E	D O F I

Block 5

A J O H	I B G P	M F C L	E N K D
J H A O	B P I G	F L M C	N D E K
O A H J	G I P B	C M L F	K E D N
H O J A	P G B I	L C F M	D K N E

Appendix 5.C. Designs PL 30–54. Extra–period balanced designs with $t = k$

Design PL30 (Parent Design PL1). $t = k = 3$, $p = 3$, $b = 2$, $n = 6$, $\text{IF}_d = 5.333$, $\text{IF}_r = 13.333$. An example is

		Block					
		1			2		
Subject →		1	2	3	4	5	6
Period	1	C	A	B	C	A	B
	2	B	C	A	A	B	C
	3	B	C	A	A	B	C

Design PL31 (Parent Design PL2). $t = k = 3$, $p = 4$, $b = 2$, $n = 6$, $\text{IF}_d = 1.5625$, $\text{IF}_r = 2.25$. An example is

		Block					
		1			2		
Subject →		1	2	3	4	5	6
Period	1	B	C	A	A	B	C
	2	A	B	C	B	C	A
	3	C	A	B	C	A	B
	4	C	A	B	C	A	B

Design PL32 (Parent Design PL3). $t = k = 4$, $p = 3$, $b = 3$, $n = 12$, $\text{IF}_d = 4.000$, $\text{IF}_r = 9.333$. An example is

		Block		
		1	2	3
Subject →		1 2 3 4	1 2 3 4	1 2 3 4
Period	1	B A D C	C D A B	D C B A
	2	C D A B	D C B A	B A D C
	3	C D A B	D C B A	B A D C

Design PL33 (Parent Design PL4). $t = k = 4$, $p = 4$, $b = 3$, $n = 12$, $\text{IF}_d = 1.544$, $\text{IF}_r = 2.184$. An example is

		Block		
		1	2	3
Subject →		1 2 3 4	1 2 3 4	1 2 3 4
Period	1	A B C D	A B C D	A B C D
	2	B A D C	C D A B	D C B A
	3	D C B A	B A D C	C D A B
	4	D C B A	B A D C	C D A B

Design PL34 (Parent Design PL5). $t = k = 4$, $p = 5$, $b = 1$, $n = 4$, $IF_d = 1.320$, $IF_r = 1.600$. An example is

```
                    Block
                      1
     Subject  →   1   2   3   4
               ┌──────────────────┐
            1  │ A   B   C   D    │
            2  │ B   D   A   C    │
    Period  3  │ C   A   D   B    │
            4  │ D   C   B   A    │
            5  │ D   C   B   A    │
               └──────────────────┘
```

Design PL35 (Parent Design PL6). $t = k = 4$, $p = 5$, $b = 3$, $n = 12$, $IF_d = 1.320$, $IF_r = 1.600$. An example is

```
                          Block
                 1          2          3
     Subject  →  1 2 3 4   1 2 3 4   1 2 3 4
               ┌─────────────────────────────┐
            1  │ A B C D   A B C D   A B C D │
            2  │ B A D C   C D A B   D C B A │
    Period  3  │ C D A B   D C B A   B A D C │
            4  │ D C B A   B A D C   C D A B │
            5  │ D C B A   B A D C   C D A B │
               └─────────────────────────────┘
```

Design PL36 (Parent Design PL7). $t = k = 5$, $p = 3$, $b = 4$, $n = 20$, $IF_d = 3.556$, $IF_r = 8.000$. An example is

```
                               Block
                  1            2            3            4
     Subject  →  1 2 3 4 5    1 2 3 4 5    1 2 3 4 5    1 2 3 4 5
               ┌──────────────────────────────────────────────────┐
            1  │ B C D E A    C D E A B    D E A B C    E A B C D │
    Period  2  │ E A B C D    D E A B C    C D E A B    B C D E A │
            3  │ E A B C D    D E A B C    C D E A B    B C D E A │
               └──────────────────────────────────────────────────┘
```

Design PL37 (Parent Design PL8). $t = k = 5$, $p = 4$, $b = 4$, $n = 20$, $IF_d = 1.534$, $IF_r = 2.148$. An example is

```
                               Block
                  1            2            3            4
     Subject  →  1 2 3 4 5    1 2 3 4 5    1 2 3 4 5    1 2 3 4 5
               ┌──────────────────────────────────────────────────┐
            1  │ B C D E A    C D E A B    D E A B C    E A B C D │
    Period  2  │ C D E A B    E A B C D    B C D E A    D E A B C │
            3  │ D E A B C    B C D E A    E A B C D    C D E A B │
            4  │ D E A B C    B C D E A    E A B C D    C D E A B │
               └──────────────────────────────────────────────────┘
```

Design PL38 (Parent Design PL9). $t = k = 5$, $p = 5$, $b = 4$, $n = 20$, $\text{IF}_d = 1.318$, $\text{IF}_r = 1.590$. An example is

		Block 1	Block 2	Block 3	Block 4
Subject →		1 2 3 4 5	1 2 3 4 5	1 2 3 4 5	1 2 3 4 5
Period	1	A B C D E	A B C D E	A B C D E	A B C D E
	2	C D E A B	E A B C D	B C D E A	D E A B C
	3	D E A B C	B C D E A	E A B C D	C D E A B
	4	E A B C D	D E A B C	C D E A B	B C D E A
	5	E A B C D	D E A B C	C D E A B	B C D E A

Design PL39 (Parent Design PL10). $t = k = 5$, $p = 6$, $b = 2$, $n = 10$, $\text{IF}_d = 1.232$, $\text{IF}_r = 1.389$. An example is

		Block 1	Block 2
Subject →		1 2 3 4 5	1 2 3 4 5
Period	1	A B C D E	A B C D E
	2	B C D E A	E A B C D
	3	E A B C D	B C D E A
	4	C D E A B	D E A B C
	5	D E A B C	C D E A B
	6	D E A B C	C D E A B

Design PL40 (Parent Design PL11). $t = k = 5$, $p = 6$, $b = 4$, $n = 20$, $\text{IF}_d = 1.231$, $\text{IF}_r = 1.389$. An example is

		Block 1	Block 2	Block 3	Block 4
Subject →		1 2 3 4 5	1 2 3 4 5	1 2 3 4 5	1 2 3 4 5
Period	1	A B C D E	A B C D E	A B C D E	A B C D E
	2	B C D E A	C D E A B	D E A B C	E A B C D
	3	C D E A B	E A B C D	B C D E A	D E A B C
	4	D E A B C	B C D E A	E A B C D	C D E A B
	5	E A B C D	D E A B C	C D E A B	B C D E A
	6	E A B C D	D E A B C	C D E A B	B C D E A

Design PL41 (Parent Design PL12). $t = k = 6$, $p = 3$, $b = 5$, $n = 30$, $IF_d = 3.333$, $IF_r = 7.333$. An example is

				Block 3		
		1	2	Subjects	4	5
	1	ABCDEF	ABCDEF	ABCDEF	ABCDEF	ABCDEF
Period	2	BCDEFA	CDEFAB	DEFABC	EFABCD	FABCDE
	3	BCDEFA	CDEFAB	DEFABC	EFABCD	FABCDE

Design PL42 (Parent Design PL13). $t = k = 6$, $p = 6$, $b = 5$, $n = 30$, $IF_d = 1.231$, $IF_r = 1.386$. An example is

				Block 3		
		1	2	Subjects	4	5
	1	ABCDEF	ABCDEF	ABCDEF	ABCDEF	ABCDEF
	2	BCDEFA	CDEFAB	DEFABC	EFABCD	FABCDE
Period	3	CDEFAB	DEFABC	EFABCD	CDEFAB	BCDEFA
	4	FABCDE	FABCDE	CDEFAB	FABCDE	EFABCD
	5	DEFABC	EFABCD	BCDEFA	BCDEFA	DEFABC
	6	DEFABC	EFABCD	BCDEFA	BCDEFA	DEFABC

Design PL43 (Parent Design PL16). $t = k = 7$, $p = 3$, $b = 6$, $n = 42$, $IF_d = 3.200$, $IF_r = 6.933$. An example is (the rows representing periods)

			Block 4		
1	2	3	Subjects	5	6
BCDEFGA	CDEFGAB	DEFGABC	EFGABCD	FGABCDE	GABCDEF
EFGABCD	BCDEFGA	FGABCDE	CDEFGAB	GABCDEF	DEFGABC
EFGABCD	BCDEFGA	FGABCDE	CDEFGAB	GABCDEF	DEFGABC

Design PL44 (Parent Design PL17). $t = k = 7$, $p = 4$, $b = 3$, $n = 21$, $IF_d = 1.523$, $IF_r = 2.109$. An example is

		Block 1	Block 2	Block 3
Subject →		1 2 3 4 5 6 7	1 2 3 4 5 6 7	1 2 3 4 5 6 7
	1	A B C D E F G	A B C D E F G	A B C D E F G
Period	2	B C D E F G A	C D E F G A B	E F G A B C D
	3	E F G A B C D	B C D E F G A	C D E F G A B
	4	E F G A B C D	B C D E F G A	C D E F G A B

Design PL45 (Parent Design PL18). $t = k = 7$, $p = 4$, $b = 6$, $n = 42$, $\text{IF}_d = 1.523$, $\text{IF}_r = 2.109$. An example is (the rows representing periods)

			Block		
1	2	3	4	5	6
			Subjects		

	1	2	3	4	5	6
1	ABCDEFG	ABCDEFG	ABCDEFG	ABCDEFG	ABCDEFG	ABCDEFG
2	EFGABCD	BCDEFGA	FGABCDE	CDEFGAB	GABCDEF	DEFGABC
3	FGABCDE	DEFGABC	BCDEFGA	GABCDEF	EFGABCD	CDEFGAB
4	FGABCDE	DEFGABC	BCDEFGA	GABCDEF	EFGABCD	CDEFGAB

Design PL46 (Parent Design PL19). $t = k = 7$, $p = 5$, $b = 2$, $n = 14$, $\text{IF}_d = 1.315$, $\text{IF}_r = 1.578$. An example is (the rows representing periods)

		Block	
		1	2
Subject →		1 2 3 4 5 6 7	1 2 3 4 5 6 7

		Block 1	Block 2
	1	A B C D E F G	A B C D E F G
	2	B C D E F G A	G A B C D E F
Period	3	D E F G A B C	E F G A B C D
	4	G A B C D E F	B C D E F G A
	5	G A B C D E F	B C D E F G A

Design PL47 (Parent Design PL20). $t = k = 7$, $p = 5$, $b = 6$, $n = 42$, $\text{IF}_d = 1.315$, $\text{IF}_r = 1.578$. An example is (the rows representing periods)

			Block		
1	2	3	4	5	6
			Subjects		

	1	2	3	4	5	6
1	ABCDEFG	ABCDEFG	ABCDEFG	ABCDEFG	ABCDEFG	ABCDEFG
2	CDEFGAB	EFGABCD	GABCDEF	BCDEFGA	DEFGABC	FGABCDE
3	DEFGABC	GABCDEF	CDEFGAB	FGABCDE	BCDEFGA	EFGABCD
4	EFGABCD	BCDEFGA	FGABCDE	CDEFGAB	GABCDEF	DEFGABC
5	EFGABCD	BCDEFGA	FGABCDE	CDEFGAB	GABCDEF	DEFGABC

Design PL48 (Parent Design PL21). $t = k = 7$, $p = 6$, $b = 3$, $n = 21$, $IF_d = 1.230$, $IF_r = 1.384$. An example is

		Block	
		1	2	3
Subject →		1 2 3 4 5 6 7	1 2 3 4 5 6 7	1 2 3 4 5 6 7

1	A B C D E F G	A B C D E F G	A B C D E F G
2	B C D E F G A	C D E F G A B	E F G A B C D
Period 3	E F G A B C D	B C D E F G A	C D E F G A B
4	G A B C D E F	F G A B C D E	D E F G A B C
5	F G A B C D E	D E F G A B C	G A B C D E F
6	F G A B C D E	D E F G A B C	G A B C D E F

Design PL49 (Parent Design PL22). $t = k = 7$, $p = 6$, $b = 6$, $n = 42$, $IF_d = 1.230$, $IF_r = 1.384$. An example is (the rows representing periods)

Block

	1	2	3	4	5	6
				Subjects		
1	ABCDEFG	ABCDEFG	ABCDEFG	ABCDEFG	ABCDEFG	ABCDEFG
2	BCDEFGA	CDEFGAB	DEFGABC	EFGABCD	FGABCDE	GABCDEF
3	CDEFGAB	EFGABCD	GABCDEF	BCDEFGA	DEFGABC	FGABCDE
4	EFGABCD	BCDEFGA	FGABCDE	CDEFGAB	GABCDEF	DEFGABC
5	FGABCDE	DEFGABC	BCDEFGA	GABCDEF	EFGABCD	CDEFGAB
6	FGABCDE	DEFGABC	BCDEFGA	GABCDEF	EFGABCD	CDEFGAB

Design PL50 (Parent Design PL24). $t = k = 11$, $p = 4$, $b = 5$, $n = 55$, $IF_d = 1.514$, $IF_r = 2.077$. An example is

Block

		1	2	3
			Subjects	
Period	1	ABCDEFGHIJK	ABCDEFGHIJK	ABCDEFGHIJK
	2	BCDEFGHIJKA	DEFGHIJKABC	EFGHIJKABCD
	3	JKABCDEFGHI	FGHIJKABCDE	DEFGHIJKABC
	4	JKABCDEFGHI	FGHIJKABCDE	DEFGHIJKABC

(continued)

Block

		4	5
		Subjects	
Period	1	ABCDEFGHIJK	ABCDEFGHIJK
	2	FGHIJKABCDE	JKABCDEFGHI
	3	BCDEFGHIJKA	EFGHIJKABCD
	4	BCDEFGHIJKA	EFGHIJKABCD

Design PL51 (Parent Design PL25). $t = k = 11$, $p = 6$, $b = 5$, $n = 55$, $IF_d = 1.230$, $IF_r = 1.379$. An example is

		Block 1	Block 2 Subjects	Block 3
Period	1	ABCDEFGHIJK	ABCDEFGHIJK	ABCDEFGHIJK
	2	BCDEFGHIJKA	DEFGHIJKABC	EFGHIJKABCD
	3	JKABCDEFGHI	FGHIJKABCDE	DEFGHIJKABC
	4	DEFGHIJKABC	JKABCDEFGHI	BCDEFGHIJKA
	5	KABCDEFGHIJ	IJKABCDEFGH	HIJKABCDEFG
	6	KABCDEFGHIJ	IJKABCDEFGH	HIJKABCDEFG

(continued)

		Block 4	Block 5
		Subjects	
Period	1	ABCDEFGHIJK	ABCDEFGHIJK
	2	FGHIJKABCDE	JKABCDEFGHI
	3	BCDEFGHIJKA	EFGHIJKABCD
	4	EFGHIJKABCD	FGHIJKABCDE
	5	GHIJKABCDEF	CDEFGHIJKAB
	6	GHIJKABCDEF	CDEFGHIJKAB

Design PL52 (Parent Design PL27). $t = k = 13$, $p = 5$, $b = 4$, $n = 52$, $IF_d = 1.312$, $IF_r = 1.566$. An example is (the rows representing periods)

	Block 1	Block 2	Block 3 Subjects	Block 4
1	ABCDEFGHIJKLM	ABCDEFGHIJKLM	ABCDEFGHIJKLM	ABCDEFGHIJKLM
2	BCDEFGHIJKLMA	IJKLMABCDEFGH	KLMABCDEFGHIJ	HIJKLMABCDEFG
3	DEFGHIJKLMABC	MABCDEFGHIJKL	GHIJKLMABCDEF	FGHIJKLMABCDE
4	JKLMABCDEFGHI	CDEFGHIJKLMAB	LMABCDEFGHIJK	EFGHIJKLMABCD
5	JKLMABCDEFGHI	CDEFGHIJKLMAB	LMABCDEFGHIJK	EFGHIJKLMABCD

Design PL53 (Parent Design PL28). $t = k = 13$, $p = 6$, $b = 3$, $n = 39$, $IF_d = 1.229$, $IF_r = 1.377$. An example is

Block

	1	2	3
		Subjects	

		1	2	3
	1	ABCDEFGHIJKLM	ABCDEFGHIJKLM	ABCDEFGHIJKLM
	2	BCDEFGHIJKLMA	DEFGHIJKLMABC	JKLMABCDEFGHI
Period	3	DEFGHIJKLMABC	JKLMABCDEFGHI	BCDEFGHIJKLMA
	4	HIJKLMABCDEFG	IJKLMABCDEFGH	LMABCDEFGHIJK
	5	CDEFGHIJKLMAB	GHIJKLMABCDEF	FGHIJKLMABCDE
	6	CDEFGHIJKLMAB	GHIJKLMABCDEF	FGHIJKLMABCDE

Design PL54 (Parent Design PL29). $t = k = 16$, $p = 5$, $b = 5$, $n = 80$, $IF_d = 1.312$, $IF_r = 1.563$. An example is (the rows representing periods)

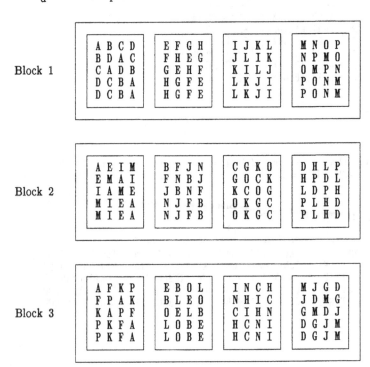

(continued)

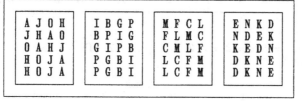

Block 4

A N G L	M B K H	E J C P	I F O D
N L A G	B H M K	J P E C	F D I O
G A L N	K M H B	C E P J	O I D F
L G N A	H K B M	P C J E	D O F I
L G N A	H K B M	P C J E	D O F I

Block 5

A J O H	I B G P	M F C L	E N K D
J H A O	B P I G	F L M C	N D E K
O A H J	G I P B	C M L F	K E D N
H O J A	P G B I	L C F M	D K N E
H O J A	P G B I	L C F M	D K N E

Appendix 5.D. Designs PL 55–75. Balanced designs with $t \geq k$

Design PL55. $t = 4$, $p = 2$, $k = 3$, $b = 4$, $n = 12$, VIF = 4.000
    Constructed by deleting any one period of Design PL56. An example is

			Block									
		1			2			3			4	
Subject →	1	2	3	4	5	6	7	8	9	10	11	12
Period 1	C	D	B	D	C	A	A	B	D	B	A	C
2	D	B	C	C	A	D	B	D	A	A	C	B

Design PL56. $t = 4$, $p = 3$, $k = 3$, $b = 4$, $n = 12$, VIF = 1.250
    An example is

			Block									
		1			2			3			4	
Subject →	1	2	3	4	5	6	7	8	9	10	11	12
Period 1	C	D	B	D	C	A	A	B	D	B	A	C
2	D	B	C	C	A	D	B	D	A	A	C	B
3	B	C	D	A	D	C	D	A	B	C	B	A

Design PL57. $t = 5$, $p = 2$, $k = 4$, $b = 5$, $n = 20$, VIF = 3.000
    Constructed by deleting any two periods of Design PL59. The periods in the design may then be randomized, provided the same randomization is used for all blocks. An example is

				Block		
		1	2	3	4	5
				Subjects		
Period 1		C A D B	D B E C	E C A D	A D B E	B E C A
2		B D A C	C E B D	D A C E	E B D A	A C E B

Design PL58. $t = 5$, $p = 3$, $k = 4$, $b = 5$, $n = 20$, VIF = 1.235
    Constructed by deleting any one period of Design PL59. The periods in the design may then be randomized, provided the same randomization is used for all blocks. An example is

			Block			
		1	2	3	4	5
Subject →		1 2 3 4	1 2 3 4	1 2 3 4	1 2 3 4	1 2 3 4
	1	B D A C	C E B D	D A C E	E B D A	A C E B
Period	2	D C B A	E D C B	A E D C	B A E D	C B A E
	3	C A D B	D B E C	E C A D	A D B E	B E C A

Design PL59. $t = 5$, $p = 4$, $k = 4$, $b = 5$, $n = 20$, VIF = 1.100
    An example is

		Block				
		1	2	3	4	5
Subject →		1 2 3 4	1 2 3 4	1 2 3 4	1 2 3 4	1 2 3 4
Period	1	A B C D	B C D E	C D E A	D E A B	E A B C
	2	B D A C	C E B D	D A C E	E B D A	A C E B
	3	C A D B	D B E C	E C A D	A D B E	B E C A
	4	D C B A	E D C B	A E D C	B A E D	C B A E

Design PL60. $t = 6$, $p = 5$, $k = 5$, $b = 6$, $n = 30$, VIF = 1.056
    An example is

		Block					
		1	2	3	4	5	6
				Subjects			
Period	1	ABCDE	BCDEF	ACDEF	ABDEF	ABCEF	ABCDF
	2	CDEAB	EFCDB	DFCAE	EDAFB	EFABC	DABFC
	3	DEABC	FDEBC	EDAFC	DFBAE	CEFAB	BCFAD
	4	EABCD	DBFCF	CEFDA	FAEBD	BCEFA	FDACB
	5	BCDEA	CEBFD	FAECD	BEFDA	FABCE	CFDBA

Design PL61. $t = 7$, $p = 2$, $k = 3$, $b = 14$, $n = 42$, VIF = 4.000
    Constructed by deleting any one period of Design PL63. An example is

		Block												
	1	2	3	4	5	6	7	8	9	10	11	12	13	14
							Subjects							
Period 1	ACD	BDE	CEF	DFG	EGA	FAB	GBC	ACD	BDE	CEF	DFG	EGA	FAB	GBC
2	DAC	EBD	FCE	GDF	AEG	BFA	CGB	CDA	DEB	EFC	FGD	GAE	ABF	BCG

Design PL62. $t = 7$, $p = 2$, $k = 6$, $b = 7$, $n = 42$, VIF = 2.500
    Constructed by deleting any four periods of Design PL68. The periods in the design may then be randomized, provided the same randomization is used for all blocks. An example is (the rows representing periods)

	Block					
1	2	3	4	5	6	7
			Subjects			
BDFACE	CEGBDF	DFACEG	EGBDFA	FACEGB	GBDFAC	ACEGBD
ECAFDB	FDBGEC	GECAFD	AFDBGE	BGECAF	CAFDBG	DBGECA

Design PL63. $t = 7$, $p = 3$, $k = 3$, $b = 14$, $n = 42$, VIF $= 1.250$
An example is

		1	2	3	4	5	6	Block 7	8	9	10	11	12	13	14
								Subjects							
	1	ACD	BDE	CEF	DFG	EGA	FAB	GBC	ACD	BDE	CEF	DFG	EGA	FAB	GBC
Period	2	CDA	DEB	EFC	FGD	GAE	ABF	BCG	DAC	EBD	FCE	GDF	AEG	BFA	CGB
	3	DAC	EBD	FCE	GDF	AEG	BFA	CGB	CDA	DEB	EFC	FGD	GAE	ABF	BCG

Design PL64. $t = 7$, $p = 3$, $k = 6$, $b = 7$, $n = 28$, VIF $= 1.222$
Constructed by deleting any three periods of Design PL68. The periods in the design may then be randomized, provided the same randomization is used for all blocks. An example is (the rows representing periods)

			Block			
1	2	3	4	5	6	7
			Subjects			
ABCDEF	BCDEFG	CDEFGA	DEFGAB	EFGABC	FGABCD	GABCDE
DAEBFC	EBFCGD	FCGDAE	GDAEBF	AEBFCG	BFCGDA	CGDAEB
FEDCBA	GFEDCB	AGFEDC	BAGFED	CBAGFE	DCBAGF	EDCBAG

Design PL65. $t = 7$, $p = 4$, $k = 4$, $b = 7$, $n = 42$, VIF $= 1.100$
An example is

		1	2	3	Block 4	5	6	7
					Subjects			
	1	ABDG	BCEA	CDFB	DEGC	EFAD	FGBE	GACF
Period	2	BGAD	CABE	DBCF	ECDG	FDEA	GEFB	AFGC
	3	DAGB	EBAC	FCBD	GDCE	AEDF	BFEG	CGFA
	4	GDBA	AECB	BFDC	CGED	DAFE	EBGF	FCAG

Design PL66. $t = 7$, $p = 4$, $k = 6$, $b = 7$, $n = 42$, VIF $= 1.097$
Constructed by deleting any two periods of Design PL68. The periods in the design may then be randomized, provided the same randomization is used for all blocks. An example is (the rows representing periods)

			Block			
1	2	3	4	5	6	7
			Subjects			
ABCDEF	BCDEFG	CDEFGA	DEFGAB	EFGABC	FGABCD	GABCDE
CFBEAD	DGCFBE	EADGCF	FBEADG	GCFBEA	ADGCFB	BEADGC
DAEBFC	EBFCGD	FCGDAE	GDAEBF	AEBFCG	BFCGDA	CGDAEB
FEDCBA	GFEDCB	AGFEDC	BAGFED	CBAGFE	DCBAGF	EDCBAG

Design PL67. $t = 7$, $p = 5$, $k = 6$, $b = 7$, $n = 42$, VIF = 1.055

Constructed by deleting any one period of Design PL68. The periods in the design may then be randomized, provided the same randomization is used for all blocks. An example is (the rows representing periods)

			Block			
1	2	3	4	5	6	7
			Subjects			
ABCDEF	BCDEFG	CDEFGA	DEFGAB	EFGABC	FGABCD	GABCDE
BDFACE	CEGBDF	DFACEG	EGBDFA	FACEGB	GBDFAC	ACEGBD
DAEBFC	EBFCGD	FCGDAE	GDAEBF	AEBFCG	BFCGDA	CGDAEB
ECAFDB	FDBGEC	GECAFD	AFDBGE	BGECAF	CAFDBG	DBGECA
FEDCBA	GFEDCB	AGFEDC	BAGFED	CBAGFE	DCBAGF	EDCBAG

Design PL68. $t = 7$, $p = 6$, $k = 6$, $b = 7$, $n = 42$, VIF = 1.036

An example is (the rows representing periods and the columns subjects)

			Block			
1	2	3	4	5	6	7
			Subjects			
ABCDEF	BCDEFG	CDEFGA	DEFGAB	EFGABC	FGABCD	GABCDE
BDFACE	CEGBDF	DFACEG	EGBDFA	FACEGB	GBDFAC	ACEGBD
CFBEAD	DGCFBE	EADGCF	FBEADG	GCFBEA	ADGCFB	BEADGC
DAEBFC	EBFCGD	FCGDAE	GDAEBF	AEBFCG	BFCGDA	CGDAEB
ECAFDB	FDBGEC	GECAFD	AFDBGE	BGECAF	CAFDBG	DBGECA
FEDCBA	GFEDCB	AGFEDC	BAGFED	CBAGFE	DCBAGF	EDCBAG

Design PL69. $t = 10$, $p = 4$, $k = 4$, $b = 15$, $n = 60$, VIF = 1.100

An example is (the rows representing periods)

A B C D	A B E F	A C G H	A D I K	A E G I
B D A C	B F A E	C H A G	D K A I	E I A G
C A D B	E A F B	G A H C	I A K D	G A I E
D C B A	F E B A	H G C A	K I D A	I G E A

A F H J	B C F I	B D G J	B E H J	B G H I
F J A H	C I B F	D J B G	E J B H	G I B H
H A J F	F B I C	G B J D	H B J E	H B I G
J H F A	I F C B	J G D B	J H E B	I H G B

C E I J	C F G J	C D E H	D E F G	D F H I
E J C I	F J C G	D H C E	E G D F	F I D H
I C J E	G C J F	E C H D	F D G E	H D I F
J I E C	J G F C	H E D C	G F E D	I H F D

Design PL70. $t = 11$, $p = 3$, $k = 5$, $b = 11$, $n = 55$, VIF = 1.227

    Constructed by deleting the first two periods of Design PL71. An example is

		Block 1	2	3	4	5	6
		Subjects					
	1	HACBE	IBDCF	JCEDG	KDFEH	AEGFI	BFHGJ
Period	2	ECBHA	FDCIB	GEDJC	HFEKD	IGFAE	JHGBF
	3	CHEAB	DIFBC	EJGCD	FKHDE	GAIEF	HBJFG

(continued)

		7	8	Block 9	10	11
				Subjects		
	1	CGIHK	DHJIA	EIKJB	FJAKC	GKBAD
Period	2	KIHCG	AJIDH	BKJEI	CAKFJ	DBAGK
	3	ICKGH	JDAHI	KEBIJ	AFCJK	BGDKA

Design PL71. $t = 11$, $p = 5$, $k = 5$, $b = 11$, $n = 55$, VIF = 1.056

    An example is

		Block 1	2	3	4	5	6
		Subjects					
	1	ABHEC	BCIFD	CDJGE	DEKHF	EFAIG	FGBJH
	2	BEACH	CFBDI	DGCEJ	EHDFK	FIEGA	GJFHB
Period	3	HACBE	IBDCF	JCEDG	KDFEH	AEGFI	BFHGJ
	4	ECBHA	FDCIB	GEDJC	HFEKD	IGFAE	JHGBF
	5	CHEAB	DIFBC	EJGCD	FKHDE	GAIEF	HBJFG

(continued)

		7	8	Block 9	10	11
				Subjects		
	1	GHCKI	HIDAJ	IJEBK	JKFCA	KAGDB
	2	HKGIC	IAHJD	JBIKE	KCJAF	ADKBG
Period	3	CGIHK	DHJIA	EIKJB	FJAKC	GKBAD
	4	KIHCG	AJIDH	BKJEI	CAKFJ	DBAGK
	5	ICKGH	JDAHI	KEBIJ	AFCJK	BGDKA

Design PL72. $t = 11$, $p = 6$, $k = 6$, $b = 11$, $n = 66$, VIF = 1.036
    An example is (the rows representing periods)

Block

1	2	3	4	5	6

Subjects

ACFGDH	BDGHEI	CEHIFJ	DFIJGK	EGJKHA	FHKAIB
CGHAFD	DHIBGE	EIJCHF	FJKDIG	GKAEJH	HABFKI
FHCDAG	GIDEBH	HJEFCI	IKFGDJ	JAGHEK	KBHIFA
GADCHF	HBEDIG	ICFEJH	JDGFKI	KEHGAJ	AFIHBK
DFAHGC	EGBIHD	FHCJIE	GIDKJF	HJEAKG	IKFBAH
HDGFCA	IEHGDB	JFIHEC	KGJIFD	AHKJGE	BIAKHF

(continued)

Block

7	8	9	10	11

Subjects

GIABJC	HJBCKD	IKCDAE	JADEBF	KBEFCG
IBCGAJ	JCDHBK	KDEICA	AEFJDB	BFGKEC
ACIJGB	BDJKHC	CEKAID	DFABJE	EGBCKF
BGJICA	CHKJDB	DIAKEC	EJBAFD	FKCBGE
JAGCBI	KBHDCJ	ACIEDK	BDJFEA	CEKGFB
CJBAIG	DKCBJH	EADCKI	FBEDAJ	GCFEBK

Design PL73. $t = 13$, $p = 4$, $k = 4$, $b = 13$, $n = 52$, VIF = 1.100
    An example is (the rows representing periods and the columns subjects)

A B D J	B C E K	C D F L	D E G M	E F H A
B J A D	C K B E	D L C F	E M D G	F A E H
D A J B	E B K C	F C L D	G D M E	H E A F
J D B A	K E C B	L F D C	M G E D	A H F E

F G I B	G H J C	H I K D	I J L E	J K M F
G B F I	H C G J	I D H K	J E I L	K F J M
I F B G	J G C H	K H D I	L I E J	M J F K
B I G F	C J H G	D K I H	E L J I	F M K J

K L A G	L M B H	M A C I
L G K A	M H L B	A I M C
A K G L	B L H M	C M I A
G A L K	H B M L	I C A M

Design PL74. $t = 16$, $p = 4$, $k = 4$, $b = 20$, $n = 80$, VIF = 1.100
   An example is

```
A B C D      E F G H      I J K L      M N O P      A E I M
B D A C      F H E G      J L I K      N P M O      E M A I
C A D B      G E H F      K I L J      O M P N      I A M E
D C B A      H G F E      L K J I      P O N M      M I E A


B F J N      C G K O      D H L P      A F K P      E B O L
F N B J      G O C K      H P D L      F P A K      B L E O
J B N F      K C O G      L D P H      K A P F      O E L B
N J F B      O K G C      P L H D      P K F A      L O B E


I N C H      M J G D      A N G L      M B K H      E J C P
N H I C      J D M G      N L A G      B H M K      J P E C
C I H N      G M D J      G A L N      K M H B      C E P J
H C N I      D G J M      L G N A      H K B M      P C J E


I F O D      A J O H      I B G P      M F C L      E N K D
F D I O      J H A O      B P I G      F L M C      N D E K
O I D F      O A H J      G I P B      C M L F      K E D N
D O F I      H O J A      P G B I      L C F M      D K N E
```

Design PL75. $t = 16$, $p = 6$, $k = 6$, $b = 16$, $n = 96$, VIF = 1.036
   An example is

			Block			
1	2	3	4	5	6	7
			Subjects			
ABCDEF	BGHIJA	CAMGKL	DHAKNO	ELNAPI	FJOMAP	GNLPOC
BCDEFA	GHIJAB	AMGKLC	HAKNOD	LNAPIE	JOMAPF	NLPOCG
FABCDE	ABGHIJ	LCAMGK	ODHAKN	IELNAP	PFJOMA	CGNLPO
CDEFAB	HIJABG	MGKLCA	AKNODH	NAPIEL	OMAPFJ	LPOCGN
EFABCD	JABGHI	KLCAMG	NODHAK	PIELNA	APFJOM	OCGNLP
DEFABC	IJABGH	GKLCAM	KNODHA	APIELN	MAPFJO	POCGNL

(continued)

			Block 11 Subjects	12	13	14
8	9	10				
HPLBDM	IOKEMB	JKFLBN	KDPCIJ	LCJOHE	MFINCH	NMEJGA
PLBDMH	OKEMBI	KFLBNJ	DPCIJK	CJOHEL	FINCHM	MEJGAN
MHPLBD	BIOKEM	NJKFLB	JKDPCI	ELCJOH	HMFINC	ANMEJG
LBDMHP	KEMBIO	FLBNJK	PCIJKD	JOHELC	INCHMF	EJGANM
DMHPLB	MBIOKE	BNJKFL	IJKDPC	HELCJO	CHMFIN	GANMEJ
BDMHPL	EMBIOK	LBNJKF	CIJKDP	OHELCJ	NCHMFI	JGANME

(continued)

Block 15	16 Subjects
OIDFLG	PEGHFK
IDFLGO	EGHFKP
GOIDFL	KPEGHF
DFLGOI	GHFKPE
LGOIDF	FKPEGH
FLGOID	HFKPEG

Appendix 5.E. Designs PL 76–93. Extra-period balanced designs with $t > k$

Design PL76 (Parent Design PL55). $t = 4$, $p = 3$, $k = 3$, $b = 4$, $n = 12$, $IF_d = 5.333$, $IF_r = 13.333$. An example is

		Block											
		1			2			3			4		
Subject →		1	2	3	4	5	6	7	8	9	10	11	12
Period	1	C	D	B	D	C	A	A	B	D	B	A	C
	2	D	B	C	C	A	D	B	D	A	A	C	B
	3	D	B	C	C	A	D	B	D	A	A	C	B

Design PL77 (Parent Design PL56). $t = 4$, $p = 4$, $k = 3$, $b = 4$, $n = 12$, $IF_d = 1.5625$, $IF_r = 2.250$. An example is

		Block											
		1			2			3			4		
Subject →		1	2	3	4	5	6	7	8	9	10	11	12
Period	1	C	D	B	D	C	A	A	B	D	B	A	C
	2	D	B	C	C	A	D	B	D	A	A	C	B
	3	B	C	D	A	D	C	D	A	B	C	B	A
	4	B	C	D	A	D	C	D	A	B	C	B	A

Design PL78 (Parent Design PL57). $t = 5$, $p = 3$, $k = 4$, $b = 5$, $n = 20$, $IF_d = 4.000$, $IF_r = 9.333$. An example is

		Block				
		1	2	3	4	5
				Subjects		
Period	1	D C B A	E D C B	A E D C	B A E D	C B A E
	2	B D A C	C E B D	D A C E	E B D A	A C E B
	3	B D A C	C E B D	D A C E	E B D A	A C E B

Design PL79 (Parent Design PL58). $t = 5$, $p = 4$, $k = 4$, $b = 5$, $n = 20$, $IF_d = 1.544$, $IF_r = 2.184$. An example is

		Block				
		1	2	3	4	5
Subject →		1 2 3 4	1 2 3 4	1 2 3 4	1 2 3 4	1 2 3 4
Period	1	A B C D	B C D E	C D E A	D E A B	E A B C
	2	B D A C	C E B D	D A C E	E B D A	A C E B
	3	D C B A	E D C B	A E D C	B A E D	C B A E
	4	D C B A	E D C B	A E D C	B A E D	C B A E

Design PL80 (Parent Design PL59). $t = 5$, $p = 5$, $k = 4$, $b = 5$, $n = 20$, $IF_d = 1.320$, $IF_r = 1.600$. An example is

		Block				
		1	2	3	4	5
Subject →		1 2 3 4	1 2 3 4	1 2 3 4	1 2 3 4	1 2 3 4
	1	A B C D	B C D E	C D E A	D E A B	E A B C
	2	B D A C	C E B D	D A C E	E B D A	A C E B
Period	3	C A D B	D B E C	E C A D	A D B E	B E C A
	4	D C B A	E D C B	A E D C	B A E D	C B A E
	5	D C B A	E D C B	A E D C	B A E D	C B A E

Design PL81 (Parent Design PL60). $t = 6$, $p = 6$, $k = 5$, $b = 6$, $n = 30$, $IF_d = 1.231$, $IF_r = 1.389$. An example is

		Block					
		1	2	3	4	5	6
				Subjects			
	1	ABCDE	BCDEF	ACDEF	ABDEF	ABCEF	ABCDF
	2	CDEAB	EFCDB	DFCAE	EDAFB	EFABC	DABFC
Period	3	DEABC	FDEBC	EDAFC	DFBAE	CEFAB	BCFAD
	4	EABCD	DBFCF	CEFDA	FAEBD	BCEFA	FDACB
	5	BCDEA	CEBFD	FAECD	BEFDA	FABCE	CFDBA
	6	BCDEA	CEBFD	FAECD	BEFDA	FABCE	CFDBA

Design PL82 (Parent Design PL61). $t = 7$, $p = 3$, $k = 3$, $b = 14$, $n = 42$ $IF_d = 5.333$, $IF_r = 13.333$. An example is

		Block													
		1	2	3	4	5	6	7	8	9	10	11	12	13	14
								Subjects							
	1	ACD	BDE	CEF	DFG	EGA	FAB	GBC	ACD	BDE	CEF	DFG	EGA	FAB	GBC
Period	2	DAC	EBD	FCE	GDF	AEG	BFA	CGB	CDA	DEB	EFC	FGD	GAE	ABF	BCG
	3	DAC	EBD	FCE	GDF	AEG	BFA	CGB	CDA	DEB	EFC	FGD	GAE	ABF	BCG

Design PL83 (Parent Design PL62). $t = 7$, $p = 3$, $k = 6$, $b = 7$, $n = 42$, $IF_d = 3.333$, $IF_r = 7.333$. An example is (the rows representing periods)

			Block			
1	2	3	4	5	6	7
			Subjects			
ECAFDB	FDBGEC	GECAFD	AFDBGE	BGECAF	CAFDBG	DBGECA
BDFACE	CEGBDF	DFACEG	EGBDFA	FACEGB	GBDFAC	ACEGBD
BDFACE	CEGBDF	DFACEG	EGBDFA	FACEGB	GBDFAC	ACEGBD

Design PL84 (Parent Design PL63). $t = 7$, $p = 4$, $k = 3$, $b = 14$, $n = 42$
IF$_d$ = 1.5625, IF$_r$ = 2.250. An example is

		1	2	3	4	5	6	Block 7	8	9	10	11	12	13	14
								Subjects							
	1	ACD	BDE	CEF	DFG	EGA	FAB	GBC	ACD	BDE	CEF	DFG	EGA	FAB	GBC
Period	2	CDA	DEB	EFC	FGD	GAE	ABF	BCG	DAC	EBD	FCE	GDF	AEG	BFA	CGB
	3	DAC	EBD	FCE	GDF	AEG	BFA	CGB	CDA	DEB	EFC	FGD	GAE	ABF	BCG
	4	DAC	EBD	FCE	GDF	AEG	BFA	CGB	CDA	DEB	EFC	FGD	GAE	ABF	BCG

Design PL85 (Parent Design PL64). $t = 7$, $p = 4$, $k = 6$, $b = 7$, $n = 42$,
IF$_d$ = 1.528, IF$_r$ = 2.125. An example is (the rows representing periods)

1	2	3	Block 4	5	6	7
			Subjects			
CFBEAD	DGCFBE	EADGCF	FBEADG	GCFBEA	ADGCFB	BEADGC
DAEBFC	EBFCGD	FCGDAE	GDAEBF	AEBFCG	BFCGDA	CGDAEB
FEDCBA	GFEDCB	AGFEDC	BAGFED	CBAGFE	DCBAGF	EDCBAG
FEDCBA	GFEDCB	AGFEDC	BAGFED	CBAGFE	DCBAGF	EDCBAG

Design PL86 (Parent Design PL65). $t = 7$, $p = 5$, $k = 4$, $b = 7$, $n = 28$,
IF$_d$ = 1.320, IF$_r$ = 1.600. An example is

		1	2	3	Block 4	5	6	7
					Subjects			
	1	ABDG	BCEA	CDFB	DEGC	EFAD	FGBE	GACF
	2	BGAD	CABE	DBCF	ECDG	FDEA	GEFB	AFGC
Period	3	DAGB	EBAC	FCBD	GDCE	AEDF	BFEG	CGFA
	4	GDBA	AECB	BFDC	CGED	DAFE	EBGF	FCAG
	5	GDBA	AECB	BFDC	CGED	DAFE	EBGF	FCAG

Design PL87 (Parent Design PL66). $t = 7$, $p = 5$, $k = 6$, $b = 7$, $n = 42$,
IF$_d$ = 1.316, IF$_r$ = 1.583. An example is (the rows representing periods)

1	2	3	Block 4	5	6	7
			Subjects			
DAEBFC	EBFCGD	FCGDAE	GDAEBF	AEBFCG	BFCGDA	CGDAEB
ECAFDB	FDBGEC	GECAFD	AFDBGE	BGECAF	CAFDBG	DBGECA
FEDCBA	GFEDCB	AGFEDC	BAGFED	CBAGFE	DCBAGF	EDCBAG
ABCDEF	BCDEFG	CDEFGA	DEFGAB	EFGABC	FGABCD	GABCDE
ABCDEF	BCDEFG	CDEFGA	DEFGAB	EFGABC	FGABCD	GABCDE

Design PL88 (Parent Design PL67). $t = 7$, $p = 6$, $k = 6$, $b = 7$, $n = 42$, $\text{IF}_d = 1.231$, IF_r 1.386. An example is (the rows representing periods)

			Block			
1	2	3	4	5	6	7
			Subjects			
FEDCBA	GFEDCB	AGFEDC	BAGFED	CBAGFE	DCBAGF	EDCBAG
DAEBFC	EBFCGD	FCGDAE	GDAEBF	AEBFCG	BFCGDA	CGDAEB
ECAFDB	FDBGEC	GECAFD	AFDBGE	BGECAF	CAFDBG	DBGECA
ABCDEF	BCDEFG	CDEFGA	DEFGAB	EFGABC	FGABCD	GABCDE
ABCDEF	BCDEFG	CDEFGA	DEFGAB	EFGABC	FGABCD	GABCDE

Design PL89 (Parent Design PL69). $t = 10$, $p = 5$, $k = 4$, $b = 15$, $n = 60$, $\text{IF}_d = 1.320$, $\text{IF}_r = 1.600$. An example is

A B C D	A B E F	A C G H	A D I K	A E G I
B D A C	B F A E	C H A G	D K A I	E I A G
C A D B	E A F B	G A H C	I A K D	G A I E
D C B A	F E B A	H G C A	K I D A	I G E A
D C B A	F E B A	H G C A	K I D A	I G E A

(continued)

A F H J	B C F I	B D G J	B E H J	B G H I
F J A H	C I B F	D J B G	E J B H	G I B H
H A J F	F B I C	G B J D	H B J E	H B I G
J H F A	I F C B	J G D B	J H E B	I H G B
J H F A	I F C B	J G D B	J H E B	I H G B

(continued)

C E I J	C F G J	C D E H	D E F G	D F H I
E J C I	F J C G	D H C E	E G D F	F I D H
I C J E	G C J F	E C H D	F D G E	H D I F
J I E C	J G F C	H E D C	G F E D	I H F D
J I E C	J G F C	H E D C	G F E D	I H F D

Design PL90 (Parent Design PL70). $t = 11$, $p = 4$, $k = 5$, $b = 11$, $n = 55$, $IF_d = 1.534$, $IF_r = 2.148$. An example is

		Block					
		1	2	3	4	5	6
				Subjects			
	1	HACBE	IBDCF	JCEDG	KDFEH	AEGFI	BFHGJ
Period	2	ECBHA	FDCIB	GEDJC	HFEKD	IGFAE	JHGBF
	3	CHEAB	DIFBC	EJGCD	FKHDE	GAIEF	HBJFG
	4	CHEAB	DIFBC	EJGCD	FKHDE	GAIEF	HBJFG

(continued)

		Block				
		7	8	9	10	11
				Subjects		
	1	CGIHK	DHJIA	EIKJB	FJAKC	GKBAD
Period	2	KIHCG	AJIDH	BKJEI	CAKFJ	DBAGK
	3	ICKGH	JDAHI	KEBIJ	AFCJK	BGDKA
	4	ICKGH	JDAHI	KEBIJ	AFCJK	BGDKA

Design PL91 (Parent Design PL71). $t = 11$, $p = 6$, $k = 5$, $b = 11$, $n = 55$, $IF_d = 1.231$, $IF_r = 1.389$. An example is

		Block					
		1	2	3	4	5	6
				Subjects			
	1	ABHEC	BCIFD	CDJGE	DEKHF	EFAIG	FGBJH
	2	BEACH	CFBDI	DGCEJ	EHDFK	FIEGA	GJFHB
Period	3	HACBE	IBDCF	JCEDG	KDFEH	AEGFI	BFHGJ
	4	ECBHA	FDCIB	GEDJC	HFEKD	IGFAE	JHGBF
	5	CHEAB	DIFBC	EJGCD	FKHDE	GAIEF	HBJFG
	6	CHEAB	DIFBC	EJGCD	FKHDE	GAIEF	HBJFG

(continued)

		Block				
		7	8	9	10	11
				Subjects		
	1	GHCKI	HIDAJ	IJEBK	JKFCA	KAGDB
	2	HKGIC	IAHJD	JBIKE	KCJAF	ADKBG
Period	3	CGIHK	DHJIA	EIKJB	FJAKC	GKBAD
	4	KIHCG	AJIDH	BKJEI	CAKFJ	DBAGK
	5	ICKGH	JDAHI	KEBIJ	AFCJK	BGDKA
	6	ICKGH	JDAHI	KEBIJ	AFCJK	BGDKA

Design PL92 (Parent Design PL73). $t = 13$, $p = 5$, $k = 4$, $b = 13$, $n = 52$, $\text{IF}_d = 1.320$, $\text{IF}_r = 1.600$. An example is (the rows representing periods and the columns subjects)

A B D J	B C E K	C D F L	D E G M	E F H A
B J A D	C K B E	D L C F	E M D G	F A E H
D A J B	E B K C	F C L D	G D M E	H E A F
J D B A	K E C B	L F D C	M G E D	A H F E
J D B A	K E C B	L F D C	M G E D	A H F E

F G I B	G H J C	H I K D	I J L E	J K M F
G B F I	H C G J	I D H K	J E I L	K F J M
I F B G	J G C H	K H D I	L I E J	M J F K
B I G F	C J H G	D K I H	E L J I	F M K J
B I G F	C J H G	D K I H	E L J I	F M K J

K L A G	L M B H	M A C I
L G K A	M H L B	A I M C
A K G L	B L H M	C M I A
G A L K	H B M L	I C A M
G A L K	H B M L	I C A M

Design PL93 (Parent Design PL74). $t = 16$, $p = 5$, $k = 4$, $b = 20$, $n = 80$, $\text{IF}_d = 1.320$, $\text{IF}_r = 1.600$. An example is (the rows representing periods and the columns subjects)

A B C D	E F G H	I J K L	M N O P	A E I M
B D A C	F H E G	J L I K	N P M O	E M A I
C A D B	G E H F	K I L J	O M P N	I A M E
D C B A	H G F E	L K J I	P O N M	M I E A
D C B A	H G F E	L K J I	P O N M	M I E A

B F J N	C G K O	D H L P	A F K P	E B O L
F N B J	G O C K	H P D L	F P A K	B L E O
J B N F	K C O G	L D P H	K A P F	O E L B
N J F B	O K G C	P L H D	P K F A	L O B E
N J F B	O K G C	P L H D	P K F A	L O B E

(continued)

```
I N C H     M J G D     A N G L     M B K H     E J C P
N H I C     J D M G     N L A G     B H M K     J P E C
C I H N     G M D J     G A L N     K M H B     C E P J
H C N I     D G J M     L G N A     H K B M     P C J E
H C N I     D G J M     L G N A     H K B M     P C J E
```

```
I F O D     A J O H     I B G P     M F C L     E N K D
F D I O     J H A O     B P I G     F L M C     N D E K
O I D F     O A H J     G I P B     C M L F     K E D N
D O F I     H O J A     P G B I     L C F M     D K N E
D O F I     H O J A     P G B I     L C F M     D K N E
```

Appendix 5.F. Designs PL 94–98. Extra–period balanced designs with $k = 2$

Design PL94. $t = 3$, $p = 3$, $k = 2$, $b = 3$, $n = 6$. An example is

		Block					
		1		2		3	
Subject →		1	2	3	4	5	6
Period	1	A	B	A	C	B	C
	2	B	A	C	A	C	B
	3	B	A	C	A	C	B

Design PL95. $t = 4$, $p = 3$, $k = 2$, $b = 6$, $n = 12$. An example is (the rows representing periods)

	Block											
	1		2		3		4		5		6	
					Subjects							
	1	2	3	4	5	6	7	8	9	10	11	12
1	A	B	A	C	A	D	B	C	B	D	C	D
2	B	A	C	A	D	A	C	B	D	B	D	C
3	B	A	C	A	D	A	C	B	D	B	D	C

Design PL96. $t = 5$, $p = 3$, $k = 2$, $b = 10$, $n = 20$. An example is (the rows representing periods)

	Block									
	1	2	3	4	5	6	7	8	9	10
					Subjects					
1	A B	A C	A D	A E	B C	B D	B E	C D	C E	D E
2	B A	C A	D A	E A	C B	D B	E B	D C	E C	E D
3	B A	C A	D A	E A	C B	D B	E B	D C	E C	E D

Design PL97. $t = 6$, $p = 3$, $k = 2$, $b = 15$, $n = 30$. An example is (the rows representing periods)

	Block									
	1	2	3	4	5	6	7	8	9	10
					Subjects					
1	A B	A C	A D	A E	A F	B C	B D	B E	B F	C D
2	B A	C A	D A	E A	F A	C B	D B	E B	F B	D C
3	B A	C A	D A	E A	F A	C B	D B	E B	F B	D C

(continued)

		Block		
	11	12	13	14	15
			Subjects		

	11	12	13	14	15
1	C E	C F	D E	D F	E F
2	E C	F C	E D	F D	F E
3	E C	F C	E D	F D	F E

Design PL98. $t = 7$, $p = 3$, $k = 2$, $b = 21$, $n = 42$. An example is (the rows representing periods)

					Block					
	1	2	3	4	5	6	7	8	9	10	11
						Subjects					

	1	2	3	4	5	6	7	8	9	10	11
1	A B	A C	A D	A E	A F	A G	B C	B D	B E	B F	B G
2	B A	C A	D A	E A	F A	G A	C B	D B	E B	F B	G B
3	B A	C A	D A	E A	F A	G A	C B	D B	E B	F B	G B

(continued)

				Block					
	12	13	14	15	16	17	18	19	20	21
				Subjects					

	12	13	14	15	16	17	18	19	20	21
1	C D	C E	C F	C G	D E	D F	D G	E F	E G	F G
2	D C	E C	F C	G C	E D	F D	G D	F E	G E	G F
3	D C	E C	F C	G C	E D	F D	G D	F E	G E	G F

Appendix 5.G. **Designs QBP 1-4.** **Quenouille–Berenblut–Patterson designs for**
3 ≤ *t* ≤ 6

Design QBP1. *t* = 3, *p* = 6, *n* = 9.

	Period					
	1	2	3	4	5	6
1	A	C	B	B	C	A
2	B	A	C	C	A	B
3	C	B	A	A	B	C
4	A	B	B	A	C	C
5	B	C	C	B	A	A
6	C	A	A	C	B	B
7	A	A	B	C	C	B
8	B	B	C	A	A	C
9	C	C	A	B	B	A

Subject (rows 4–6)

Design QBP2. *t* = 4, *p* = 8, *n* = 16.

	Period							
	1	2	3	4	5	6	7	8
1	A	D	C	B	B	C	D	A
2	B	A	D	C	C	D	A	B
3	C	B	A	D	D	A	B	C
4	D	C	B	A	A	B	C	D
5	A	C	C	A	B	B	D	D
6	B	D	D	B	C	C	A	A
7	C	A	A	C	D	D	B	B
8	D	B	B	D	A	A	C	C
9	A	B	C	D	B	A	D	C
10	B	C	D	A	C	B	A	D
11	C	D	A	B	D	C	B	A
12	D	A	B	C	A	D	C	B
13	A	A	C	C	B	D	D	B
14	B	B	D	D	C	A	A	C
15	C	C	A	A	D	B	B	D
16	D	D	B	B	A	C	C	A

Subject

Design QBP3. $t = 5$, $p = 10$, $n = 25$.

					Period					
	1	2	3	4	5	6	7	8	9	10
1	A	E	D	C	B	B	C	D	E	A
2	B	A	E	D	C	C	D	E	A	B
3	C	B	A	E	D	D	E	A	B	C
4	D	C	B	A	E	E	A	B	C	D
5	E	D	C	B	A	A	B	C	D	E
6	A	D	D	B	B	A	C	C	E	E
7	B	E	E	C	C	B	D	D	A	A
8	C	A	A	D	D	C	E	E	B	B
9	D	B	B	E	E	D	A	A	C	C
10	E	C	C	A	A	E	B	B	D	D
11	A	C	D	A	B	E	C	B	E	D
12	B	D	E	B	C	A	D	C	A	E
13	C	E	A	C	D	B	E	D	B	A
14	D	A	B	D	E	C	A	E	C	B
15	E	B	C	E	A	D	B	A	D	C
16	A	B	D	E	B	D	C	A	E	C
17	B	C	E	A	C	E	D	B	A	D
18	C	D	A	B	D	A	E	C	B	E
19	D	E	B	C	E	B	A	D	C	A
20	E	A	C	D	A	C	B	E	D	B
21	A	A	D	D	B	C	C	E	E	B
22	B	B	E	E	C	D	D	A	A	C
23	C	C	A	A	D	E	E	B	B	D
24	D	D	B	B	E	A	A	C	C	E
25	E	E	C	C	A	B	B	D	D	A

Subject

Design QBP4. $t = 6$, $p = 12$, $n = 36$.

	\multicolumn{12}{c}{Period}										
	1	2	3	4	5	6	7	8	9	10	11	12
1	A	F	E	D	C	B	B	C	D	E	F	A
2	B	A	F	E	D	C	C	D	E	F	A	B
3	C	B	A	F	E	D	D	E	F	A	B	C
4	D	C	B	A	F	F	F	A	B	C	D	E
5	E	D	C	B	A	F	F	A	B	C	D	E
6	F	E	D	C	B	A	A	B	C	D	E	F
7	A	E	E	C	C	A	B	B	D	D	F	F
8	B	F	F	D	D	B	C	C	E	E	A	A
9	C	A	A	E	E	C	D	D	F	F	B	B
10	D	B	B	F	F	D	E	E	A	A	C	C
11	E	C	C	A	A	E	F	F	B	B	D	D
12	F	D	D	B	B	F	A	A	C	C	E	E
13	A	D	E	B	C	F	B	A	D	C	F	E
14	B	E	F	C	D	A	C	B	E	D	A	F
15	C	F	A	D	E	B	D	C	F	E	B	A
16	D	A	B	E	F	C	E	D	A	F	C	B
17	E	B	C	F	A	D	F	E	B	A	D	C
18	F	C	D	A	B	E	A	F	C	B	E	D
19	A	C	E	A	C	E	B	F	D	B	F	D
20	B	D	F	B	D	F	C	A	E	C	A	E
21	C	E	A	C	E	A	D	B	F	D	B	F
22	D	F	B	D	F	B	E	C	A	E	C	A
23	E	A	C	E	A	C	F	D	B	F	D	B
24	F	B	D	F	B	D	A	E	C	A	E	C
25	A	B	E	F	C	D	B	E	D	A	F	C
26	B	C	F	A	D	E	C	F	E	B	A	D
27	C	D	A	B	E	F	D	A	F	C	B	E
28	D	E	B	C	F	A	E	B	A	D	C	F
29	E	F	C	D	A	B	F	C	B	E	D	A
30	F	A	D	E	B	C	A	D	C	F	E	B
31	A	A	E	E	C	C	B	D	D	F	F	B
32	B	B	F	F	D	D	C	E	E	A	A	C
33	C	C	A	A	E	E	D	F	F	B	B	D
34	D	D	B	B	F	F	E	A	A	C	C	E
35	E	E	C	C	A	A	F	B	B	D	D	F
36	F	F	D	D	B	B	A	C	C	E	E	A

Subject (row label for the table)

6

Cross-over Designs Lacking Variance Balance

The present chapter catalogues a large number of designs that do not possess the property of variance balance. Variance–balanced designs were the topic of Chapter 5. Unlike variance–balanced designs, where all contrasts among the treatment effects and all contrasts among the carryover effects are of equal precision, the designs in this chapter lack that particular type of balance. For some or most purposes, users would wish designs that approximate variance–balance, and this condition is fulfilled by the cyclic designs of Davis and Hall (1969) and the majority of the partially balanced incomplete block designs of Patterson and Lucas (1962). For other purposes, users might wish some comparisons to be much more sensitive than others, and a few such designs may be found among the 62 designs listed here of Patterson and Lucas (1962).

6.1. The Partially–Balanced Cross–over Designs of Patterson and Lucas

The paper of Patterson and Lucas (1962) listed many designs not possessing variance balance, in addition to those designs possessing that condition. We present an index in Appendix 6.A of all of the Patterson–Lucas designs in this chapter. In further appendices, we list examples of each of these designs, using alphabetical letters to indicate the treatments. In some cases, there are other designs that are of the same type, that cannot be constructed merely by a randomization of the letters. In these cases, that information will be given.

For all the designs in this chapter, the treatments are to be assigned randomly to the alphabetical letters. In addition, some other forms of randomization may be required. For example, in some designs, the rows may be permuted, provided that the permutation is done for the same rows in each of the blocks. This will be indicated wherever such a randomization is permissible.

Appendix 6.B lists 26 partially balanced designs whose method of construction is based upon combining a balanced incomplete block design with two associate classes of Bose *et al.* (1954) with one of the balanced cross–over designs of Patterson and Lucas (1962) which appeared in Appendix 5.B of the previous chapter. Only a few of the latter designs were used, these being PL1, PL2, PL5, PL10 and PL14. The resulting designs are designated PL99–124, and a typical representative of each design is given. In each of these designs, no treatment appears more than once in each block. Over the total design, each ordered pair of treatments occur sequentially (that is, in consecutive periods) either n_1 or n_2 times.

In all of the designs, an ordered treatment pair, such as AB, occurs with the same frequency as its reverse–ordered pair BA. Treatments whose pairs occur sequentially n_1 times are said to be first associates, while treatments whose pairs occur sequentially n_2 times are said to be second associates. Either n_1 or n_2 may be zero, but neither has a theoretical upper limit, although in Appendix 6.B the highest value for either n_1 or n_2 is four. Because n_1 and n_2 are never equal (else the design would be balanced), there are two efficiencies for treatment comparisons, depending upon whether the comparison involves treatments in the same or different associate classes. The average overall efficiency of the design is reported in Appendix 6.A, and is based on a comparison of the partially–balanced cross–over design with a balanced incomplete block design for k $(= t)$ treatments in blocks of k. The two individual efficiencies (depending upon whether one is comparing first associates or second associates) can be determined from information tabulated in Bose *et al.* (1954), and these are presented in Appendix 6.B along with an example of each design. This information should help users decide how to allocate their treatments to the associate classes, depending upon the degree of precision they may wish to have in each treatment comparison. In addition to these efficiencies, a variance inflation factor (VIF) is presented for each of these designs. The closer the VIF is to 1.0, the more separable are the treatment and carryover effects. Designs with high VIF's (say, greater than 1.5) are designs of relatively low power for detecting treatment effects while carryover parameters are in the model. Designs with VIF = 4 or greater are of doubtful value.

Appendix 6.C contains six additional partially–balanced designs (PL 125–130) that were derived by Patterson and Lucas (1962) from partially balanced incomplete block designs not listed by Bose *et al.* (1954). First and second associate classes are indicated for each treatment. In addition, VIF's are given for each of these designs. As is the case with the designs in Appendix 6.B, those with high VIF's are of doubtful utility in cross–over trials, the power of rejecting the null hypothesis of no treatment and/or carryover effects being low.

Appendix 6.D lists 22 extra–period partially–balanced cross–over designs obtained by adding an extra period to designs of Appendices 6.B and 6.C. The treatment in the extra period, for each experimental unit, is identical to that used in the last period of the parent design. The extra period guarantees orthogonality, that is, the covariances of treatments and carryover effects are zero, with resulting complete separability of those effects. Since the variance inflation factors (VIF's) for these designs are all 1.0, they are not listed individually in this appendix. The first and second associate classes for each treatment is identical to those of the parent design and are not given. Instead of VIF's, "improvement factors" IF_d and IF_r for treatment and carryover effects, respectively, are given. These improvement factors represent the ratio of the variance of the effect in question in the parent design (with $p - 1$ periods) to the variance of the same effect in the extra–period design (p periods).

Thus, for Design PL131, obtained by adding an extra period having the same treatments as the previous final period of the parent design, Design PL125, to each subject, the improvement factors are $IF_d = 5.333$ and $IF_r = 13.333$, respectively. Therefore, compared to the two–period Design PL125, the variance of the treatment effects in Design PL131 is less by a factor of 5.333, and the variance of the carryover effects is less by a factor of 13.333, than the corresponding variances for those same effects in Design PL125. This increase in precision is dramatic, and is attributable to the fact that the parent design had only two periods. Two–period designs are notoriously inefficient, having high variances and high covariances between the estimators of the treatment and carryover effects. The result of the high covariance is poor separability between treatment and carryover effects. Therefore, two period designs should be avoided as much as possible. The main reason for wishing to use a two–period design is the high dropout rate that often accompanies the use of multi–period designs. However, the price one pays for restricting the number of periods to two is poor power in detecting treatment and carryover effects.

The last appendix listing designs of Patterson and Lucas (1962) is Appendix 6.E, which concerns extra–period partially–balanced cross–over designs with treatments always occurring in block of two (that is, $k = 2$), the "parent" designs having been tabulated by Clatworthy (1955). These designs all have three periods (that is, $p = 3$). For each design, the first and second associate classes for each treatment are identified.

6.2. Analysis of Partially–Balanced Designs of Patterson and Lucas

The analyses of all the designs catalogued in Appendix 6.A, examples of which are given in Appendices 6.B – 6.E, are carried out in an identical fashion in a manner similar to that described in Section 1.1 of Chapter 1. To illustrate, consider the data set that appeared in Table 3.1 of Patterson and Lucas (1962), reproduced in the previous chapter as Table 5.1. Patterson and Lucas rearranged those data into three blocks of four experimental units (cows) each, according to performance in a preliminary period, the results of which were not reported. The data for the rearranged design are shown in Table 6.1. Readers can verify that the plan conforms to Design PL101 given in Appendix 6.B, with Blocks 2 and 3 interchanged and the subjects allocated at random within blocks (see top of next page).

The analysis using SAS involves first the creation of a data step, such as:

```
DATA FCM;
 INPUT BLOCK UNIT PERIOD TREAT$ CARRY$ FCM @@;
IF CARRY = '0' THEN CARRY = 'F';
CARDS;
1 1 1 A 0 38.7     1 1 2 D A 37.4     1 1 3 B D 34.3     1 1 4 E B 31.3
1 2 1 B 0 48.9     1 2 2 A B 46.9     1 2 3 E A 42.0     1 2 4 D E 39.6
1 3 1 E 0 34.6     1 3 2 B E 32.3     1 3 3 D B 28.5     1 3 4 A D 27.1
1 4 1 D 0 35.2     1 4 2 E D 33.5     1 4 3 A E 28.4     1 4 4 B A 25.1
2 1 1 D 0 32.9     2 1 2 A D 33.1     2 1 3 F A 27.5     2 1 4 C F 25.1
2 2 1 F 0 30.4     2 2 2 D F 29.5     2 2 3 C D 26.7     2 2 4 A C 23.1
2 3 1 C 0 30.8     2 3 2 F C 29.3     2 3 3 A F 26.4     2 3 4 D A 23.2
2 4 1 A 0 25.7     2 4 2 C A 26.1     2 4 3 D C 23.4     2 4 4 F D 18.7
3 1 1 E 0 25.4     3 1 2 F E 26.0     3 1 3 B F 23.9     3 1 4 C B 19.9
3 2 1 B 0 21.8     3 2 2 E B 23.9     3 2 3 C E 21.7     3 2 4 F C 17.6
```

(continued on next page)

Table 6.1. Data from Patterson and Lucas (1962), illustrating analysis of data from a partially–balanced cross–over trial. (Reproduced here with the permission of the North Carolina State University.) Treatments given in parentheses.

Block 1

		Cow 1	Cow 2	Cow 3	Cow 4
	1	38.7(A)	48.9(B)	34.6(E)	35.2(D)
Period	2	37.4(D)	46.9(A)	32.3(B)	33.5(E)
	3	34.3(B)	42.0(E)	28.5(D)	28.4(A)
	4	31.3(E)	39.6(D)	27.1(A)	25.1(B)

Block 2

		Cow 1	Cow 2	Cow 3	Cow 4
	1	32.9(D)	30.4(F)	30.8(C)	25.7(A)
Period	2	33.1(A)	29.5(D)	29.3(F)	26.1(C)
	3	27.5(F)	26.7(C)	26.4(A)	23.4(D)
	4	25.1(C)	23.1(A)	23.2(D)	18.7(F)

Block 3

		Cow 1	Cow 2	Cow 3	Cow 4
	1	25.4(E)	21.8(B)	21.4(F)	22.8(C)
Period	2	26.0(F)	23.9(E)	22.0(C)	21.0(B)
	3	23.9(B)	21.7(C)	19.4(E)	18.6(F)
	4	19.9(C)	17.6(F)	16.6(B)	16.1(E)

```
3 3 1 F 0 21.4     3 3 2 C F 22.0     3 3 3 E C 19.4     3 3 4 B E 16.6
3 4 1 C 0 22.8     3 4 2 B C 21.0     3 4 3 F B 18.6     3 4 4 E F 16.1
RUN;
```

The IF statement is needed to ensure estimability of the treatment and carryover effects using the LSMEANS statement in the procedure which follows. We assume, as in Section 5.2, that the time frame may be different in the three blocks, or that there may be a difference in the effect of time from block to block, and therefore choose to use a nested term, periods within blocks.

```
PROC GLM;
 CLASS BLOCK UNIT PERIOD TREAT CARRY;
MODEL FCM = BLOCK UNIT(BLOCK) PERIOD(BLOCK) TREAT CARRY/
     SOLUTION SS1 SS2 E1 E2;
RANDOM UNIT(BLOCK);
LSMEANS TREAT CARRY / STDERR PDIFF;
RUN;
```

The above code results in the following edited output:

Dependent Variable: FCM

Source	DF	Sum of Squares	Mean Square	F Value	Pr > F
Model	30	2650.1330	88.3378	161.14	0.0001
Error	17	9.3195	0.5482		
Corrected Total	47	2659.4525			

Source	DF	Type I SS	Mean Square	F Value	Pr > F
BLOCK	2	1607.011250	803.50563	1465.71	0.0001
UNIT(BLOCK)	9	628.706250	69.85625	127.43	0.0001
PERIOD(BLOCK)	9	408.031250	45.33681	82.70	0.0001
TREAT	5	2.500000	0.50000	0.91	0.4964
CARRY	5	3.884292	0.77686	1.42	0.2680

Source	DF	Type II SS	Mean Square	F Value	Pr > F
TREAT	5	3.139595	0.62792	1.15	0.3750
CARRY	5	3.884292	0.77686	1.42	0.2680

Below is presented a portion of the output of PROC GLM which gives estimates of the various parameters of the model; only the output relating to the treatment and carryover parameters is presented. It can be seen that the standard errors of the individual treatment parameters are not all identical, and the same is true of the standard errors of the carryover parameters. The standard error associated with Treatment C is smaller (more precise) than that of the other parameters, both for treatment and carryover. Since all estimates are contrasted against Treatment F in the set-to-zero convention employed by PROC GLM, this implies that the contrast of Treatment C against Treatment F has less variability than the contrast of any other treatment with Treatment F. Examining the association scheme for Design PL101 in Appendix 6.B reveals that Treatments C and F occur sequentially more frequently with each other than with other treatments, explaining the lower variance of their difference.

Parameter		Estimate	T for H0: Parameter=0	Pr > |T|	Std Error of Estimate
TREAT	A	0.40770833	0.97	0.3446	0.41938280
	B	0.18791667	0.45	0.6598	0.41938280
	C	0.74625000	1.92	0.0715	0.38827303
	D	0.74020833	1.76	0.0955	0.41938280
	E	0.51166667	1.22	0.2391	0.41938280
	F	0.00000000	.	.	.

(continued on next page)

Parameter		Estimate	T for H0: Parameter=0	Pr > |T|	Std Error of Estimate
CARRY	A	−0.24416667	−0.48	0.6354	0.50579469
	B	0.07666667	0.15	0.8813	0.50579469
	C	0.18500000	0.40	0.6977	0.46827489
	D	0.93583333	1.85	0.0817	0.50579469
	E	0.22166667	0.44	0.6667	0.50579469
	F	0.00000000	.	.	.

Similarly, one can show that the differences of Treatments B and E, and of Treatments A and D, the other first associates for Design PL101, have smaller variances than those of "second associate" treatment differences. One way of showing this is to set up another SAS DATA step, or modify the existing one, so that treatment label B is changed to label G (say). Since G occurs last in the alphabet in the set of six treatments, it will appear last in the set of treatments, so all other treatments will be contrasted against it. This will show that Treatment E will have a smaller standard error than any of the other contrasts with G, which is really Treatment B. Similarly, a further data step in which all occurrences of Treatment A is denoted H (say), so that that treatment is listed last, will reveal that the contrast of Treatments A and D is more precise than those of their second associates.

Neither the carryover effect nor the treatment effect is significant. Since the design used (PL101) has a relatively small variance inflation factor (VIF = 1.10), there is no reason to suspect that when carryover is dropped from the model the treatment effect might become significant. This was confirmed by a further run with carryover not in the model, which showed that the P-value for treatment was P = 0.3209. A further run using "period" instead of "period within blocks" reduced the explained sum of squares for periods from 125.802 to 109.962, a non-significant decrease. Hence, it is permissible to consider that there is no interaction between blocks and periods; that is, the same time frame applies to all periods.

6.3. The Designs of Davis and Hall

Davis and Hall (1969) proposed a class of cyclic cross-over designs which exist for any number of treatments and periods, by extending cyclic incomplete block designs. The word cyclic derives from the fact that given the sequence of treatments in the different time periods for the first subject of each block, the sequence for the second subject is derived by going to the next letter in the alphabet, the one exception being that when the letter corresponding to the last treatment is reached, one reverts back to the first letter, A. Thus, when there are six treatments ($t = 6$), three periods ($p = 3$), two blocks ($b = 2$), the starting sequences given by Davis and Hall (1969) for the two blocks are ADE and AFB, respectively. One simply goes to the next letter following A, D and E, respectively, to determine the treatments for the second subject, and so on for the remaining subjects in the block. The result is shown as Design DH1 in Appendix 6.F, and is also reproduced below.

Design DH1. $t = k = 6$, $p = 3$, $b = 2$, $n = 12$. $E_t = 79$, $E_d = 58$, $E_r = 34$.

An example is

		Block	
		1	2
		Subjects	
	1	ABCDEF	ABCDEF
Period	2	DEFABC	FABCDE
	3	EFABCD	BCDEFA

Note that $k = t$, so that block size is always equal to the number of treatments in any of the Davis–Hall designs. Also listed with each of the 45 Davis–Hall designs reported in Appendix 6.F are the efficiencies E_t, E_d and E_r. These are consistent with, and correspond to, the efficiencies for the designs of Patterson and Lucas (1962). E_t refers to the efficiency of the design, compared to a balanced incomplete block design, when carryover effects are not in the model. E_d and E_r refer to the efficiencies of the treatment and carryover effects, respectively, when both terms are in the model. Values reported in Appendix 6.F are average efficiencies of treatment and carryover contrasts, respectively. Corresponding to each average value is a range of efficiencies that are not reported in Appendix 6.F, but interested readers can find the extremes listed in Table 4 of Davis and Hall (1969). In most cases, this range of efficiencies is rather narrow, reflecting the fact that although, in general, the Davis–Hall designs are not variance–balanced, the degree of imbalance is rather small. For example, in Design DH1, shown above, it can be seen that Treatment A is followed by Treatments B, C, D and F, but not by Treatment E. Therefore, the contrast of Treatment A vs. E, both for treatment and carryover effects, may be expected to be more variable than other contrasts involving those treatments.

However, it is not as simple as that, since Treatment A is preceded by Treatments B, D, E and F, which is not exactly the same list of treatments as those that followed Treatment A, whereas in the Patterson–Lucas designs the same treatments both precede and follow each treatment equally. The range of E_t values for Design DH1 is 77 to 83 (average 79), the range of E_d is 56 to 63 (average 58), and the range of E_r is 31 to 45 (average 34).

The advantage of the Davis–Hall designs in comparison with other existing designs such as the Patterson and Lucas (1962) designs is that their efficiencies are comparable to those designs but tend to require fewer subjects. The other advantage is that these designs exist for any number of treatments and periods. Although Appendix 6.F is restricted to a selection of designs for 6 to 20 treatments in 3, 4 or 5 periods, similar designs may be constructed for other numbers of treatments and periods. In most cases, the efficiencies of these designs are reasonably high and the degree of imbalance is small. Only one design in Appendix 6.F, Design DH17, with $t = 7$, $p = 4$ and $b = 2$, is variance balanced and is identical with Design PL19 of Appendix 5.B of Chapter 5.

Analysis of the Davis–Hall cyclic designs is carried out in exactly the same way as the analysis of other cross–over designs, balanced or unbalanced, using the approach described in Section 1.1. Let us consider the set of data reported in Table 2 of Davis and Hall (1969), reproduced here as Table 6.2.

Table 6.2. Data of Davis and Hall (1969, Table 2) corresponding to Design DH1, and illustrating the analysis of a cyclic cross–over design.

Block 1
Subjects

		1	2	3	4	5	6
	1	38.7(A)	48.9(B)	35.2(C)	34.6(D)	32.9(E)	30.4(F)
Period	2	37.4(D)	46.9(E)	33.5(F)	32.3(A)	33.1(B)	29.5(C)
	3	34.3(E)	42.0(F)	28.4(A)	28.5(B)	27.5(C)	26.7(D)

Block 2
Subjects

		1	2	3	4	5	6
	1	25.7(A)	30.8(B)	25.4(C)	21.8(D)	21.4(E)	22.8(F)
Period	2	26.1(F)	29.3(A)	26.0(B)	23.9(C)	22.0(D)	21.0(E)
	3	23.4(B)	26.4(C)	23.9(D)	21.7(E)	19.4(F)	18.6(A)

The analysis using SAS involves first the creation of a data step, such as:

```
DATA DAVHALL;
  INPUT BLOCK SUBJ PERIOD TREAT$ CARRY$ RESP @@;
IF CARRY = '0' THEN CARRY = 'F';
CARDS;
1 1 1 A 0 38.7      1 1 2 D A 37.4      1 1 3 E D 34.3
1 2 1 B 0 48.9      1 2 2 E B 46.9      1 2 3 F E 42.0
1 3 1 C 0 35.2      1 3 2 F C 33.5      1 3 3 A F 28.4
1 4 1 D 0 34.6      1 4 2 A D 32.3      1 4 3 B A 28.5
1 5 1 E 0 32.9      1 5 2 B E 33.1      1 5 3 C B 27.5
1 6 1 F 0 30.4      1 6 2 C F 29.5      1 6 3 D C 26.7
2 1 1 A 0 25.7      2 1 2 F A 26.1      2 1 3 B F 23.4
2 2 1 B 0 30.8      2 2 2 A B 29.3      2 2 3 C A 26.4
2 3 1 C 0 25.4      2 3 2 B C 26.0      2 3 3 D B 23.9
2 4 1 D 0 21.8      2 4 2 C D 23.9      2 4 3 E C 21.7
2 5 1 E 0 21.4      2 5 2 D E 22.0      2 5 3 F D 19.4
2 6 1 F 0 22.8      2 6 2 E F 21.0      2 6 3 A E 18.6
RUN;
```

The IF statement is needed to ensure estimability of the treatment and carryover effects using the LSMEANS statement in the procedure which follows. We assume, as in Section 5.2, that the time frame may be different in the three blocks, or that there may be a difference in the effect of time from block to block, and therefore we choose to use a nested term, periods within blocks.

```
PROC GLM;
  CLASS BLOCK UNIT PERIOD TREAT CARRY;
  MODEL RESP = BLOCK UNIT(BLOCK) PERIOD(BLOCK) TREAT CARRY
      / SOLUTION SS1 SS2 E1 E2;
  RANDOM UNIT(BLOCK);
  LSMEANS TREAT CARRY / STDERR PDIFF;
  RUN;
```

This code results in the following edited output:

Dependent Variable: FCM

Source	DF	Sum of Squares	Mean Square	F Value	Pr > F
Model	25	1863.0940	74.5238	112.43	0.0001
Error	10	6.6282	0.6628		
Corrected Total	35	1869.7222			

Source	DF	Type I SS	Mean Square	F Value	Pr > F
BLOCK	1	1015.4844	1015.4844	1535.06	0.0001
UNIT(BLOCK)	10	714.8844	71.4884	107.85	0.0001
PERIOD(BLOCK)	4	124.7444	31.1861	47.05	0.0001
TREAT	5	3.1357	0.6272	0.95	0.4925
CARRY	5	4.8449	0.9690	1.46	0.2844

(continued on next page)

Source	DF	Type II SS	Mean Square	F Value	Pr > F
TREAT	5	4.750124	0.9500	1.43	0.2931
CARRY	5	4.844919	0.9690	1.46	0.2844

A portion of the output of PROC GLM giving the estimates of the treatment and carryover parameters of the model is presented below. Inspection of the standard errors shows that they are not all identical, and the same is true of the standard errors of the carryover parameters. The comparison of C vs. F is the most precise of any contrast with F, with those of B vs. F and D vs. F being the least precise. The contrasts of A vs. F and E vs. F are intermediate in precision.

Parameter		Estimate	T for H0: Parameter=0	Pr > |T|	Std Error of Estimate
TREAT	A	−0.54777098	−0.89	0.3921	0.61238016
	B	1.00871212	1.60	0.1406	0.63033002
	C	0.36378205	0.61	0.5526	0.59205598
	D	0.45094697	0.72	0.4907	0.63033002
	E	0.29749417	0.49	0.6376	0.61238016
	F	0.00000000	.	.	.
CARRY	A	0.22261072	0.27	0.7923	0.82319111
	B	1.27291667	1.50	0.1648	0.84923161
	C	1.40480769	2.00	0.0738	0.70355615
	D	1.50946970	1.78	0.1059	0.84923161
	E	1.03189103	1.25	0.2385	0.82319111
	F	0.00000000	.	.	.

As neither treatment nor carryover is significant, the analysis can end there. It is clear from the Type I sums of squares and subsequent F–test that there is a very strong period effect, with the response falling off with time.

6.4. The Designs of Federer and Atkinson

Federer and Atkinson (1964) introduced what they called tied–double change–over designs. Appendix 6.G lists the pattern of these designs for 3 and 4 treatments. The basis of construction is a set of $t-1$ orthogonal Latin squares. Thus, for $t = 3$, Design FA1 is constructed from the pair of squares given as Design PL2 in Appendix 5.B. For $t = 4$, Design FA2 has the set of 3 mutually orthogonal Latin squares given as Design PL6 in Appendix 5.B as the basis for the design. The aim of the tied–double designs is to approximate as close as possible a situation where the variances of differences among treatments are nearly equal to the variances of differences among carryover effects. The usual situation with other designs is that the precision of treatment effect differences is greater than that of carryover effect differences, generally because there is no carryover into the first period.

In general, Federer–Atkinson designs have t treatments, p periods, c subjects, and two further integer constants q and s, defined respectively as

$$q = (p - 1)/t$$

and

$$s = c/t.$$

Although these designs do not in general possess variance balance, there are some combinations of s and q for which the design is balanced. For example, for $t = 3$, balanced designs are obtained with the 3 subject, 7 period design with $s = 1$ and $q = 2$, or the 6 subject, 10 period design with $s = 2$, $q = 3$, or the 9 subject, 13 period design with $s = 3$, $q = 4$, and, in general, when $q - s = 1$, for $s \geq 1$. To decide whether or not a design is balanced, one constructs a contingency table for treatment vs. carryover, as was done in Section 1.2. For the design with $t = 3$, $q = s = 2$, where there are 7 periods and 6 subjects, we get the following frequencies of occurrence:

		CARRYOVER			
		0	A	B	C
	A	2	0	6	6
TREATMENT	B	2	6	0	6
	C	2	6	6	0

Since all of the cells involving carryover by A, B and C into treatments other than themselves have the same frequency, viz. 6, the design is variance balanced. However, the contingency table reveals a major weakness of the Federer–Atkinson designs, in that a treatment never carries over into itself. This will result in a considerable variance inflation factor for these designs, as will be demonstrated later in this section.

In the allocation of subjects to the experimental units, one assigns the c subjects to the columns of the design at random, if there is no stratification amongst the subjects. The t treatments are further assigned to the letters A, B, C, etc. at random. If the columns are stratified into blocks of $k = t$, then groups of t units are first allocated at random to the blocks, and then there is a separate randomization to decide which of the t subjects in a block are assigned to which of the c columns.

Analysis of data from Federer–Atkinson designs is carried out in an identical fashion to the analysis of all other designs in this book, using procedures described in Section 1.1. Consider the data used in the example given by Federer and Atkinson (1964) for a 6 column, 7 row design ($s = q = 2$). It was a design of this type that resulted in a contingency table (see above) which showed this design to be variance balanced. The data are presented in Table 6.3.

Table 6.3. Data from Federer and Atkinson (1964), illustrating analysis of data from a tied–double cross–over trial. (Reproduced with the permission of The Biometric Society) Treatments given in parentheses.

		Column					
		1	2	3	4	5	6
	1	21 (C)	21 (C)	11 (A)	12 (A)	7 (B)	18 (B)
	2	16 (A)	20 (B)	25 (C)	16 (B)	20 (C)	23 (A)
	3	18 (B)	15 (A)	19 (B)	26 (C)	11 (A)	31 (C)
Period	4	26 (C)	25 (C)	16 (A)	17 (A)	11 (B)	25 (B)
	5	19 (B)	19 (A)	18 (B)	26 (C)	11 (A)	33 (C)
	6	19 (A)	18 (B)	28 (C)	22 (B)	21 (C)	24 (A)
	7	27 (C)	28 (C)	19 (A)	17 (A)	15 (B)	26 (B)

The analysis of the data by SAS requires a "data" step, such as:

```
DATA FEDATK;
 INPUT SUBJECT PERIOD TREAT$ CARRY$ RESP @@;
IF CARRY = '0' THEN CARRY = 'C';
CARDS;
1 1 C 0 21    1 2 A C 16    1 3 B A 18    1 4 C B 26    1 5 B C 19
         1 6 A B 19    1 7 C A 27
2 1 C 0 21    2 2 B C 20    2 3 A B 15    2 4 C A 25    2 5 A C 19
         2 6 B A 18    2 7 C B 28
3 1 A 0 11    3 2 C A 25    3 3 B C 19    3 4 A B 16    3 5 B A 18
         3 6 C B 28    3 7 A C 19
4 1 A 0 12    4 2 B A 16    4 3 C B 26    4 4 A C 17    4 5 C A 26
         4 6 B C 22    4 7 A B 17
5 1 B 0 7     5 2 C B 20    5 3 A C 11    5 4 B A 11    5 5 A B 11
```

(continued on next page)

```
         5 6 C A 21    5 7 B C 15
6 1 B 0 18     6 2 A B 23     6 3 C A 31     6 4 B C 25     6 5 C B 33
         6 6 A C 24    6 7 B A 26
RUN;
```

In common with previous examples given in this book of the analysis of cross–over designs, the IF statements needs to be present to ensure estimability of treatment and carryover effects in subsequent analysis. Since there is no block structure indicated for the six subjects, one must assume that these subjects were not classified in any way and were allocated to the six sequences at random. The code for the mixed–model analysis of variance is as follows:

```
PROC GLM;
  CLASS SUBJECT PERIOD TREAT CARRY;
  MODEL RESP=SUBJECT PERIOD TREAT CARRY/SOLUTION SS1 SS2 E1
    E2;
  RANDOM SUBJECT;
  LSMEANS TREAT CARRY / PDIFF STDERR;
  RUN;
```

The above code results in the following edited output:

Dependent Variable: FCM

Source	DF	Sum of Squares	Mean Square	F Value	Pr > F
Model	15	1396.0000	93.0667	151.23	0.0001
Error	26	16.0000	0.6154		
Corrected Total	41	1412.0000			

Source	DF	Type I SS	Mean Square	F Value	Pr > F
SUBJECT	5	520.000000	104.00000	169.00	0.0001
PERIOD	6	204.000000	34.00000	55.25	0.0001
TREAT	2	653.250000	326.62500	530.77	0.0001
CARRY	2	18.750000	9.37500	15.23	0.0001

Source	DF	Type II SS	Mean Square	F Value	Pr > F
TREAT	2	600.000000	300.00000	487.50	0.0001
CARRY	2	18.750000	9.37500	15.23	0.0001

Below is presented a portion of the output of PROC GLM giving the estimates of the treatment and carryover parameters. The standard errors for each treatment parameter and for each carryover parameter are the same, as expected since this design was shown earlier to be a balanced one.

Parameter		Estimate	T for H0: Parameter=0	Pr > |T|	Std Error of Estimate
TREAT	A	−10.0000000	−29.50	0.0001	0.33892742
	B	−8.0000000	−23.60	0.0001	0.33892742
	C	0.0000000	.	.	.
CARRY	A	−2.0000000	−5.52	0.0001	0.36232865
	B	−1.0000000	−2.76	0.0105	0.36232865
	C	0.0000000	.	.	.

Since both treatment and carryover effects are significant, it is of interest to examine which levels of the treatments are significantly different from which others. This is achieved using the least squares means that are shown as follows:

Least Squares Means

TREAT	RESP LSMEAN	Pr > |T| i/j	H0: LSMEAN(i)=LSMEAN(j)	
			1 2 3	
A	15.8571429	1	. 0.0001 0.0001	
B	17.8571429	2	0.0001 . 0.0001	
C	25.8571429	3	0.0001 0.0001 .	

CARRY	RESP LSMEAN	Pr > |T| i/j	H0: LSMEAN(i)=LSMEAN(j)	
			1 2 3	
A	18.8571429	1	. 0.0105 0.0001	
B	19.8571429	2	0.0105 . 0.0105	
C	20.8571429	3	0.0001 0.0105 .	

It is clear that in this contrived set of data all levels of treatment are highly significantly different, and that differential carryover involving Treatments A and C are also highly significant, whereas differential carryover for Treatments A vs. B and for Treatments B vs. C is not as strongly significantly different.

We now examine this problem using matrix operations to see what the variance inflation factor is when carryover parameters are dropped from the model. The model under consideration is virtually identical to that used throughout this book,

$$Y_{ijk} = \mu + \gamma_i + \pi_j + \tau_t + \delta_j \alpha_r + \epsilon_{ijk}$$

where the term γ_i refers to the sequences, $i=1,...,6$, each column being a unique sequence, π_j refers to the periods, $j=1,...,7$, τ_t refers to treatments, $t=1,...,3$, α_r is the carryover effect from the rth treatment having been applied in the previous period, $r=1,...,3$, and δ_j is an indicator variable whose value is zero in the first period and unity in subsequent periods, and ϵ_{ijk} is a random experimental error corresponding to square i, period j, and subject k. Note that there is no error term in the Between Subjects stratum, since each sequence is unique; a subjects within sequences term arises when the sequences are replicated. The five degrees of freedom in that stratum are all taken up by differences between sequences.

Using matrix operations similar to the illustration in Section 1.1.2, one obtains the following portion of the inverse of the $\mathbf{X}'\mathbf{X}$ matrix pertaining to the treatment and carryover parameters.

$$\text{Cov}(\hat{\tau}_1, \hat{\tau}_2, \hat{\lambda}_1, \hat{\lambda}_2) = \begin{pmatrix} 0.06222 & -0.03111 & 0.03111 & -0.01556 \\ -0.03111 & 0.06222 & -0.01556 & 0.03111 \\ 0.03111 & -0.01556 & 0.07111 & -0.03556 \\ -0.01556 & 0.03111 & -0.03556 & 0.07111 \end{pmatrix}.$$

The objective of the Federer–Atkinson designs is to make the variance of the carryover parameters almost equal to that of the treatment parameters. the above results show that the objective has not been fully realized, although the variances get closer together as s and q increase. Eliminating the carryover parameters leads to the following portion of the inverse of the $\mathbf{X}'\mathbf{X}$ matrix pertaining to the treatment parameters.

$$(\mathbf{X}'\mathbf{X})^{-1} = \begin{pmatrix} 0.04861 & -0.02431 \\ -0.02431 & 0.04861 \end{pmatrix}.$$

Thus, the variance inflation factor is:

$$\text{VIF} = \frac{0.06222}{0.04861} = 1.280,$$

which is not an excessively high value. Nevertheless, the Federer–Atkinson designs are not orthogonal designs like the extra–period designs of Patterson and Lucas (1962), studied in Sections 5.1 and 6.1.

Exercises

6.1. Consider the "parent" of Design PL153 in Appendix 6.E, viz.

	Block			
	1	2	3	4
		Subjects		
Period 1	AB	BC	CD	DA
2	BA	CB	DC	AD

Set up matrix equations for this design using an approach similar to that of Section 1.1.2. Note, however, that the subjects are in blocks, so that in addition to the mean (μ), there are three parameters (γ_1, γ_2, γ_3) representing block differences, and there is a need for four parameters (ζ_{11}, ζ_{21}, ζ_{31}, ζ_{41}) to represent the different sequences within blocks (see the approach in Section 5.5). With a single parameter for periods (π_1), three for treatments (τ_1, τ_2, τ_3) and three for carryover (λ_1, λ_2, λ_3), set up the 16 x 15 design matrix for this design, and show that $X'X$ is singular. It is for this reason that the parent design of PL 153 cannot be used as a cross-over design.

6.2. Consider the extra–period partially–balanced Design PL 153, with treatments occurring in blocks of $k = 2$, viz.

		Block			
		1	2	3	4
			Subjects		
	1	AB	BC	CD	DA
Period	2	BA	CB	DC	AD
	3	BA	CB	DC	AD

(a) As in Exercise 6.1, set up the design matrix X for this design. Note that as there are eight subjects and three periods, or 24 "cells", and two period parameters (π_1, π_2), the dimensions of X will be 24 x 16.

(b) Find the elements of the inverse of the $X'X$ matrix, and show that the treatment parameters are orthogonal to the carryover parameters. It is the use of an extra–period that results in orthogonality, as well as low variances for these parameters.

Consider the parent of Design PL153 in Exercise 6.1. Precede the first nent for each subject with a baseline measurement to give the following design:

```
                      Block
                1   2   3   4
                  Subjects
```

		Block
Period	1	— — — —
	2	AB BC CD DA
	3	BA CB DC AD

(a) Using matrix operations as in Section 1.1.2 or the previous exercises in this chapter, determine the elements of the inverse of the $X'X$ matrix for this design.

(b) Comparing these results with the results of Exercise 6.2, where an extra period was added to the same "parent" design, show that obtaining a baseline measurement prior to the commencement of active treatments is not as efficient as the addition of an extra period.

Appendix 6.A. Index to Partially–Balanced Cross–over Designs of Patterson and Lucas

(t = no. of treatments, p = no. of periods, k = no. of units per block, b = minimum no. of blocks, n = total number of subjects. Efficiencies of designs are compared to a complete Latin square: E_t = efficiency when carryover parameters are not in the model, E_d = efficiency of treatment effects in the presence of carryover, E_r = efficiency of carryover effects in the presence of treatment).

t	p	k	b	n	E_t	E_d	E_r	Design No.	Appendix No.
4	3	2	4	8	53	53	40	PL153	6.E
5	2	3	10	30	61	15	5	PL125	6.C
5	3	2	5	10	44	44	33	PL154	6.E
5	3	3	10	30	81	65	36	PL126	6.C
5	3	3	10	30	54	54	45	PL131	6.D
5	4	3	10	30	76	76	61	PL132	6.D
6	2	3	8	24	58	14	5	PL99	6.B
6	3	2	9	18	49	49	37	PL155	6.E
6	3	3	8	24	77	62	34	PL100	6.B
6	3	3	8	24	51	51	43	PL133	6.D

(continued on following page)

t	p	k	b	n	E_t	E_d	E_r	Design No.	Appendix No.
6	4	3	8	24	72	72	58	PL134	6.D
6	4	4	3	12	88	80	55	PL101	6.B
6	5	4	3	12	85	85	71	PL135	6.D
6	5	4	3	12	85	85	71	PL135	6.D
8	2	3	16	48	56	14	5	PL102	6.B
8	3	2	16	32	48	48	36	PL156	6.E
8	3	3	16	48	75	60	33	PL103	6.B
8	3	3	16	48	50	50	41	PL136	6.D
8	4	3	16	48	70	70	56	PL137	6.D
8	4	4	6	24	82	75	51	PL104	6.B
8	5	4	6	24	79	79	66	PL138	6.D
8	6	6	4	24	95	92	74	PL105	6.B
9	2	3	12	36	50	12	4	PL127	6.C
9	3	2	18	36	44	44	33	PL157	6.E
9	3	3	12	36	67	53	30	PL128	6.C
9	3	3	12	36	44	44	37	PL139	6.D
9	4	3	12	36	62	62	50	PL140	6.D
9	4	4	9	36	83	76	52	PL106	6.B
9	5	4	9	36	80	80	67	PL141	6.D
9	6	6	3	18	92	89	72	PL107	6.B
10	2	3	20	60	53	13	4	PL108	6.B
10	3	2	15	30	40	40	30	PL158	6.E
10	3	3	20	60	70	56	31	PL109	6.B
10	3	3	20	60	47	47	39	PL142	6.D
10	4	3	20	60	66	66	53	PL143	6.D
10	4	4	5	20	79	72	49	PL110	6.B
10	5	4	5	20	76	76	63	PL144	6.D
10	5	5	12	60	88	83	63	PL111	6.B
10	6	5	12	60	85	85	73	PL145	6.D
10	6	6	5	30	92	89	71	PL112	6.B
12	3	2	36	72	47	47	35	PL159	6.E
12	4	4	9	36	80	73	50	PL113	6.B
12	5	4	9	36	77	77	64	PL146	6.D
12	6	6	6	36	88	85	68	PL114	6.B
13	2	3	26	78	50	12	4	PL115	6.B
13	3	2	39	78	44	44	33	PL160	6.E
13	3	3	26	78	67	53	30	PL116	6.B
13	3	3	26	78	44	44	37	PL147	6.D
13	4	3	26	78	62	62	50	PL148	6.D
14	4	4	14	56	80	73	50	PL117	6.B
14	5	4	14	56	77	77	64	PL149	6.D
14	6	6	7	42	88	85	69	PL118	6.B
15	4	4	15	60	80	72	50	PL119	6.B
15	5	4	15	60	76	76	64	PL150	6.D
15	5	5	12	60	81	77	58	PL120	6.B
15	6	5	12	60	79	79	67	PL151	6.D
15	6	6	10	60	89	86	69	PL121	6.B
16	4	4	8	32	71	65	45	PL129	6.C
16	5	4	8	32	69	69	57	PL152	6.D
18	6	6	12	72	86	83	67	PL122	6.B
21	6	6	7	42	82	80	64	PL123	6.B
26	6	6	13	78	84	82	66	PL124	6.B
36	6	6	12	72	78	75	60	PL130	6.C

Appendix 6.B. Designs PL 99–124. Partially–balanced designs based on Bose *et al.*

Design PL99. $t = 6$, $p = 2$, $k = 3$, $b = 8$, $n = 24$, $n_1 = 0$, $n_2 = 1$, VIF $= 4.0$

First associates are A D, B E, C F. First associates never occur sequentially, but occur once with second associates. $E_{d1} = 12$, $E_{d2} = 15$.

An example is

		Block							
		1	2	3	4	5	6	7	8
					Subjects				
Period	1	ABC	BCA	AEF	EFA	BDF	DFB	CDE	DEC
	2	BCA	ABC	EFA	AEF	DFB	BDF	DEC	CDE

Design PL100. $t = 6$, $p = 3$, $k = 3$, $b = 8$, $n = 24$, $n_1 = 0$, $n_2 = 2$, VIF $= 1.25$

First associates are A D, B E, C F. First associates never occur sequentially, but occur twice with second associates. $E_{d1} = 54$, $E_{d2} = 64$.

An example is

		Block							
		1	2	3	4	5	6	7	8
					Subjects				
	1	ABC	ABC	AEF	AEF	BDF	BDF	CDE	CDE
Period	2	BCA	CAB	EFA	FAE	DFB	FBD	DEC	ECD
	3	CAB	BCA	FAE	EFA	FBD	DFB	ECD	DEC

Design PL101. $t = 6$, $p = 4$, $k = 4$, $b = 3$, $n = 12$, $n_1 = 2$, $n_2 = 1$, VIF $= 1.1$

First associates are A D, B E, C F; these treatments occur sequentially twice, while second associates occur once only. $E_{d1} = 91$, $E_{d2} = 78$.

An example is

		Block		
		1	2	3
Subject →		1 2 3 4	1 2 3 4	1 2 3 4
	1	A D B E	B E C F	C F A D
Period	2	D E A B	E F B C	F D C A
	3	B A E D	C B F E	A C D F
	4	E B D A	F C E B	D A F C

Design PL102. $t = 8$, $p = 2$, $k = 3$, $b = 16$, $n = 48$, $n_1 = 0$, $n_2 = 1$, VIF = 4.0 First associates are A E, B F, C G, D H. First associates never occur sequentially, but occur once with second associates. $E_{d1} = 13$, $E_{d2} = 14$.

An example is

		Block						
	1	2	3	4	5	6	7	8
				Subjects				
Period 1	ABD	ABD	BCE	BCE	CDF	CDF	DEG	DEG
2	BDA	DAB	CEB	EBC	DFC	FCD	EGD	GDE

(continued)

	Block							
	9	10	11	12	13	14	15	16
				Subjects				
Period 1	EFH	EFH	FGA	FGA	GHB	GHB	HAC	HAC
2	FHE	HEF	GAF	AFG	HBG	BGH	ACH	CHA

Design PL103. $t = 8$, $p = 3$, $k = 3$, $b = 16$, $n = 48$, $n_1 = 0$, $n_2 = 2$, VIF = 1.25. First associates are A E, B F, C G, D H. First associates never occur sequentially, but occur twice with second associates. $E_{d1} = 54$, $E_{d2} = 61$.

An example is

	Block													
	1	2	3	4	5	6	7	8	9	10	11	12	13	14
								Subjects						
Period 1	ABD	ABD	BCE	BCE	CDF	CDF	DEG	DEG	EFH	EFH	FGA	FGA	GHB	GHB
2	BDA	DAB	CEB	EBC	DFC	FCD	EGD	GDE	FHE	HEF	GAF	AFG	HBG	BGH
3	DAB	BDA	EBC	CEB	FCD	DFC	GDE	EGD	HEF	FHE	AFG	GAF	BGH	HBG

(continued)

	Block	
	15	16
	Subjects	
Period 1	HAC	HAC
2	ACH	CHA
3	CHA	ACH

Design PL104. $t = 8$, $p = 4$, $k = 4$, $b = 6$, $n = 24$, $n_1 = 3$, $n_2 = 1$, VIF = 1.10
First associates are A E, B F, C G, D H. First associates occur sequentially thrice compared to once only for second associates. $E_{d1} = 91$, $E_{d2} = 73$.

An example is (the columns representing subjects)

	1	A E B F	C G D H	A E C G
Period	2	E F A B	G H C D	E G A C
	3	B A F E	D C H G	C A G E
	4	F B E A	H D G C	G C E A

(continued)

	1	B F D H	A E D H	B F C G
Period	2	F H B D	E H A D	F G B C
	3	D B H F	D A H E	C B G F
	4	H D F B	H D E A	G C F B

Design PL105. $t = 8$, $p = 6$, $k = 6$, $b = 4$, $n = 24$, $n_1 = 3$, $n_2 = 2$, VIF = 1.036. First associates are A E, B F, C G, D H. First associates occur sequentially thrice compared to twice for second associates. $E_{d1} = 97$, $E_{d2} = 91$.

An example is

		Block 1	2	3	4
		Subjects
	1	AEBGCF	BFAEDH	CGDHAE	DHCFBG
	2	EBGCFA	FAEDHB	GDHAEC	HCFBGD
Period	3	FAEBGC	HBFAED	ECGDHA	GDHCFB
	4	BGCFAE	AEDHBF	DHAECG	CFBGDH
	5	CFAEBG	DHBFAE	AECGDH	BGDHCF
	6	GCFAEB	EDHBFA	HAECGD	FBGDHC

Design PL106. $t = 9$, $p = 4$, $k = 4$, $b = 9$, $n = 36$, $n_1 = 1$, $n_2 = 2$, VIF = 1.10
First associates occur in the same row or column in the following scheme:

$$\begin{array}{ccc} A & B & C \\ D & E & F \\ G & H & I \end{array}$$

First associates occur sequentially once, compared to twice for second associates. $E_{d1} = 73$, $E_{d2} = 80$.

An example is

(continued)

Subjects

	1	A F I B	F H B D	H A D I	I B E G	B D G C
Period	2	F B A I	H D F B	A I H D	B G I E	D C B G
	3	I A B F	B F D H	D H I A	E I G B	G B C D
	4	B I F A	D B H F	I D A H	G E B I	C G D B

(continued)

Subjects

	1	D I C E	E G A F	G C F H	C E H A
Period	2	I E D C	G F E A	C H G F	E A C H
	3	C D E I	A E F G	F G H C	H C A E
	4	E C I D	F A G E	H F C G	A H E C

Design PL107. $t = 9$, $p = 6$, $k = 6$, $b = 3$, $n = 18$, $n_1 = 2$, $n_2 = 1$, VIF = 1.036. First associates occur in the same row in the following scheme:

$$
\begin{array}{ccc}
A & D & G \\
B & E & H \\
C & F & I
\end{array}
$$

First associates occur sequentially twice, compared to once for second associates. $E_{d1} = 97$, $E_{d2} = 87$.

An example is

		1	Block 2	3
			Subjects	
	1	ADGBEH	BEHCFI	CFIADG
	2	DGBEHA	EHCFIB	FIADGC
Period	3	HADGBE	IBEHCF	GCFIAD
	4	GBEHAD	HCFIBE	IADGCF
	5	EHADGB	FIBEHC	DGCFIA
	6	BEHADG	CFIBEH	ADGCFI

Design PL108. $t = 10$, $p = 2$, $k = 3$, $b = 20$, $n = 60$, $n_1 = 1$, $n_2 = 0$, VIF = 4.0 First associates occur in the same row or the same column in the following scheme:

$$
\begin{array}{ccccc}
x & A & B & C & D \\
A & x & E & F & G \\
B & E & x & H & I \\
C & F & H & x & J \\
D & G & I & J & x
\end{array}
$$

(continued)

First associates occur sequentially once, whereas second associates never occur. $E_{d1} = 14$, $E_{d2} = 12$.

An example is:

		1	2	3	4	5	6	7	8	9	10	11	12	13	14

Block
Subjects

Period	1	ABE	ABE	HIJ	HIJ	BHC	BHC	GEI	GEI	IDB	IDB	EFH	EFH	CJD	CJD
	2	BEA	EAB	IJH	JHI	HCB	CBH	EIG	IGE	DBI	BID	FHE	HEF	JDC	DCJ

(continued)

Block

		15	16	17	18	19	20

Subjects

Period	1	JGF	JGF	DAG	DAG	FCA	FCA
	2	GFJ	FJG	AGD	GDA	CAF	AFC

Design PL109. $t = 10$, $p = 3$, $k = 3$, $b = 20$, $n = 60$, $n_1 = 2$, $n_2 = 0$. VIF = 1.25. First associates occur in the same row or the same column in the following scheme:

$$
\begin{array}{ccccc}
x & A & B & C & D \\
A & x & E & F & G \\
B & E & x & H & I \\
C & F & H & x & J \\
D & G & I & J & x
\end{array}
$$

First associates occur sequentially twice each, whereas second associates never occur. $E_{d1} = 59$, $E_{d2} = 50$.

An example is:

		1	2	3	4	5	6	7	8	9	10	11	12	13	14

Block
Subjects

	1	ABE	ABE	HIJ	HIJ	BHC	BHC	GEI	GEI	IDB	IDB	EFH	EFH	CJD	CJD
Period	2	BEA	EAB	IJH	JHI	HCB	CBH	EIG	IGE	DBI	BID	FHE	HEF	JDC	DCJ
	3	EAB	BEA	JHI	IJH	CBH	HBC	IGE	EIG	BID	DBI	HEF	FHE	DCJ	JDC

(continued)

Block

		15	16	17	18	19	20

Subjects

	1	JGF	JGF	DAG	DAG	FCA	FCA
Period	2	GFJ	FJG	AGD	GDA	CAF	AFC
	3	FJG	GFJ	GDA	AGD	AFC	CAF

Design PL110. $t = 10$, $p = 4$, $k = 4$, $b = 5$, $n = 20$, $n_1 = 1$, $n_2 = 0$, VIF = 1.1
First associates occur in the same row or the same column in the following scheme:

```
x  A  B  C  D
A  x  E  F  G
B  E  x  H  I
C  F  H  x  J
D  G  I  J  x
```

First associates occur sequentially once, whereas second associates never occur.
$E_{d1} = 76$, $E_{d2} = 65$.

An example is:

		1	2	Block 3 Subjects	4	5
Period	1	A B C D	E F G A	H I B E	J C F H	D G I J
	2	B D A C	F A E G	I E H B	C H J F	G J D I
	3	C A D B	G E A F	B H E I	F J H C	I D J G
	4	D C B A	A G F E	E B I H	H F C J	J I G D

Design PL111. $t = 10$, $p = 5$, $k = 5$, $b = 12$, $n = 60$, $n_1 = 2$, $n_2 = 4$, VIF = 1.056. First associates occur in the same row or the same column in the following scheme:

```
x  A  B  C  D
A  x  E  F  G
B  E  x  H  I
C  F  H  x  J
D  G  I  J  x        (continued
```

First associates occur sequentially twice, whereas second associates occur four times. $E_{d1} = 81$, $E_{d2} = 86$.

An example is:

		1	2	Block 3 Subjects	4	5	6
	1	AHIGC	AHIGC	AHDJE	AHDJE	HDFGB	HDFGB
	2	HIGCA	CAHIG	HDJEA	EAHDJ	DFGBH	BHDFG
Period	3	CAHIG	HIGCA	EAHDJ	HDJEA	BHDFG	DFGBH
	4	IGCAH	GCAHI	DJEAH	JEAHD	FGBHD	GBHDF
	5	GCAHI	IGCAH	JEAHD	DJEAH	GBHDF	FGBHD

(continued)

Block

	7	8	9	10	11	12

Subjects

Period		7	8	9	10	11	12
	1	DFIEC	DFIEC	AFIJB	AFIJB	JGEBC	JGEBC
	2	FIECD	CDFIE	FIJBA	BAFIJ	GEBCJ	CJGEB
	3	CDFIE	FIECD	BAFIJ	FIJBA	CJGEB	GEBCJ
	4	IECDF	ECDFI	IJBAF	JBAFI	EBCJG	BCJGE
	5	ECDFI	IECDF	JBAFI	IJBAF	BCJGE	EBCJG

Design PL112. $t = 10$, $p = 6$, $k = 6$, $b = 5$, $n = 30$, $n_1 = 2$, $n_2 = 1$, VIF = 1.036. First associates occur in the same row or the same column in the following scheme:

$$
\begin{array}{ccccc}
x & A & B & C & D \\
A & x & E & F & G \\
B & E & x & H & I \\
C & F & H & x & J \\
D & G & I & J & x \\
\end{array}
$$

First associates occur sequentially twice, whereas second associates occur once. $E_{d1} = 91$, $E_{d2} = 86$.

An example is:

Block

	1	2	3	4	5

Subjects

Period		1	2	3	4	5
	1	EFGHIJ	JIDBCH	DCAJGF	GBIEAD	AHCFEB
	2	FGHIJE	IDBCHJ	CAJGFD	BIEADG	HCFEBA
	3	JEFGHI	HJIDBC	FDCAJG	DGBIEA	BAHCFE
	4	GHIJEF	DBCHJI	AJGFDC	IEADGB	CFEBAH
	5	IJEFGH	CHJIDB	GFDCAJ	ADGBIE	EBAHCF
	6	HIJEFG	BCHJID	JGFDCA	EADGBI	FEBAHC

Design PL113. $t = 12$, $p = 4$, $k = 4$, $b = 9$, $n = 36$, $n_1 = 0$, $n_2 = 1$, VIF = 1.1 First associates occur in the same row in the following scheme:

$$
\begin{array}{ccc}
A & E & I \\
B & F & J \\
C & G & K \\
D & H & L \\
\end{array}
$$

First associates never occur sequentially, whereas second associates occur once. $E_{d1} = 68$, $E_{d2} = 75$.

An example is:

Subjects

Period																				
1	A B C D		G J E D		F K I D		A G F H		K E B H	
2	B D A C		J D G E		K D F I		G H A F		E H K B	
3	C A D B		E G D J		I F D K		F A H G		B K H E	
4	D C B A		D E J G		D I K F		H F G A		H B E K	

(continued)

Subjects

Period								
1	J I C H	A K J L	I B G L	E C F L
2	I H J C	K L A J	B L I G	C L E F
3	C J H I	J A L K	G I L B	F E L C
4	H C I J	L J K A	L G B I	L F C E

Design PL114. $t = 12$, $p = 6$, $k = 6$, $b = 6$, $n = 36$, $n_1 = 3$, $n_2 = 1$, VIF = 1.036. First associates occur in the same row in the following scheme:

$$
\begin{array}{ccc}
A & E & I \\
B & F & J \\
C & G & K \\
D & H & L \\
\end{array}
$$

These treatments occur sequentially thrice, whereas second associates occur once. $E_{d1} = 97$, $E_{d2} = 83$.

(continued)

An example is (the rows representing periods)

Block

1	2	3	4	5	6

Subjects

ABEFIJ	CDGHKL	IKACEG	JLBDFH	EHILAD	FGJKBC
BEFIJA	DGHKLC	KACEGI	LBDFHJ	HILADE	GJKBCF
JABEFI	LCDGHK	GIKACE	HJLBDF	DEHILA	CFGJKB
EFIJAB	GHKLCD	ACEGIK	BDFHJL	ILADEH	JKBCFG
IJABEF	KLCDGH	EGIKAC	FHJLBD	ADEHIL	BCFGJK
FIJABE	HKLCDG	CEGIKA	DFHJLB	LADEHI	KBCFGJ

Design PL115. $t = 13$, $p = 2$, $k = 3$, $b = 26$, $n = 78$, $n_1 = 1$, $n_2 = 0$, VIF = 4.0. Each treatment has six first associates with which it occurs sequentially once, while second associates never occur. $E_{d1} = 13$, $E_{d2} = 11$.

An example is

(continued)

Block
1   2   3   4   5   6   7   8   9   10   11   12   13   14
Subjects

Period	1	ACI	ACI	BDJ	BDJ	CEK	CEK	DFL	DFL	EGM	EGM	FHA	FHA	GIB	GIB
	2	CIA	IAC	DJB	JBD	EKC	KCE	FLD	LDF	GME	MEG	HAF	AFH	IBG	BGI

(continued)

Block
15   16   17   18   19   20   21   22   23   24   25   26
Subjects

Period	1	HJC	HJC	IKD	IKD	JLE	JLE	KMF	KMF	LAG	LAG	MBH	MBH
	2	JCH	CHJ	KDI	DIK	LEJ	EJL	MFK	FKM	AGL	GLA	BHM	HMB

Design PL116. $t = 13$, $p = 3$, $k = 3$, $b = 26$, $n = 78$, $n_1 = 2$, $n_2 = 0$, VIF = 1.250. Each treatment has six first associates with which it occurs sequentially twice, while second associates never occur. $E_{d1} = 57$, $E_{d2} = 49$.

An example is

Block
1   2   3   4   5   6   7   8   9   10   11   12   13   14
Subjects

	1	ACI	ACI	BDJ	BDJ	CEK	CEK	DFL	DFL	EGM	EGM	FHA	FHA	GIB	GIB
Period	2	CIA	IAC	DJB	JBD	EKC	KCE	FLD	LDF	GME	MEG	HAF	AFH	IBG	BGI
	3	IAC	CIA	JBD	DJB	KCE	EKC	LDF	FLD	MEG	GME	AFH	HAF	BGI	IBG

(continued)

Block
15   16   17   18   19   20   21   22   23   24   25   26
Subjects

	1	HJC	HJC	IKD	IKD	JLE	JLE	KMF	KMF	LAG	LAG	MBH	MBH
Period	2	JCH	CHJ	KDI	DIK	LEJ	EJL	MFK	FKM	AGL	GLA	BHM	HMB
	3	CHJ	JCH	DIK	KDI	EJL	LEJ	FKM	MFK	GLA	AGL	HMB	BHM

Design PL117. $t = 14$, $p = 4$, $k = 4$, $b = 14$, $n = 56$, $n_1 = 0$, $n_2 = 1$, VIF = 1.100. First associates are A H, B I, C J, D K, E L, F M, G N. First associates never occur sequentially, whereas second associates occur once. $E_{d1} = 68$, $E_{d2} = 74$.

An example is

(continued)

Subjects

	1	B C E H	C D F I	D E G J	E F A K	F G B L
Period	2	C H B E	D I C F	E J D G	F K E A	G L F B
	3	E B H C	F C I D	G D J E	A E K F	B F L G
	4	H E C B	I F D C	J G E D	K A F E	L B G F

(continued)

Subjects

	1	G A C M	A B D N	I J L A	J K M B	K L N C
Period	2	A M G C	B N A D	J A I L	K B J M	L C K N
	3	C G M A	D A N B	L I A J	M J B K	N K C L
	4	M C A G	N D B A	A L J I	B M K J	C N L K

(continued)

Subjects

	1	L M H D	M N I E	N H J F	H I K G
Period	2	M D L H	N E M I	H F N J	I G H K
	3	H L D M	I M E N	J N F H	K H G I
	4	D H M L	E I N M	F J H N	G K I H

Design PL118. $t = 14$, $p = 6$, $k = 6$, $b = 7$, $n = 42$, $n_1 = 3$, $n_2 = 1$, VIF = 1.036. First asssociates are A H, B I, C J, D K, E L, F M, G N. First associates occur sequentially thrice, while second associates occur once. $E_{d1} = 97$, $E_{d2} = 85$.

An example is (the rows representing periods)

Block

1	2	3	4	5	6	7

Subjects

AHBIDK	BICJEL	CJDKFM	DKELGN	ELFMAH	FMGNBI	GNAHCJ
HBIDKA	ICJELB	JDKFMC	KELGND	LFMAHE	MGNBIF	NAHCJG
KAHBID	LBICJE	MCJDKF	NDKELG	HELFMA	IFMGNB	JGNAHC
BIDKAH	CJELBI	DKFMCJ	ELGNDK	FMAHEL	GNBIFM	AHCJGN
DKAHBI	ELBICJ	FMCJDK	GNDKEL	AHELFM	BIFMGN	CJGNAH
IDKAHB	JELBIC	KFMCJD	LGNDKE	MAHELF	NBIFMG	HCJGNA

Design PL119. $t = 15$, $p = 4$, $k = 4$, $b = 15$, $n = 60$, $n_1 = 0$, $n_2 = 1$, VIF = 1.100. First associates occur in the same row in the following scheme:

$$
\begin{array}{ccc}
A & F & K \\
B & G & L \\
C & H & M \\
D & I & N \\
E & J & O \\
\end{array}
$$

First associates never occur sequentially, whereas second associates occur once. $E_{d1} = 67$, $E_{d2} = 72$.

An example is

Subjects in Blocks

	1	B D E M	C E F N	D F G O	E G H A	F H I B
Period	2	D M B E	E N C F	F O D G	G A E H	H B F I
	3	E B M D	F C N E	G D O F	H E A G	I F B H
	4	M E D B	N F E C	O G F D	A H G E	B I H F

(continued)

Subjects in Blocks

	1	G I J C	H J K D	I K L E	J L M F	K M N G
Period	2	I C G J	J D H K	K E I L	L F J M	M G K N
	3	J G C I	K H D J	L I E K	M J F L	N K G M
	4	C J I G	D K J H	E L K I	F M L J	G N M K

(continued)

Subjects in Blocks

	1	L N O H	M O A I	N A B J	O B C K	A C D L
Period	2	N H L O	O I M A	A J N B	B K O C	C L A D
	3	O L H N	A M I O	B N J A	C O K B	D A L C
	4	H O N L	I A O M	J B A N	K C B O	L D C A

Design PL120. $t = 15$, $p = 5$, $k = 5$, $b = 12$, $n = 60$, $n_1 = 2$, $n_2 = 0$, VIF = 1.056. First associates occur in the same row or the same column in the following scheme:

$$
\begin{array}{cccccc}
x & A & B & C & D & E \\
A & x & F & G & H & I \\
B & F & x & J & K & L \\
C & G & J & x & M & N \\
D & H & K & M & x & O \\
E & I & L & N & O & x \\
\end{array}
$$

(continued)

First associates occur sequentially twice, while second associates never occur. $E_{d1} = 82$, $E_{d2} = 71$.

An example is

		Block					
		1	2	3	4	5	6
				Subjects			
	1	ABCDE	ABCDE	FGHIA	FGHIA	JKLBF	JKLBF
	2	BCDEA	EABCD	GHIAF	AFGHI	KLBFJ	FJKLB
Period	3	EABCD	BCDEA	AFGHI	GHIAF	FJKLB	KLBFJ
	4	CDEAB	DEABC	HIAFG	IAFGH	LBFJK	BFJKL
	5	DEABC	CDEAB	IAFGH	HIAFG	BFJKL	LBFJK

(continued)

		Block					
		7	8	9	10	11	12
				Subjects			
	1	MNCGJ	MNCGJ	ODHKM	ODHKM	EILNO	EILNO
	2	NCGJM	JMNCG	DHKMO	MODHK	ILNOE	OEILN
Period	3	JMNCG	NCGJM	MODHK	DHKMO	OEILN	ILNOE
	4	CGJMN	GJMNC	HKMOD	KMODH	LNOEI	NOEIL
	5	GJMNC	CGJMN	KMODH	HKMOD	NOEIL	LNOEI

Design PL121. $t = 15$, $p = 6$, $k = 6$, $b = 10$, $n = 60$, $n_1 = 1$, $n_2 = 2$, VIF = 1.036. First associates occur in the same row or the same column in the following scheme:

```
x  A  B  C  D  E
A  x  F  G  H  I
B  F  x  J  K  L
C  G  J  x  M  N
D  H  K  M  x  O
E  I  L  N  O  x
```

First associates occur sequentially once, whereas second associates occur twice. $E_{d1} = 84$, $E_{d2} = 88$.

An example is (the rows representing periods)

	Block					
1	2	3	4	5	6	
		Subjects				
HDJLNA	DOJFGE	OHIBCJ	FBMONA	BLEHGM	LFMDCI	
DJLNAH	OJFGED	HIBCJO	BMONAF	LEHGMB	FMDCIL	
AHDJLN	EDOJFG	JOHIBC	AFBMON	MBLEHG	ILFMDC	
JLNAHD	JFGEDO	IBCJOH	MONAFB	EHGMBL	MDCILF	
NAHDJL	GEDOJF	CJOHIB	NAFBMO	GMBLEH	CILFMD	
LNAHDJ	FGEDOJ	BCJOHI	ONAFBM	HGMBLE	DCILFM	

(continued)

Block

7	8	9	10

Subjects

GCAOLK	CNKHFE	NGKDBI	AEIJMK
CAOLKG	NKHFEC	GKDBIN	EIJMKA
KGCAOL	ECNKHF	INGKDB	KAEIJM
AOLKGC	KHFECN	KDBING	IJMKAE
LKGCAO	FECNKH	BINGKD	MKAEIJ
OLKGCA	HFECNK	DBINGK	JMKAEI

Design PL122. $t = 18$, $p = 6$, $k = 6$, $b = 12$, $n = 72$, $n_1 = 4$, $n_2 = 1$, VIF = 1.036. First associates are A J, B K, C L, D M, E N, F O, G P, H Q, I R. First associates occur sequentially four times, while second associates occur once. $E_{d1} = 97$, $E_{d2} = 83$.

An example is (the rows representing periods)

Block

1	2	3	4	5	6

Subjects

AJDMGP	BKENHQ	CLFOIR	BKCLDM	ENFOGP	HQIRAJ
JDMGPA	KENHQB	LFOIRC	KCLDMB	NFOGPE	QIRAJH
PAJDMG	QBKENH	RCLFOI	MBKCLD	PENFOG	JHQIRA
DMGPAJ	ENHQBK	FOIRCL	CLDMBK	FOGPEN	IRAJHQ
GPAJDM	HQBKEN	IRCLFO	DMBKCL	GPENFO	AJHQIR
MGPAJD	NHQBKE	OIRCLF	LDMBKC	OGPENF	RAJHQI

(continued)

Block

1	2	3	4	5	6

Subjects

CLAJEN	FODMHQ	IRGPBK	AJBKFO	DMENIR	GPHQCL
LAJENC	ODMHQF	RGPBKI	JBKFOA	MENIRD	PHQCLG
NCLAJE	QFODMH	KIRGPB	OAJBKF	RDMENI	LGPHQC
AJENCL	DMHQFO	GPBKIR	BKFOAJ	ENIRDM	HQCLGP
ENCLAJ	HQFODM	BKIRGP	FOAJBK	IRDMEN	CLGPHQ
JENCLA	MHQFOD	PBKIRG	KFOAJB	NIRDME	QCLGPH

Design PL123. $t = 21$, $p = 6$, $k = 6$, $b = 7$, $n = 42$, $n_1 = 1$, $n_2 = 0$, VIF = 1.036. First associates occur in the same row or the same column in the following scheme:

x	A	B	C	D	E	F
A	x	G	H	I	J	K
B	G	x	L	M	N	O
C	H	L	x	P	Q	R
D	I	M	P	x	S	T
E	J	N	Q	S	x	U
F	K	O	R	T	U	x

These treatments occur sequentially once, while second associates never occur. $E_{d1} = 86$, $E_{d2} = 76$.

An example is (the rows representing periods)

Block

	1	2	3	4	5	6	7
				Subjects			
	ABCDEF	GHIJKA	LMNOBG	PQRCHL	STDIMP	UEJNQS	FKORTU
	BCDEFA	HIJKAG	MNOBGL	QRCHLP	TDIMPS	EJNQSU	KORTUF
	FABCDE	AGHIJK	GLMNOB	LPQRCH	PSTDIM	SUEJNQ	UFKORT
	CDEFAB	IJKAGH	NOBGLM	RCHLPQ	DIMPST	JNQSUE	ORTUFK
	EFABCD	KAGHIJ	BGLMNO	HLPQRC	MPSTDI	QSUEJN	TUFKOR
	DEFABC	JKAGHI	OBGLMN	CHLPQR	IMPSTD	NQSUEJ	RTUFKO

Design PL124. $t = 26$, $p = 6$, $k = 6$, $b = 13$, $n = 78$, $n_1 = 1$, $n_2 = 0$, VIF = 1.036. Each treatment has 15 other treatments with which it occurs once sequentially, and 10 additional treatments with which it never occurs. $E_{d1} = 85$, $E_{d2} = 79$.

An example is (the rows representing periods)

Block

	1	2	3	4	5	6	7
				Subjects			
	AVFWLZ	BWGXMN	CXHYAO	DYIZBP	EZJNCQ	FNKODR	GOLPES
	VFWLZA	WGXMNB	XHYAOC	YIZBPD	ZJNCQE	NKODRF	OLPESG
	ZAVFWL	NBWGXM	OCXHYA	PDYIZB	QEZJNC	RFNKOD	SGOLPE
	FWLZAV	GXMNBW	HYAOCX	IZBPDY	JNCQEZ	KODRFN	LPESGO
	LZAVFW	MNBWGX	AOCXHY	BPDYIZ	CQEZJN	RFNKOD	ESGOLP
	WLZAVF	XMNBWG	YAOCXH	ZBPDYI	NCQEZJ	DRFNKO	PESGOL

(continued)

Block

8	9	10	11	12	13

Subjects

HPMQFT	IQARGU	JRBSHV	KSCTIW	LTDUJX	MUEVKY
PMQFTH	QARGUI	RBSHVJ	SCTIWK	TDUJXL	UEVKYM
THPMQF	UIQARG	VJRBSH	WKSCTI	XLTDUJ	YMUEVK
MQFTHP	ARGUIQ	BSHVJR	CTIWKS	DUJXLT	EVKYMU
FTHPMQ	GUIQAR	HVJRBS	IWKSCT	JXLTDU	KYMUEV
QFTHPM	RGUIQA	SHVJRB	TIWKSC	UJXLTD	VKYMUE

Appendix 6.C. Designs PL 125–130. Additional partially–balanced cross–over designs

Design PL125. $t = 5$, $p = 2$, $k = 3$, $b = 10$, $n = 30$, $n_1 = 2$, $n_2 = 1$, VIF = 4.0
First associates are as follows: A : C D   D : A B   B : D E
   E : B C   C : E A

First associates occur sequentially twice, second associates once. The design is constructed by deleting any one period of Design PL126.
An example is

		Block									
		1	2	3	4	5	6	7	8	9	10
						Subjects					
Period	1	ABD	ABD	BCE	BCE	CDA	CDA	DEB	DEB	EAC	EAC
	2	BDA	DAB	CEB	EBC	DAC	ACD	EBD	BDE	ACE	CEA

Design PL126. $t = 5$, $p = 3$, $k = 3$, $b = 10$, $n = 30$, $n_1 = 4$, $n_2 = 2$. VIF = 1.250. First associates are as follows: A : C D   D : A B
   B : D E   E : B C   C : E A

First associates occur sequentially four times; second associates twice.
An example is

		Block									
		1	2	3	4	5	6	7	8	9	10
						Subjects					
	1	ABD	ABD	BCE	BCE	CDA	CDA	DEB	DEB	EAC	EAC
Period	2	BDA	DAB	CEB	EBC	DAC	ACD	EBD	BDE	ACE	CEA
	3	DAB	BDA	EBC	CEB	ACD	DAC	BDE	EBD	CEA	ACE

Design PL127. $t = 9$, $p = 2$, $k = 3$, $b = 12$, $n = 36$, $n_1 = 1$, $n_2 = 0$, VIF = 4.0
First associates are in the same row or column of the asociation scheme:

$$
\begin{array}{ccc}
A & B & C \\
D & E & F \\
G & H & I
\end{array}
$$

First associates occur sequentially once; second associates never. The design is constructed by deleting any one period of Design PL128.
An example is

		Block											
		1	2	3	4	5	6	7	8	9	10	11	12
							Subjects						
Period	1	ABC	ABC	DEF	DEF	GHI	GHI	ADG	ADG	BEH	BEH	CFI	CFI
	2	CAB	BCA	FDE	EFD	IGH	HIG	GAD	DGA	HBE	EHB	ICF	FIC

Design PL128. $t = 9$, $p = 3$, $k = 3$, $b = 12$, $n = 36$, $n_1 = 2$, $n_2 = 0$, VIF = 1.250. First associates are in the same row or column of the association scheme:

$$
\begin{array}{ccc}
A & B & C \\
D & E & F \\
G & H & I
\end{array}
$$

First associates occur sequentially twice; second associates never.
An example is

<table>
<tr><td></td><td></td><td colspan="12">Block</td></tr>
<tr><td></td><td></td><td>1</td><td>2</td><td>3</td><td>4</td><td>5</td><td>6</td><td>7</td><td>8</td><td>9</td><td>10</td><td>11</td><td>12</td></tr>
<tr><td></td><td></td><td colspan="12">Subjects</td></tr>
<tr><td></td><td>1</td><td>ABC</td><td>ABC</td><td>DEF</td><td>DEF</td><td>GHI</td><td>GHI</td><td>ADG</td><td>ADG</td><td>BEH</td><td>BEH</td><td>CFI</td><td>CFI</td></tr>
<tr><td>Period</td><td>2</td><td>BCA</td><td>CAB</td><td>EFD</td><td>FDE</td><td>HIG</td><td>IGH</td><td>DGA</td><td>GAD</td><td>EHB</td><td>HBE</td><td>FIC</td><td>ICF</td></tr>
<tr><td></td><td>3</td><td>CAB</td><td>BCA</td><td>FDE</td><td>EFD</td><td>IGH</td><td>HIG</td><td>GAD</td><td>DGA</td><td>HBE</td><td>EHB</td><td>ICF</td><td>FIC</td></tr>
</table>

Design PL129. $t = 16$, $p = 4$, $k = 4$, $b = 8$, $n = 32$, $n_1 = 1$, $n_2 = 0$, VIF = 1.1
First associates are in the same row or column of the association scheme:

$$
\begin{array}{cccc}
A & B & C & D \\
E & F & G & H \\
I & J & K & L \\
M & N & O & P
\end{array}
$$

First associates occur sequentially once; second associates never.
An example is

Subjects in Blocks

Period	1	A	B	C	D		E	F	G	H		I	J	K	L		M	N	O	P
	2	B	D	A	C		F	H	E	G		J	L	I	K		N	P	M	O
	3	C	A	D	B		G	E	H	F		K	I	L	J		O	M	P	N
	4	D	C	B	A		H	G	F	E		L	K	J	I		P	O	N	M

(continued)

Subjects in Blocks

Period	1	A	E	I	M		B	F	J	N		C	G	K	O		D	H	L	P
	2	E	M	A	I		F	N	B	J		G	O	C	K		H	P	D	L
	3	I	A	M	E		J	B	N	F		K	C	O	G		L	D	P	H
	4	M	I	E	A		N	J	F	B		O	K	G	C		P	L	H	D

Design PL130. $t = 36$, $p = 6$, $k = 6$, $b = 12$, $n = 72$, $n_1 = 1$, $n_2 = 0$, VIF = 1.036. First associates are in the same row or column of the association scheme:

$$
\begin{array}{cccccc}
A & B & C & D & E & F \\
G & H & I & J & K & L \\
M & N & O & P & Q & R \\
S & T & U & V & W & X \\
Y & Z & 1 & 2 & 3 & 4 \\
5 & 6 & 7 & 8 & 9 & 0 \\
\end{array}
$$

First associates occur sequentially once; second associates never.
An example is

			Block		
1	2	3	4	5	6
			Subjects		

ABCDEF	GHIJKL	MNOPQR	STUVWX	YZ1234	567890
BCDEFA	HIJKLG	NOPQRM	TUVWXS	Z1234Y	678905
FABCDE	LGHIJK	RMNOPQ	XSTUVW	4YZ123	056789
CDEFAB	IJKLGH	OPQRMN	UVWXST	1234YZ	789056
EFABCD	KLGHIJ	QRMNOP	WXSTUV	34YZ12	905678
DEFABC	JKLGHI	PQRMNO	VWXSTU	234YZ1	890567

(continued)

			Block		
7	8	9	10	11	12
			Subjects		

AGMSY5	BHNTZ6	CIOU17	DJPV28	EKQW39	FLRX40
GMSY5A	HNTZ6B	IOU17C	JPV28D	KQW39E	LRX40F
5AGMSY	6BHNTZ	7CIOU1	8DJPV2	9EKQW3	0FLRX4
MSY5AG	NTZ6BH	OU17CI	PV28DJ	QW39EK	RX40FL
Y5AGMS	Z6BHNT	17CIOU	28DJPV	39EKQW	40FLRX
SY5AGM	TZ6BHN	U17CIO	V28DJP	W39EKQ	X40FLR

Appendix 6.D. Designs PL 131–152.   Extra–period partially–balanced cross–over designs

Design PL131 (Parent Design PL125). $t = 5$, $p = 3$, $k = 3$, $b = 10$, $n = 30$, $\text{IF}_d = 5.333$, $\text{IF}_r = 13.333$. An example is

Block

| | 1 | 2 | 3 | 4 | 5 | 6 | 7 | 8 | 9 | 10 |

Subjects

	1	ABD	ABD	BCE	BCE	CDA	CDA	DEB	DEB	EAC	EAC
Period	2	BDA	DAB	CEB	EBC	DAC	ACD	EBD	BDE	ACE	CEA
	3	BDA	DAB	CEB	EBC	DAC	ACD	EBD	BDE	ACE	CEA

Design PL132 (Parent Design PL126). $t = 5$, $p = 4$, $k = 3$, $b = 10$, $n = 30$, $\text{IF}_d = 1.5625$, $\text{IF}_r = 2.250$. An example is

Block

| | | 1 | 2 | 3 | 4 | 5 | 6 | 7 | 8 | 9 | 10 |

Subjects

	1	ABD	ABD	BCE	BCE	CDA	CDA	DEB	DEB	EAC	EAC
Period	2	BDA	DAB	CEB	EBC	DAC	ACD	EBD	BDE	ACE	CEA
	3	DAB	BDA	EBC	CEB	ACD	DAC	BDE	EBD	CEA	ACE
	4	DAB	BDA	EBC	CEB	ACD	DAC	BDE	EBD	CEA	ACE

Design PL133 (Parent Design PL99). $t = 6$, $p = 3$, $k = 3$, $b = 8$, $n = 24$, $\text{IF}_d = 5.333$, $\text{IF}_r = 13.333$. An example is

Block

| | | 1 | 2 | 3 | 4 | 5 | 6 | 7 | 8 |

Subjects

	1	ABC	BCA	AEF	EFA	BDF	DFB	CDE	DEC
Period	2	BCA	ABC	EFA	AEF	DFB	BDF	DEC	CDE
	3	BCA	ABC	EFA	AEF	DFB	BDF	DEC	CDE

Design PL134 (Parent Design PL100). $t = 6$, $p = 4$, $k = 3$, $b = 8$, $n = 24$, $\text{IF}_d = 1.5625$, $\text{IF}_r = 2.250$. An example is

Block

| | | 1 | 2 | 3 | 4 | 5 | 6 | 7 | 8 |

Subjects

	1	ABC	ABC	AEF	AEF	BDF	BDF	CDE	CDE
Period	2	BCA	CAB	EFA	FAE	DFB	FBD	DEC	ECD
	3	CAB	BCA	FAE	EFA	FBD	DFB	ECD	DEC
	4	CAB	BCA	FAE	EFA	FBD	DFB	ECD	DEC

Design PL135 (Parent Design PL101). $t = 6$, $p = 5$, $k = 4$, $b = 3$, $n = 12$, $IF_d = 1.320$, $IF_r = 1.600$. An example is

		Block	
	1	2	3
Subject →	1 2 3 4	1 2 3 4	1 2 3 4

		Block 1	Block 2	Block 3
	1	A D B E	B E C F	C F A D
	2	D E A B	E F B C	F D C A
Period	3	B A E D	C B F E	A C D F
	4	E B D A	F C E B	D A F C
	5	E B D A	F C E B	D A F C

Design PL136 (Parent Design PL102). $t = 8$, $p = 3$, $k = 3$, $b = 16$, $n = 48$, $IF_d = 5.333$, $IF_r = 13.333$. An example is

		Block							
		1	2	3	4	5	6	7	8
					Subjects				
	1	ABD	ABD	BCE	BCE	CDF	CDF	DEG	DEG
Period	2	BDA	DAB	CEB	EBC	DFC	FCD	EGD	GDE
	3	BDA	DAB	CEB	EBC	DFC	FCD	EGD	GDE

(continued)

		Block							
		9	10	11	12	13	14	15	16
					Subjects				
	1	EFH	EFH	FGA	FGA	GHB	GHB	HAC	HAC
Period	2	FHE	HEF	GAF	AFG	HBG	BGH	ACH	CHA
	3	FHE	HEF	GAF	AFG	HBG	BGH	ACH	CHA

Design PL137 (Parent Design PL103). $t = 8$, $p = 4$, $k = 3$, $b = 16$, $n = 48$, $IF_d = 1.5625$, $IF_r = 2.250$. An example is

		Block												
		1	2	3	4	5	6	7	8	9	10	11	12	13	14
								Subjects						
	1	ABD	ABD	BCE	BCE	CDF	CDF	DEG	DEG	EFH	EFH	FGA	FGA	GHB	GHB
Period	2	BDA	DAB	CEB	EBC	DFC	FCD	EGD	GDE	FHE	HEF	GAF	AFG	HBG	BGH
	3	DAB	BDA	EBC	CEB	FCD	DFC	GDE	EGD	HEF	FHE	AFG	GAF	BGH	HBG
	4	DAB	BDA	EBC	CEB	FCD	DFC	GDE	EGD	HEF	FHE	AFG	GAF	BGH	HBG

(continued)

```
                    Block
                    15   16
                    Subjects

           1    | HAC HAC
Period     2    | ACH CHA
           3    | CHA ACH
           4    | CHA ACH
```

Design PL138 (Parent Design PL104). $t = 8$, $p = 5$, $k = 4$, $b = 6$, $n = 24$, $IF_d = 1.320$, $IF_r = 1.600$. An example is

```
           1    | A E B F  | C G D H  | A E C G
           2    | E F A B  | G H C D  | E G A C
Period     3    | B A F E  | D C H G  | C A G E
           4    | F B E A  | H D G C  | G C E A
           5    | F B E A  | H D G C  | G C E A
```

(continued)

```
           1    | B F D H  | A E D H  | B F C G
           2    | F H B D  | E H A D  | F G B C
Period     3    | D B H F  | D A H E  | C B G F
           4    | H D F B  | H D E A  | G C F B
           5    | H D F B  | H D E A  | G C F B
```

Design PL139 (Parent Design PL127). $t = 9$, $p = 3$, $k = 3$, $b = 12$, $n = 36$, $IF_d = 5.333$, $IF_r = 13.333$. An example is

```
                                    Block
           1   2   3   4   5   6   7   8   9  10  11  12
                                   Subjects

        1 | ABC ABC DEF DEF GHI GHI ADG ADG BEH BEH CFI CFI
Period  2 | CAB BCA FDE EFD IGH HIG GAD DGA HBE EHB ICF FIC
        3 | CAB BCA FDE EFD IGH HIG GAD DGA HBE EHB ICF FIC
```

Design PL140 (Parent Design PL128). $t = 9$, $p = 4$, $k = 3$, $b = 12$, $n = 36$, $IF_d = 1.5625$, $IF_r = 2.250$. An example is

		Block											
	1	2	3	4	5	6	7	8	9	10	11	12
						Subjects						
Period 1	ABC	ABC	DEF	DEF	GHI	GHI	ADG	ADG	BEH	BEH	CFI	CFI
2	BCA	CAB	EFD	FDE	HIG	IGH	DGA	GAD	EHB	HBE	FIC	ICF
3	CAB	BCA	FDE	EFD	IGH	HIG	GAD	DGA	HBE	EHB	ICF	FIC
4	CAB	BCA	FDE	EFD	IGH	HIG	GAD	DGA	HBE	EHB	ICF	FIC

Design PL141 (Parent Design PL106). $t = 9$, $p = 5$, $k = 4$, $b = 9$, $n = 36$, $IF_d = 1.320$, $IF_r = 1.600$. An example is

		Subjects			
Period 1	A F I B	F H B D	H A D I	I B E G	B D G C
2	F B A I	H D F B	A I H D	B G I E	D C B G
3	I A B F	B F D H	D H I A	E I G B	G B C D
4	B I F A	D B H F	I D A H	G E B I	C G D B
5	B I F A	D B H F	I D A H	G E B I	C G D B

(continued)

		Subjects		
Period 1	D I C E	E G A F	G C F H	C E H A
2	I E D C	G F E A	C H G F	E A C H
3	C D E I	A E F G	F G H C	H C A E
4	E C I D	F A G E	H F C G	A H E C
5	E C I D	F A G E	H F C G	A H E C

Design PL142 (Parent Design PL108). $t = 10$, $p = 3$, $k = 3$, $b = 20$, $n = 60$, $IF_d = 5.333$, $IF_r = 13.333$. An example is

		Block												
	1	2	3	4	5	6	7	8	9	10	11	12	13	14
							Subjects							
Period 1	ABE	ABE	HIJ	HIJ	BHC	BHC	GEI	GEI	IDB	IDB	EFH	EFH	CJD	CJD
2	BEA	EAB	IJH	JHI	HCB	CBH	EIG	IGE	DBI	BID	FHE	HEF	JDC	DCJ
3	BEA	EAB	IJH	JHI	HCB	CBH	EIG	IGE	DBI	BID	FHE	HEF	JDC	DCJ

(continued)

Block
	15	16	17	18	19	20
			Subjects			

	1	JGF	JGF	DAG	DAG	FCA	FCA
Period	2	GFJ	FJG	AGD	GDA	CAF	AFC
	3	GFJ	FJG	AGD	GDA	CAF	AFC

Design PL143 (Parent Design PL109). $t = 10$, $p = 4$, $k = 3$, $b = 20$, $n = 60$, $IF_d = 1.5625$, $IF_r = 2.250$. An example is

Block

		1	2	3	4	5	6	7	8	9	10	11	12	13	14
								Subjects							
	1	ABE	ABE	HIJ	HIJ	BHC	BHC	GEI	GEI	IDB	IDB	EFH	EFH	CJD	CJD
Period	2	BEA	EAB	IJH	JHI	HCB	CBH	EIG	IGE	DBI	BID	FHE	HEF	JDC	DCJ
	3	EAB	BEA	JHI	IJH	CBH	HBC	IGE	EIG	BID	DBI	HEF	FHE	DCJ	JDC
	4	EAB	BEA	JHI	IJH	CBH	HBC	IGE	EIG	BID	DBI	HEF	FHE	DCJ	JDC

(continued)

Block
	15	16	17	18	19	20
			Subjects			

	1	JGF	JGF	DAG	DAG	FCA	FCA
Period	2	GFJ	FJG	AGD	GDA	CAF	AFC
	3	FJG	GFJ	GDA	AGD	AFC	CAF
	4	FJG	GFJ	GDA	AGD	AFC	CAF

Design PL144 (Parent Design PL110). $t = 10$, $p = 5$, $k = 4$, $b = 5$, $n = 20$, $IF_d = 1.320$, $IF_r = 1.600$. An example is

Block

		1	2	3	4	5
				Subjects		
	1	A B C D	E F G A	H I B E	J C F H	D G I J
	2	B D A C	F A E G	I E H B	C H J F	G J D I
Period	3	C A D B	G E A F	B H E I	F J H C	I D J G
	4	D C B A	A G F E	E B I H	H F C J	J I G D
	5	D C B A	A G F E	E B I H	H F C J	J I G D

Design PL145 (Parent Design PL111). $t = 10$, $p = 6$, $k = 5$, $b = 12$, $n = 60$, $\text{IF}_d = 1.231$, $\text{IF}_r = 1.389$. An example is

		Block				
		1	2	3	4	5	6
				Subjects			

		1	2	3	4	5	6
	1	AHIGC	AHIGC	AHDJE	AHDJE	HDFGB	HDFGB
	2	HIGCA	CAHIG	HDJEA	EAHDJ	DFGBH	BHDFG
Period	3	CAHIG	HIGCA	EAHDJ	HDJEA	BHDFG	DFGBH
	4	IGCAH	GCAHI	DJEAH	JEAHD	FGBHD	GBHDF
	5	GCAHI	IGCAH	JEAHD	DJEAH	GBHDF	FGBHD
	6	GCAHI	IGCAH	JEAHD	DJEAH	GBHDF	FGBHD

(continued)

		Block				
		7	8	9	10	11	12
				Subjects			

		7	8	9	10	11	12
	1	DFIEC	DFIEC	AFIJB	AFIJB	JGEBC	JGEBC
	2	FIECD	CDFIE	FIJBA	BAFIJ	GEBCJ	CJGEB
Period	3	CDFIE	FIECD	BAFIJ	FIJBA	CJGEB	GEBCJ
	4	IECDF	ECDFI	IJBAF	JBAFI	EBCJG	BCJGE
	5	ECDFI	IECDF	JBAFI	IJBAF	BCJGE	EBCJG
	6	ECDFI	IECDF	JBAFI	IJBAF	BCJGE	EBCJG

Design PL146 (Parent Design PL113). $t = 12$, $p = 5$, $k = 4$, $b = 9$, $n = 36$, $\text{IF}_d = 1.320$, $\text{IF}_r = 1.600$. An example is

Subjects

	1	A	B	C	D		G	J	E	D		F	K	I	D		A	G	F	H		K	E	B	H
	2	B	D	A	C		J	D	G	E		K	D	F	I		G	H	A	F		E	H	K	B
Period	3	C	A	D	B		E	G	D	J		I	F	D	K		F	A	H	G		B	K	H	E
	4	D	C	B	A		D	E	J	G		D	I	K	F		H	F	G	A		H	B	E	K
	5	D	C	B	A		D	E	J	G		D	I	K	F		H	F	G	A		H	B	E	K

(continued)

Subjects

	1	J	I	C	H		A	K	J	L		I	B	G	L		E	C	F	L
	2	I	H	J	C		K	L	A	J		B	L	I	G		C	L	E	F
Period	3	C	J	H	I		J	A	L	K		G	I	L	B		F	E	L	C
	4	H	C	I	J		L	J	K	A		L	G	B	I		L	F	C	E
	5	H	C	I	J		L	J	K	A		L	G	B	I		L	F	C	E

Design PL147 (Parent Design PL115). $t = 13$, $p = 3$, $k = 3$, $b = 26$, $n = 78$, $\text{IF}_d = 5.333$, $\text{IF}_r = 13.333$. An example is

		Block												
		1	2	3	4	5	6	7	8	9	10	11	12	13	14
								Subjects							
Period	1	ACI	ACI	BDJ	BDJ	CEK	CEK	DFL	DFL	EGM	EGM	FHA	FHA	GIB	GIB
	2	CIA	IAC	DJB	JBD	EKC	KCE	FLD	LDF	GME	MEG	HAF	AFH	IBG	BGI
	3	CIA	IAC	DJB	JBD	EKC	KCE	FLD	LDF	GME	MEG	HAF	AFH	IBG	BGI

(continued)

		Block											
		15	16	17	18	19	20	21	22	23	24	25	26
							Subjects						
Period	1	HJC	HJC	IKD	IKD	JLE	JLE	KMF	KMF	LAG	LAG	MBH	MBH
	2	JCH	CHJ	KDI	DIK	LEJ	EJL	MFK	FKM	AGL	GLA	BHM	HMB
	3	JCH	CHJ	KDI	DIK	LEJ	EJL	MFK	FKM	AGL	GLA	BHM	HMB

Design PL148 (Parent Design PL116). $t = 13$, $p = 4$, $k = 3$, $b = 26$, $n = 78$, $\text{IF}_d = 1.5625$, $\text{IF}_r = 2.250$. An example is

		Block												
		1	2	3	4	5	6	7	8	9	10	11	12	13	14
								Subjects							
Period	1	ACI	ACI	BDJ	BDJ	CEK	CEK	DFL	DFL	EGM	EGM	FHA	FHA	GIB	GIB
	2	CIA	IAC	DJB	JBD	EKC	KCE	FLD	LDF	GME	MEG	HAF	AFH	IBG	BGI
	3	IAC	CIA	JBD	DJB	KCE	EKC	LDF	FLD	MEG	GME	AFH	HAF	BGI	IBG
	4	IAC	CIA	JBD	DJB	KCE	EKC	LDF	FLD	MEG	GME	AFH	HAF	BGI	IBG

(continued)

		Block											
		15	16	17	18	19	20	21	22	23	24	25	26
							Subjects						
Period	1	HJC	HJC	IKD	IKD	JLE	JLE	KMF	KMF	LAG	LAG	MBH	MBH
	2	JCH	CHJ	KDI	DIK	LEJ	EJL	MFK	FKM	AGL	GLA	BHM	HMB
	3	CHJ	JCH	DIK	KDI	EJL	LEJ	FKM	MFK	GLA	AGL	HMB	BHM
	4	CHJ	JCH	DIK	KDI	EJL	LEJ	FKM	MFK	GLA	AGL	HMB	BHM

Design PL149 (Parent Design PL117). $t = 14$, $p = 5$, $k = 4$, $b = 14$, $n = 56$, $IF_d = 1.320$, $IF_r = 1.600$. An example is

Subjects in Blocks

	1	B C E H	C D F I	D E G J	E F A K	F G B L
	2	C H B E	D I C F	E J D G	F K E A	G L F B
Period	3	E B H C	F C I D	G D J E	A E K F	B F L G
	4	H E C B	I F D C	J G E D	K A F E	L B G F
	5	H E C B	I F D C	J G E D	K A F E	L B G F

(continued)

Subjects in Blocks

	1	G A C M	A B D N	I J L A	J K M B	K L N C
	2	A M G C	B N A D	J A I L	K B J M	L C K N
Period	3	C G M A	D A N B	L I A J	M J B K	N K C L
	4	M C A G	N D B A	A L J I	B M K J	C N L K
	5	M C A G	N D B A	A L J I	B M K J	C N L K

(continued)

Subjects in Blocks

	1	L M H D	M N I E	N H J F	H I K G
	2	M D L H	N E M I	H F N J	I G H K
Period	3	H L D M	I M E N	J N F H	K H G I
	4	D H M L	E I N M	F J H N	G K I H
	5	D H M L	E I N M	F J H N	G K I H

Design PL150 (Parent Design PL119). $t = 15$, $p = 5$, $k = 4$, $b = 15$, $n = 60$, $IF_d = 1.320$, $IF_r = 1.600$. An example is

Subjects in Blocks

	1	B D E M	C E F N	D F G O	E G H A	F H I B
	2	D M B E	E N C F	F O D G	G A E H	H B F I
Period	3	E B M D	F C N E	G D O F	H E A G	I F B H
	4	M E D B	N F E C	O G F D	A H G E	B I H F
	5	M E D B	N F E C	O G F D	A H G E	B I H F

(continued)

Subjects in Blocks

	1	2	3	4	5
Period 1	G I J C	H J K D	I K L E	J L M F	K M N G
2	I C G J	J D H K	K E I L	L F J M	M G K N
3	J G C I	K H D J	L I E K	M J F L	N K G M
4	C J I G	D K J H	E L K I	F M L J	G N M K
5	C J I G	D K J H	E L K I	F M L J	G N M K

(continued)

Subjects in Blocks

	1	2	3	4	5
Period 1	L N O H	M O A I	N A B J	O B C K	A C D L
2	N H L O	O I M A	A J N B	B K O C	C L A D
3	O L H N	A M I O	B N J A	C O K B	D A L C
4	H O N L	I A O M	J B A N	K C B O	L D C A
5	H O N L	I A O M	J B A N	K C B O	L D C A

Design PL151 (Parent Design PL120). $t = 15$, $p = 6$, $k = 5$, $b = 12$, $n = 60$, $IF_d = 1.231$, $IF_r = 1.389$. An example is

Block

	1	2	3	4	5	6
			Subjects			
Period 1	ABCDE	ABCDE	FGHIA	FGHIA	JKLBF	JKLBF
2	BCDEA	EABCD	GHIAF	AFGHI	KLBFJ	FJKLB
3	EABCD	BCDEA	AFGHI	GHIAF	FJKLB	KLBFJ
4	CDEAB	DEABC	HIAFG	IAFGH	LBFJK	BFJKL
5	DEABC	CDEAB	IAFGH	HIAFG	BFJKL	LBFJK
6	DEABC	CDEAB	IAFGH	HIAFG	BFJKL	LBFJK

(continued)

Block

	7	8	9	10	11	12
			Subjects			
Period 1	MNCGJ	MNCGJ	ODHKM	ODHKM	EILNO	EILNO
2	NCGJM	JMNCG	DHKMO	MODHK	ILNOE	OEILN
3	JMNCG	NCGJM	MODHK	DHKMO	OEILN	ILNOE
4	CGJMN	GJMNC	HKMOD	KMODH	LNOEI	NOEIL
5	GJMNC	CGJMN	KMODH	HKMOD	NOEIL	LNOEI
6	GJMNC	CGJMN	KMODH	HKMOD	NOEIL	LNOEI

Design PL152 (Parent Design PL129). $t = 16$, $p = 5$, $k = 4$, $b = 8$, $n = 32$, $IF_d = 1.320$, $IF_r = 1.600$. An example is

Subjects in Blocks

	1	A B C D	E F G H	I J K L	M N O P
	2	B D A C	F H E G	J L I K	N P M O
Period	3	C A D B	G E H F	K I L J	O M P N
	4	D C B A	H G F E	L K J I	P O N M
	5	D C B A	H G F E	L K J I	P O N M

(continued)

Subjects in Blocks

	1	A E I M	B F J N	C G K O	D H L P
	2	E M A I	F N B J	G O C K	H P D L
Period	3	I A M E	J B N F	K C O G	L D P H
	4	M I E A	N J F B	O K G C	P L H D
	5	M I E A	N J F B	O K G C	P L H D

Appendix 6.E. Designs PL 153–160. Extra–period partially–balanced designs with $k = 2$

Design PL153. $t = 4$, $p = 3$, $k = 2$, $b = 4$, $n = 8$, $n_1 = 0$, $n_2 = 1$. First associates are treatments not occurring in the same block. First associates never occur sequentially, whereas second associates occur once.

An example is

		Block			
		1	2	3	4
		Subjects			
	1	AB	BC	CD	DA
Period	2	BA	CB	DC	AD
	3	BA	CB	DC	AD

Design PL154. $t = 5$, $p = 3$, $k = 2$, $b = 5$, $n = 10$, $n_1 = 0$, $n_2 = 1$. First associates are treatments not occurring in the same block. First associates never occur sequentially, whereas second associates occur once.

An example is

		Block				
		1	2	3	4	5
		Subjects				
	1	AC	BD	CE	DA	EB
Period	2	CA	DB	EC	AD	BE
	3	CA	DB	EC	AD	BE

Design PL155. $t = 6$, $p = 3$, $k = 2$, $b = 9$, $n = 18$, $n_1 = 0$, $n_2 = 1$. First associates are treatments which occur in the same row in the association scheme

$$\begin{array}{ccc} A & B & C \\ D & E & F \end{array}$$

First associates never occur sequentially, whereas second associates occur once.

An example is

		Block								
		1	2	3	4	5	6	7	8	9
		Subjects								
	1	AD	AE	AF	BD	BE	BF	CD	CE	CF
Period	2	DA	EA	FA	DB	EB	FB	DC	EC	FC
	3	DA	EA	FA	DB	EB	FB	DC	EC	FC

Design PL156. $t = 8$, $p = 3$, $k = 2$, $b = 16$, $n = 32$, $n_1 = 0$, $n_2 = 1$. First associates are treatments which occur in the same row in the association scheme

$$
\begin{array}{cccc}
A & C & E & G \\
B & D & F & H
\end{array}
$$

First associates never occur sequentially, whereas second associates occur once. An example is (the rows representing periods)

							Block								
1	2	3	4	5	6	7	8	9	10	11	12	13	14	15	16
							Subjects								

	1	2	3	4	5	6	7	8	9	10	11	12	13	14	15	16
1	AB	AD	BC	BE	CD	CF	DE	DG	EF	EH	FG	FA	GH	GB	HA	HC
2	BA	DA	CB	EB	DC	FC	ED	GD	FE	HE	GF	AF	HG	BG	AH	CH
3	BA	DA	CB	EB	DC	FC	ED	GD	FE	HE	GF	AF	HG	BG	AH	CH

Design PL157. $t = 9$, $p = 3$, $k = 2$, $b = 18$, $n = 36$, $n_1 = 1$, $n_2 = 0$. First associates are treatments which occur either in the same row or same column in the association scheme

$$
\begin{array}{ccc}
A & B & C \\
D & E & F \\
G & H & I
\end{array}
$$

First associates occur once sequentially, while second associates never occur. An example is

				Block					
	1	2	3	4	5	6	7	8	9
				Subjects					

		1	2	3	4	5	6	7	8	9
	1	AB	CA	BC	DE	FD	EF	GH	IG	HI
Period	2	BA	AC	CB	ED	DF	FE	HG	GI	IH
	3	BA	AC	CB	ED	DF	FE	HG	GI	IH

(continued)

			Block						
10	11	12	13	14	15	16	17	18
			Subjects						

		10	11	12	13	14	15	16	17	18
	1	AD	GA	DG	BE	HB	EH	CF	IC	FI
Period	2	DA	AG	GD	EB	BH	HE	FC	CI	IF
	3	DA	AG	GD	EB	BH	HE	FC	CI	IF

Design PL158. $t = 10$, $p = 3$, $k = 2$, $b = 15$, $n = 30$, $n_1 = 1$, $n_2 = 0$. First associates are treatments which occur in the same block. First associates occur once sequentially, while second associates never occur.

An example is

		Block								
		1	2	3	4	5	6	7	8	9
					Subjects					
	1	AH	AI	AJ	BF	BG	BJ	CE	CG	CI
Period	2	HA	IA	JA	FB	GB	JB	EC	GC	IC
	3	HA	IA	JA	FB	GB	JB	EC	GC	IC

(continued)

		Block					
		1	2	3	4	5	6
			Subjects				
	1	DE	DF	DH	EJ	FI	GH
Period	2	ED	FD	HD	JE	IF	HG
	3	ED	FD	HD	JE	IF	HG

Design PL159. $t = 12$, $p = 3$, $k = 2$, $b = 36$, $n = 72$, $n_1 = 0$, $n_2 = 1$. First associates are treatments which occur in the same row in the association scheme

$$
\begin{array}{cccccc}
A & C & E & G & I & K \\
B & D & F & H & J & L
\end{array}
$$

First associates never occur sequentially, whereas second associates occur once.

An example is

		Block								
		1	2	3	4	5	6	7	8	9
					Subjects					
	1	AB	AD	AF	BC	BE	BG	CD	CF	CH
Period	2	BA	DA	FA	CB	EB	GB	DC	FC	HC
	3	BA	DA	FA	CB	EB	GB	DC	FC	HC

(continued)

		Block								
		10	11	12	13	14	15	16	17	18
					Subjects					
	1	DE	DG	DI	EF	EH	EJ	FG	FI	FK
Period	2	ED	GD	ID	FE	HE	JE	GF	IF	KF
	3	ED	GD	ID	FE	HE	JE	GF	IF	KF

(continued)

		Block								
		19	20	21	22	23	24	25	26	27
					Subjects					
	1	GH	GJ	GL	HI	HK	HA	IJ	IL	IB
Period	2	HG	JG	LG	IH	KH	AH	JI	LI	BI
	3	HG	JG	LG	IH	KH	AH	JI	LI	BI

(continued)

		Block								
		28	29	30	31	32	33	34	35	36
					Subjects					
	1	JK	JA	JC	KL	KB	KD	LA	LC	LE
Period	2	KJ	AJ	CJ	LK	BK	DK	AL	CL	EL
	3	KJ	AJ	CJ	LK	BK	DK	AL	CL	EL

Design PL160. $t = 13$, $p = 3$, $k = 2$, $b = 39$, $n = 78$, $n_1 = 1$, $n_2 = 0$. First associates are treatments which occur in the same block. First associates occur once sequentially, whereas second associates never occur.

An example is

		Block												
		1	2	3	4	5	6	7	8	9	10	11	12	13
							Subjects							
	1	AC	AF	AG	BD	BG	BH	CE	CH	CI	DF	DI	DJ	EG
Period	2	CA	FA	GA	DB	GB	HB	EC	HC	IC	FD	ID	JD	GE
	3	CA	FA	GA	DB	GB	HB	EC	HC	IC	FD	ID	JD	GE

(continued)

		Block												
		14	15	16	17	18	19	20	21	22	23	24	25	26
							Subjects							
	1	EJ	EK	FH	FK	FL	GI	GL	GM	HJ	HM	HC	IK	IA
Period	2	JE	KE	HF	KF	LF	IG	LG	MG	JH	MH	CH	KI	AI
	3	JE	KE	HF	KF	LF	IG	LG	MG	JH	MH	CH	KI	AI

(continued)

		Block												
		27	28	29	30	31	32	33	34	35	36	37	38	39
							Subjects							
	1	IB	JL	JB	JC	KM	KC	KD	LA	LD	LE	MB	ME	MF
Period	2	BI	LJ	BJ	CJ	MK	CK	DK	AL	DL	EL	BM	EM	FM
	3	BI	LJ	BJ	CJ	MK	CK	DK	AL	DL	EL	BM	EM	FM

Appendix 6.F. Designs DH 1–45. Cyclic change–over designs of Davis and Hall

Design DH1. $t = k = 6$, $p = 3$, $b = 2$, $n = 12$. $E_t = 79$, $E_d = 58$, $E_r = 34$.
An example is

Block
1   2
Subjects

		1	2
	1	ABCDEF	ABCDEF
Period	2	DEFABC	FABCDE
	3	EFABCD	BCDEFA

Design DH2. $t = k = 7$, $p = 3$, $b = 2$, $n = 14$. $E_t = 78$, $E_d = 62$, $E_r = 35$.
An example is

Block
1   2
Subjects

		1	2
	1	ABCDEFG	ABCDEFG
Period	2	DEFGABC	EFGABCD
	3	BCDEFGA	FGABCDE

Design DH3. $t = k = 8$, $p = 3$, $b = 2$, $n = 16$. $E_t = 76$, $E_d = 54$, $E_r = 31$.
An example is

Block
1   2
Subjects

		1	2
	1	ABCDEFGH	ABCDEFGH
Period	2	EFGHABCD	GHABCDEF
	3	BCDEFGHA	FGHABCDE

Design DH4. $t = k = 9$, $p = 3$, $b = 2$, $n = 18$. $E_t = 74$, $E_d = 55$, $E_r = 31$.
An example is

Block
1   2
Subjects

		1	2
	1	ABCDEFGHI	ABCDEFGHI
Period	2	DEFGHIABC	GHIABCDEF
	3	IABCDEFGH	FGHIABCDE

Design DH5. $t = k = 10$, $p = 3$, $b = 2$, $n = 20$. $E_t = 73$, $E_d = 51$, $E_r = 30$.
An example is

Block
1   2
Subjects

		Block 1	Block 2
Period	1	ABCDEFGHIJ	ABCDEFGHIJ
	2	BCDEFGHIJA	FGHIJABCDE
	3	DEFGHIJABC	EFGHIJABCD

Design DH6. $t = k = 11$, $p = 3$, $b = 2$, $n = 22$. $E_t = 73$, $E_d = 52$, $E_r = 30$.
An example is

Block
1   2
Subjects

		Block 1	Block 2
Period	1	ABCDEFGHIJK	ABCDEFGHIJK
	2	BCDEFGHIJKA	KABCDEFGHIJ
	3	HIJKABCDEFG	CDEFGHIJKAB

Design DH7. $t = k = 12$, $p = 3$, $b = 2$, $n = 24$. $E_t = 71$, $E_d = 52$, $E_r = 30$.
An example is

Block
1   2
Subjects

		Block 1	Block 2
Period	1	ABCDEFGHIJKL	ABCDEFGHIJKL
	2	BCDEFGHIJKLA	LABCDEFGHIJK
	3	HIJKLABCDEFG	DEFGHIJKLABC

Design DH8. $t = k = 13$, $p = 3$, $b = 2$, $n = 26$. $E_t = 72$, $E_d = 45$, $E_r = 28$.
An example is

Block
1   2
Subjects

		Block 1	Block 2
Period	1	ABCDEFGHIJKLM	ABCDEFGHIJKLM
	2	BCDEFGHIJKLMA	LMABCDEFGHIJK
	3	EFGHIJKLMABCD	GHIJKLMABCDEF

Design DH9. $t = k = 14$, $p = 3$, $b = 2$, $n = 28$. $E_t = 70$, $E_d = 50$, $E_r = 30$.
An example is

Block
<table>
<tr><td></td><td></td><td>1</td><td>2</td></tr>
<tr><td></td><td></td><td colspan="2">Subjects</td></tr>
<tr><td></td><td>1</td><td>ABCDEFGHIJKLMN</td><td>ABCDEFGHIJKLMN</td></tr>
<tr><td>Period</td><td>2</td><td>BCDEFGHIJKLMNA</td><td>NABCDEFGHIJKLM</td></tr>
<tr><td></td><td>3</td><td>JKLMNABCDEFGHI</td><td>DEFGHIJKLMNABC</td></tr>
</table>

Design DH10. $t = k = 15$, $p = 3$, $b = 2$, $n = 30$. $E_t = 71$, $E_d = 44$, $E_r = 28$.
An example is

Block
<table>
<tr><td></td><td></td><td>1</td><td>2</td></tr>
<tr><td></td><td></td><td colspan="2">Subjects</td></tr>
<tr><td></td><td>1</td><td>ABCDEFGHIJKLMNO</td><td>ABCDEFGHIJKLMNO</td></tr>
<tr><td>Period</td><td>2</td><td>IJKLMNOABCDEFGH</td><td>OABCDEFGHIJKLMN</td></tr>
<tr><td></td><td>3</td><td>CDEFGHIJKLMNOAB</td><td>EFGHIJKLMNOABCD</td></tr>
</table>

Design DH11. $t = k = 16$, $p = 3$, $b = 2$, $n = 32$. $E_t = 70$, $E_d = 45$, $E_r = 28$.
An example is

Block
<table>
<tr><td></td><td></td><td>1</td><td>2</td></tr>
<tr><td></td><td></td><td colspan="2">Subjects</td></tr>
<tr><td></td><td>1</td><td>ABCDEFGHIJKLMNOP</td><td>ABCDEFGHIJKLMNOP</td></tr>
<tr><td>Period</td><td>2</td><td>GHIJKLMNOPABCDEF</td><td>EFGHIJKLMNOPABCD</td></tr>
<tr><td></td><td>3</td><td>JKLMNOPABCDEFGHI</td><td>PABCDEFGHIJKLMNO</td></tr>
</table>

Design DH12. $t = k = 17$, $p = 3$, $b = 2$, $n = 34$. $E_t = 70$, $E_d = 45$, $E_r = 28$.
An example is

Block
<table>
<tr><td></td><td></td><td>1</td><td>2</td></tr>
<tr><td></td><td></td><td colspan="2">Subjects</td></tr>
<tr><td></td><td>1</td><td>ABCDEFGHIJKLMNOPQ</td><td>ABCDEFGHIJKLMNOPQ</td></tr>
<tr><td>Period</td><td>2</td><td>BCDEFGHIJKLMNOPQA</td><td>PQABCDEFGHIJKLMNO</td></tr>
<tr><td></td><td>3</td><td>FGHIJKLMNOPQABCDE</td><td>JKLMNOPQABCDEFGHI</td></tr>
</table>

Design DH13. $t = k = 18$, $p = 3$, $b = 2$, $n = 36$. $E_t = 69$, $E_d = 44$, $E_r = 27$.
An example is

		Block	
		1	2
		Subjects	
Period	1	ABCDEFGHIJKLMNOPQR	ABCDEFGHIJKLMNOPQR
	2	BCDEFGHIJKLMNOPQRA	MNOPQRABCDEFGHIJKL
	3	FGHIJKLMNOPQRABCDE	CDEFGHIJKLMNOPQRAB

Design DH14. $t = k = 19$, $p = 3$, $b = 2$, $n = 38$. $E_t = 69$, $E_d = 45$, $E_r = 28$.
An example is

		Block	
		1	2
		Subjects	
Period	1	ABCDEFGHIJKLMNOPQRS	ABCDEFGHIJKLMNOPQRS
	2	EFGHIJKLMNOPQRSABCD	RSABCDEFGHIJKLMNOPQ
	3	SABCDEFGHIJKLMNOPQR	HIJKLMNOPQRSABCDEFG

Design DH15. $t = k = 20$, $p = 3$, $b = 2$, $n = 40$. $E_t = 68$, $E_d = 44$, $E_r = 27$.
An example is

		Block	
		1	2
		Subjects	
Period	1	ABCDEFGHIJKLMNOPQRST	ABCDEFGHIJKLMNOPQRST
	2	CDEFGHIJKLMNOPQRSTAB	TABCDEFGHIJKLMNOPQRS
	3	NOPQRSTABCDEFGHIJKLM	EFGHIJKLMNOPQRSTABCD

Design DH16. $t = k = 6$, $p = 4$, $b = 2$, $n = 12$. $E_t = 90$, $E_d = 81$, $E_r = 57$.
An example is

		Block	
		1	2
		Subjects	
Period	1	ABCDEF	ABCDEF
	2	BCDEFA	DEFABC
	3	DEFABC	BCDEFA
	4	CDEFAB	EFABCD

Design DH17. $t = k = 7$, $p = 4$, $b = 2$, $n = 14$. $E_t = 88$, $E_d = 80$, $E_r = 57$.

An example is (Note: this design is variance–balanced and is the same as Design PL19 of Chapter 5)

Block
1   2
Subjects

		1	2
	1	ABCDEFG	ABCDEFG
Period	2	BCDEFGA	GABCDEF
	3	DEFGABC	EFGABCD
	4	GABCDEF	BCDEFGA

Design DH18. $t = k = 8$, $p = 4$, $b = 2$, $n = 16$. $E_t = 85$, $E_d = 77$, $E_r = 55$.

An example is

Block
1   2
Subjects

		1	2
	1	ABCDEFGH	ABCDEFGH
Period	2	CDEFGHAB	BCDEFGHA
	3	BCDEFGHA	FGHABCDE
	4	EFGHABCD	DEFGHABC

Design DH19. $t = k = 9$, $p = 4$, $b = 2$, $n = 18$. $E_t = 84$, $E_d = 74$, $E_r = 54$.

An example is

Block
1   2
Subjects

		1	2
	1	ABCDEFGHI	ABCDEFGHI
Period	2	BCDEFGHIA	FGHIABCDE
	3	EFGHIABCD	CDEFGHIAB
	4	CDEFGHIAB	GHIABCDEF

Design DH20. $t = k = 10$, $p = 4$, $b = 2$, $n = 20$. $E_t = 83$, $E_d = 73$, $E_r = 51$.

An example is

Block
1   2
Subjects

		1	2
	1	ABCDEFGHIJ	ABCDEFGHIJ
Period	2	EFGHIJABCD	FGHIJABCDE
	3	CDEFGHIJAB	HIJABCDEFG
	4	BCDEFGHIJA	EFGHIJABCD

Design DH21. $t = k = 11$, $p = 4$, $b = 2$, $n = 22$. $E_t = 82$, $E_d = 73$, $E_r = 50$.
An example is

Block

		1	2
		Subjects	
Period	1	ABCDEFGHIJK	ABCDEFGHIJK
	2	FGHIJKABCDE	GHIJKABCDEF
	3	BCDEFGHIJKA	FGHIJKABCDE
	4	CDEFGHIJKAB	CDEFGHIJKAB

Design DH22. $t = k = 12$, $p = 4$, $b = 2$, $n = 24$. $E_t = 82$, $E_d = 71$, $E_r = 51$.
An example is

Block

		1	2
		Subjects	
Period	1	ABCDEFGHIJKL	ABCDEFGHIJKL
	2	LABCDEFGHIJK	KLABCDEFGHIJ
	3	BCDEFGHIJKLA	DEFGHIJKLABC
	4	EFGHIJKLABCD	EFGHIJKLABCD

Design DH23. $t = k = 13$, $p = 4$, $b = 2$, $n = 26$. $E_t = 81$, $E_d = 69$, $E_r = 50$.
An example is

Block

		1	2
		Subjects	
Period	1	ABCDEFGHIJKLM	ABCDEFGHIJKLM
	2	BCDEFGHIJKLMA	IJKLMABCDEFGH
	3	DEFGHIJKLMABC	GHIJKLMABCDEF
	4	JKLMABCDEFGHI	JKLMABCDEFGHI

Design DH24. $t = k = 14$, $p = 4$, $b = 2$, $n = 28$. $E_t = 81$, $E_d = 69$, $E_r = 50$.
An example is

Block

		1	2
		Subjects	
Period	1	ABCDEFGHIJKLMN	ABCDEFGHIJKLMN
	2	HIJKLMNABCDEFG	GHIJKLMNABCDEF
	3	BCDEFGHIJKLMNA	EFGHIJKLMNABCD
	4	FGHIJKLMNABCDE	DEFGHIJKLMNABC

Design DH25. $t = k = 15$, $p = 4$, $b = 2$, $n = 30$. $E_t = 80$, $E_d = 69$, $E_r = 51$.
An example is

Block

1   2

Subjects

		1	2
	1	ABCDEFGHIJKLMNO	ABCDEFGHIJKLMNO
Period	2	HIJKLMNOABCDEFG	FGHIJKLMNOABCDE
	3	BCDEFGHIJKLMNOA	LMNOABCDEFGHIJK
	4	FGHIJKLMNOABCDE	MNOABCDEFGHIJKL

Design DH26. $t = k = 16$, $p = 4$, $b = 2$, $n = 32$. $E_t = 79$, $E_d = 68$, $E_r = 49$.
An example is

Block

1   2

Subjects

		1	2
	1	ABCDEFGHIJKLMNOP	ABCDEFGHIJKLMNOP
Period	2	FGHIJKLMNOPABCDE	OPABCDEFGHIJKLMN
	3	HIJKLMNOPABCDEFG	GHIJKLMNOPABCDEF
	4	BCDEFGHIJKLMNOPA	NOPABCDEFGHIJKLM

Design DH27. $t = k = 17$, $p = 4$, $b = 2$, $n = 34$. $E_t = 79$, $E_d = 67$, $E_r = 49$.
An example is

Block

1   2

Subjects

		1	2
	1	ABCDEFGHIJKLMNOPQ	ABCDEFGHIJKLMNOPQ
Period	2	BCDEFGHIJKLMNOPQA	DEFGHIJKLMNOPQABC
	3	HIJKLMNOPQABCDEFG	CDEFGHIJKLMNOPQAB
	4	FGHIJKLMNOPQABCDE	JKLMNOPQABCDEFGHI

Design DH28. $t = k = 18$, $p = 4$, $b = 2$, $n = 36$. $E_t = 79$, $E_d = 66$, $E_r = 48$.
An example is

Block

1   2

Subjects

		1	2
	1	ABCDEFGHIJKLMNOPQR	ABCDEFGHIJKLMNOPQR
Period	2	BCDEFGHIJKLMNOPQRA	LMNOPQRABCDEFGHIJK
	3	IJKLMNOPQRABCDEFGH	CDEFGHIJKLMNOPQRAB
	4	FGHIJKLMNOPQRABCDE	MNOPQRABCDEFGHIJKL

Design DH29. $t = k = 19$, $p = 4$, $b = 2$, $n = 38$. $E_t = 79$, $E_d = 65$, $E_r = 47$.
An example is

		Block	
		1	2
		Subjects	
Period	1	ABCDEFGHIJKLMNOPQRS	ABCDEFGHIJKLMNOPQRS
	2	HIJKLMNOPQRSABCDEFG	MNOPQRSABCDEFGHIJKL
	3	GHIJKLMNOPQRSABCDEF	QRSABCDEFGHIJKLMNOP
	4	JKLMNOPQRSABCDEFGHI	LMNOPQRSABCDEFGHIJK

Design DH30. $t = k = 20$, $p = 4$, $b = 2$, $n = 40$. $E_t = 78$, $E_d = 65$, $E_r = 47$.
An example is

		Block	
		1	2
		Subjects	
Period	1	ABCDEFGHIJKLMNOPQRST	ABCDEFGHIJKLMNOPQRST
	2	BCDEFGHIJKLMNOPQRSTA	TABCDEFGHIJKLMNOPQRS
	3	STABCDEFGHIJKLMNOPQR	EFGHIJKLMNOPQRSTABCD
	4	GHIJKLMNOPQRSTABCDEF	HIJKLMNOPQRSTABCDEFG

Design DH31. $t = k = 6$, $p = 5$, $b = 1$, $n = 6$. $E_t = 96$, $E_d = 86$, $E_r = 66$.
An example is

		Subjects
Period	1	ABCDEF
	2	BCDEFA
	3	DEFABC
	4	CDEFAB
	5	FABCDE

Design DH32. $t = k = 7$, $p = 5$, $b = 1$, $n = 7$. $E_t = 93$, $E_d = 80$, $E_r = 61$.
An example is

		Subjects
Period	1	ABCDEFG
	2	CDEFGAB
	3	DEFGABC
	4	BCDEFGA
	5	FGABCDE

Design DH33. $t = k = 8$, $p = 5$, $b = 1$, $n = 8$. $E_t = 91$, $E_d = 79$, $E_r = 59$.

An example is

Subjects

	1	ABCDEFGH
	2	BCDEFGHA
Period	3	DEFGHABC
	4	CDEFGHAB
	5	FGHABCDE

Design DH34. $t = k = 9$, $p = 5$, $b = 1$, $n = 9$. $E_t = 90$, $E_d = 76$, $E_r = 56$.

An example is

Subjects

	1	ABCDEFGHI
	2	BCDEFGHIA
Period	3	DEFGHIABC
	4	CDEFGHIAB
	5	FGHIABCDE

Design DH35. $t = k = 10$, $p = 5$, $b = 1$, $n = 10$. $E_t = 90$, $E_d = 76$, $E_r = 56$.

An example is

Subjects

	1	ABCDEFGHIJ
	2	DEFGHIJABC
Period	3	BCDEFGHIJA
	4	IJABCDEFGH
	5	HIJABCDEFG

Design DH36. $t = k = 11$, $p = 5$, $b = 1$, $n = 11$. $E_t = 88$, $E_d = 67$, $E_r = 52$.

An example is

Subjects

	1	ABCDEFGHIJK
	2	EFGHIJKABCD
Period	3	HIJKABCDEFG
	4	BCDEFGHIJKA
	5	CDEFGHIJKAB

Design DH37. $t = k = 12$, $p = 5$, $b = 1$, $n = 12$. $E_t = 87$, $E_d = 71$, $E_r = 55$.
An example is

Subjects

	1	ABCDEFGHIJKL
	2	BCDEFGHIJKLA
Period	3	FGHIJKLABCDE
	4	EFGHIJKLABCD
	5	HIJKLABCDEFG

Design DH38. $t = k = 13$, $p = 5$, $b = 1$, $n = 13$. $E_t = 86$, $E_d = 71$, $E_r = 56$.
An example is

Subjects

	1	ABCDEFGHIJKLM
	2	CDEFGHIJKLMAB
Period	3	DEFGHIJKLMABC
	4	HIJKLMABCDEFG
	5	EFGHIJKLMABCD

Design DH39. $t = k = 14$, $p = 5$, $b = 1$, $n = 14$. $E_t = 86$, $E_d = 71$, $E_r = 55$.
An example is

Subjects

	1	ABCDEFGHIJKLMN
	2	DEFGHIJKLMNABC
Period	3	EFGHIJKLMNABCD
	4	LMNABCDEFGHIJK
	5	FGHIJKLMNABCDE

Design DH40. $t = k = 15$, $p = 5$, $b = 1$, $n = 15$. $E_t = 85$, $E_d = 68$, $E_r = 51$.
An example is

Subjects

	1	ABCDEFGHIJKLMNO
	2	BCDEFGHIJKLMNOA
Period	3	FGHIJKLMNOABCDE
	4	EFGHIJKLMNOABCD
	5	HIJKLMNOABCDEFG

Design DH41. $t = k = 16$, $p = 5$, $b = 1$, $n = 16$. $E_t = 85$, $E_d = 68$, $E_r = 53$.

   An example is

Subjects

	1	ABCDEFGHIJKLMNOP
	2	IJKLMNOPABCDEFGH
Period	3	FGHIJKLMNOPABCDE
	4	GHIJKLMNOPABCDEF
	5	BCDEFGHIJKLMNOPA

Design DH42. $t = k = 17$, $p = 5$, $b = 1$, $n = 17$. $E_t = 84$, $E_d = 66$, $E_r = 52$.

   An example is

Subjects

	1	ABCDEFGHIJKLMNOPQ
	2	CDEFGHIJKLMNOPQAB
Period	3	JKLMNOPQABCDEFGHI
	4	DEFGHIJKLMNOPQABC
	5	EFGHIJKLMNOPQABCD

Design DH43. $t = k = 18$, $p = 5$, $b = 1$, $n = 18$. $E_t = 84$, $E_d = 64$, $E_r = 49$.

   An example is

Subjects

	1	ABCDEFGHIJKLMNOPQR
	2	JKLMNOPQRABCDEFGHI
Period	3	MNOPQRABCDEFGHIJKL
	4	IJKLMNOPQRABCDEFGH
	5	HIJKLMNOPQRABCDEFG

Design DH44. $t = k = 19$, $p = 5$, $b = 1$, $n = 19$. $E_t = 84$, $E_d = 62$, $E_r = 49$.

   An example is

Subjects

	1	ABCDEFGHIJKLMNOPQRS
	2	HIJKLMNOPQRSABCDEFG
Period	3	DEFGHIJKLMNOPQRSABC
	4	IJKLMNOPQRSABCDEFGH
	5	JKLMNOPQRSABCDEFGHI

Design DH45. $t = k = 20$, $p = 5$, $b = 1$, $n = 20$. $E_t = 83$, $E_d = 63$, $E_r = 47$.
An example is

Subjects

	1	ABCDEFGHIJKLMNOPQRST
	2	BCDEFGHIJKLMNOPQRSTA
Period	3	EFGHIJKLMNOPQRSTABCD
	4	CDEFGHIJKLMNOPQRSTAB
	5	JKLMNOPQRSTABCDEFGHI

Appendix 6.G. Designs FA 1-2. Tied-Double-Change-Over Designs of Federer and Atkinson

Design FA1. $t = 3$. The design can be written as follows:

Period or row number		$s = 1$ 1 2 3	$s = 2$ 4 5 6	$s = 3$ 7 8 9	$s = 4$ 10 12	.
$q = 1$	1	A B C	A B C	A B C	A B C	A
	2	B C A	C A B	B C A	C A B	B
	3	C A B	B C A	C A B	B C A	C
	4	A B C	A B C	A B C	A B C	A
$q = 2$	5	C A B	B C A	C A B	B C A	C
	6	B C A	C A B	B C A	C A B	B
	7	A B C	A B C	A B C	A B C	A
$q = 3$	8	B C A	C A B	B C A	C A B	B
	9	C A B	B C A	C A B	B C A	C
	10	A B C	A B C	A B C	A B C	A
$q = 4$	11	C A B	B C A	C A B	B C A	C
	12	B C A	C A B	B C A	C A B	B
	13	A B C	A B C	A B C	A B C	A
	14	B C A	C A B	B C A	C A B	B

Design FA2. $t = 4$. The design can be written as follows:

Period or row number		$s = 1$ 1 2 3 4	$s = 2$ 5 6 7 8	$s = 3$ 9 11	$s = 4$ 13 15	.
	1	A B C D	A B C D	A B C D	A B C D	A
	2	B A D C	D C B A	C D A B	B A D C	D
$q = 1$	3	C D A B	B A D C	D C B A	C D A B	B
	4	D C B A	C D A B	B A D C	D C B A	C
	5	C D A B	B A D C	D C B A	C D A B	B
	6	C D A B	B A D C	D C B A	C D A B	B
$q = 2$	7	D C B A	C D A B	B A D C	D C B A	C
	8	B A D C	D C B A	C D A B	B A D C	D
	9	A B C D	A B C D	A B C D	A B C D	A
	10	D C B A	C D A B	B A D C	D C B A	C
$q = 3$	11	B A D C	D C B A	C D A B	B A D C	D
	12	C D A B	B A D C	D C B A	C D A B	B
	13	A B C D	A B C D	A B C D	A B C D	A
	14	B A D C	D C B A	C D A B	B A D C	D
$q = 4$	15	C D A B	B A D C	D C B A	C D A B	B
	16	D C B A	C D A B	B A D C	D C B A	C
	17	A B C D	A B C D	A B C D	A B C D	A

7
The Analysis of Categorical
Data from Cross-over Designs

In Chapter 3 we discussed the treatment of continuous data collected from the 2-treatment, 2-period, 2-sequence cross-over design. As previously indicated, this design is possibly both the most widely used and abused cross-over design in the current literature. Not surprisingly, the problems associated with this design carry over to the analysis for categorical data. It is unfortunate that past and present literature describing the analysis of categorical data from cross-over designs has not clarified the situation. In Chapter 3 these problems were, to some degree, resolved for continuous data. We hope to show how these results carry over to the analysis of categorical data from cross-over designs.

Some authors have developed procedures for testing treatment differences by conditional approaches (Mainland, 1963, Gart, 1969, and Le and Gomez–Marin 1984). These methods condition on the number of unlike responses for the two treatments. A subject responding favorably to one treatment, but not the other, indicates this preference by unlike response. The information contained in the like or no preference responses is believed to be limited in value and is discarded. That is, a subject that responds favorably to Treatment A and favorably to Treatment B is thought to contribute little to the comparison of treatments. The same is said to be true of a subject who responds unfavorably to both treatments.

Other authors have constructed models which incorporate the information contained in the non-preference responses with subjects showing a preference (Dunsmore, 1981, Prescott, 1981, and Le and Cary, 1984). Although incorporating the non-preference information, these models do not distinguish between favorable and nonfavorable responses. These methods do not account for the correlation between responses because no distinction is made between the favorable and nonfavorable non-preference responses.

These conditional testing procedures are not without limitations. First, they have only been developed for the 2-treatment, 2-period, 2-sequence cross-over design. No conditional testing procedures are available for discrete data from more general cross-over designs. Secondly, there is no need for conditional testing methods when unconditional testing procedures are available.

Recently, Kenward and Jones (1987), and Jones and Kenward (1989) proposed unconditional testing procedures based on log-linear models. These are the first models to incorporate effects for treatment, period, sequence and carryover, and to include parameters for the average correlation between responses made on the same subject. Although this is a big step forward in the analysis of categorical data from cross-over experiments, the parameters of these models do not have the clear meaning that the authors suggest.

It was clearly stated in Chapter 3 that parameter estimation for the 2-treatment, 2-period, 2-sequence cross-over design cannot proceed without making strong assumptions about the model structure. It is intuitive that a qualitative response cannot contain more information than a quantitative response. Thus, for this design, one cannot expect the analysis of categorical data to provide more information than the continuous case. Surprisingly, the models proposed by Kenward and Jones (1987), and Jones and Kenward (1989) appear to do exactly this. Simultaneous estimation of sequence, period, treatment and carryover parameters is possible under their model representation. This is clearly not a possibility. The reasons for this were presented in Chapter 3 for the case of a continuous response.

Jones and Kenward (1987) extended their 2-treatment, 2-period, 2-sequence cross-over design with binary response to a more general 3-treatment, 3-period, 6-sequence design. These authors also suggest how one could extend this work to even more complicated designs for binary data. This work is a natural extension of their earlier work, and is likely to suffer some of the same problems as the 2–sequence design.

Up to this point, the discussion has centered on a binary response. Generalizations to categorical data with multiple response levels have also been proposed (Le, 1984, and Kenward and Jones, 1991). The work of Le (1984) is an extension of Gart's (1969) model for the 2–treatment, 2–period, 2–sequence design, and as such, carries with it the same problems. Kenward and Jones (1991) extend their models for a binary response to the more general case of a multilevel response. As with the binary case, their multilevel response model includes parameters for sequence, period, treatment and carryover, along with the average correlation between successive observations on the same subject. Again, the parameters for these models do not have the simple interpretation indicated by the authors. As such, the models for continuous data and the models proposed by Kenward and Jones (1991) for categorical data are similar, but not of an analogous nature.

In this chapter, we will develop a logit regression model for the analysis of categorical data from cross–over designs. First we will consider a logit model for the 2–treatment, 2–period, 2–sequence cross–over design with binary response. It will become quite clear from this development that a one–to–one correspondence exists between the continuous and categorical cases, the only difference being that the parameters for correlation will be included in the models having a categorical response. Other than this, the logical development for the analysis of continuous data will be carried over to the analysis of categorical data.

Later, we will extend the methods for binary data to designs having more than two treatments, periods or sequences. Finally, with the binary case explored, extensions to a response with 2 or more categories are developed.

7.1. The 2–Treatment, 2–Period, 2–Sequence Binary Cross–over Design

Consider the 2–treatment, 2–period, 2–sequence design of Chapter 3. When the response is binary, say success or failure, the method of analysis outlined in Chapter 3 does not apply. One very important reason for this stems from the nature of the data. Categorical random variables have variances that are typically functions of the means. Therefore, the usual model assumptions of normality and constant variance are violated.

Let N_1 and N_2 subjects be selected for the AB and BA sequences, respectively. The lth subject from the ith sequence can be cross–classified according to the joint response in Periods 1 and 2. Let Y_{ijkl} denote this joint response, with j indexing the first period and k indexing the second period. Both j and k can take on the values 1 or 2, with 1 indicating a success and 2 a failure. Y_{ijkl} will take on the value 1 or 0 such that

$$\sum_{j=1}^{2} \sum_{k=1}^{2} Y_{ijkl} = 1.$$

We will make the usual assumption that $(Y_{i11l}\ Y_{i12l}\ Y_{i21l}\ Y_{i22l})$ follow a multinomial distribution and that subjects respond independently. The observable random frequency, F_{ijk}, is obtained by summing the Y_{ijkl} over all subjects in the ith sequence. Letting F_i denote the vector obtained by stacking F_{i11}, F_{i12}, F_{i21} and F_{i22}, the joint distribution of each frequency vector is multinomial. These frequencies are neatly presented in the following 2^2 contingency tables.

AB Sequence

Period 2

		1	2
Period 1	1	F_{111}	F_{112}
	2	F_{121}	F_{122}

BA Sequence

Period 2

		1	2
Period 1	1	F_{211}	F_{212}
	2	F_{221}	F_{222}

Mainland (1963) presented a conditional test for treatment effect, and Gart (1969) extended this work. The test is conditionally based on the frequencies having dissimilar joint response (F_{ijk}; $j \neq k$). These are typically represented in the form of the following 2^2 contingency table:

	Joint Response		
	12	21	Row Totals
AB Sequence	F_{112}	F_{121}	$n_{1.}$
BA Sequence	F_{212}	F_{221}	$n_{2.}$
Column Totals	$n_{.12}$	$n_{.21}$	$n_{..}$

Based on Gart's model, a Pearson–type χ^2 statistic can be computed for testing the null hypothesis of no treatment effect against the alternative hypothesis that a treatment effect is present. Under the null hypothesis, this test statistic is asymptotically distributed as chi–square with 1 degree of freedom. The test statistic is given by:

$$\chi_G^2 = \frac{(F_{112}F_{221} - F_{121}F_{212})^2 \, n_{..}}{n_{1.} \, n_{2.} \, n_{.12} \, n_{.21}} .$$

Prescott (1981) developed a test which also makes use of the no preference frequencies. A useful method of presenting the data for this test is made in the following table:

	Joint Response			
	12	11 and 22	21	Row Totals
AB Sequence	F_{112}	$F_{111} + F_{122}$	F_{121}	N_1
BA Sequence	F_{212}	$F_{211} + F_{222}$	F_{221}	N_2
Column Totals	$n_{.12}$	$n_{.11} + n_{.22}$	$n_{.21}$	$N_{.}$

If a trend exists in the above table, then a treatment effect is assumed present. Prescott (1981) developed a test for a linear trend in the above table. The test is based on a Pearson–type chi–square statistic. Under the null hypothesis of no linear trend, Prescott's chi–square statistic has an asymptotic chi–square distribution with 1 degree of freedom. Prescott's chi–square statistic is given by

$$\chi_P^2 = \frac{[(F_{112} - F_{121})N_. - (n_{.12} - n_{.21})N_1]^2 N_.}{N_1 N_2 [(n_{.12} + n_{.21})N_. - (n_{.12} - n_{.21})^2]} \quad .$$

Both Gart's test and Prescott's test are limited to the 2–treatment, 2–period, 2–sequence cross–over design. Gart's test has been extended from the binary response to the multi–level categorical response (Le, 1984), but no simple test statistic exists for this more general cross–over design. Typically, a relationship exists between temporal observations for a single subject, and so, more complex designs will require more complex models.

Observations from cross–over experiments are assumed to be correlated or associated. For the binary case, this correlation can be seen at the subject level as a continuing tendency towards success or failure in response irrespective of treatment. This is analogous to the continuous case where a subject would have a natural level of response prior to treatment which is carried through the experiment.

For multiway contingency tables, a common analysis is the assessment of association between cross–classified factors. Log–linear models are commonly used for assessing association in multiway contingency tables (Bishop *et al.* 1975). Given that data from binary cross–over experiments can be cross–classified, researchers in this area have tended to rely on log–linear models as a pathway of analysis for this type of data. Logit models, which are closely related to log–linear models, garner the majority of the remaining methods.

The parameters of interest for these log–linear and logit models are the usual parameters for sequence, period, treatment and carryover. These are location parameters, which we will distinguish from the association parameters. Although somewhat analogous to the scale parameters of the continuous case, association parameters are necessarily included in the model definition. Leaving the association parameters out of the model specification without good reason is likely to produce erroneous results. This differs from the continuous case where only location parameters are modeled and the covariance matrix is estimated secondarily.

Although it is widely accepted that dependencies exist for binary cross–over data, the methods first developed (Gart, 1969) for the analysis of this type of data did not model the association component. Kenward and Jones (1987) developed models by first defining logits of the probabilities in terms of the location parameters under the assumption of independence. Assuming independence, the joint probabilities for the cross–classified frequencies are easily constructed from the logits, in terms of the location parameters. Subsequently, Kenward and Jones add the components for the association among observations. This type of two stage process would be acceptable for a linear model, but these models are intrinsically nonlinear and are not strictly additive.

The reason for this approach appears to have been one of theoretical and computational ease. Such models can utilize existing software packages. The focus has been more on adapting existing theory, as opposed to the development of new ideas for analysis. A second reason, and intimately connected with the first, stems from the usual convention for linear models: that of using the marginal results and avoiding the probability structure of the cells from the cross–classified factors.

7.2. The Logit Model for Binary Data

Because cross–over designs involve observations taken over time, a discrete Markov process is a convenient way of expressing the expected cell frequencies. Let P_A and P_B define the unconditional probability of a success for Treatments A and B, respectively. Further, let $P_{A|B}$ denote the probability of a success for Treatment A given that prior Treatment B was a success, and let $P_{A|\overline{B}}$ denote the probability of a success for Treatment A given that prior Treatment B was a failure. Similar definitions apply to $P_{B|A}$ and $P_{B|\overline{A}}$. Given these definitions, the contingency table of observed frequencies has expectation of the following form:

AB Sequence

Period 2

		1	2
Period 1	1	$N_1 P_A P_{B\mid A}$	$N_1 P_A (1 - P_{B\mid A})$
	2	$N_1 (1 - P_A) P_{B\mid \overline{A}}$	$N_1 (1 - P_A)(1 - P_{B\mid \overline{A}})$

BA Sequence

Period 2

		1	2
Period 1	1	$N_2 P_B P_{A\mid B}$	$N_2 P_B (1 - P_{A\mid B})$
	2	$N_2 (1 - P_B) P_{A\mid \overline{B}}$	$N_2 (1 - P_B)(1 - P_{A\mid \overline{B}})$

Clearly, one would like to test hypotheses concerning the probability of success for Treatment A as opposed to Treatment B, while controlling the effects of any association. Extending the lead of Gart (1969), we consider the following definitions:

$$\text{logit}(P_A) = \ln\left(\frac{P_A}{1 - P_A}\right) = \mu + \gamma_1 + \pi_1 + \tau_1$$

$$\text{logit}(P_{B\mid A}) = \mu + \gamma_1 + \pi_2 + \tau_2 + \lambda_1 + \epsilon_{112}$$

$$\text{logit}(P_{B\mid \overline{A}}) = \mu + \gamma_1 + \pi_2 + \tau_2 + \lambda_1 + \epsilon_{1\overline{1}2}$$

$$\text{logit}(P_B) = \mu + \gamma_2 + \pi_1 + \tau_2$$

$$\text{logit}(P_{A\mid B}) = \mu + \gamma_2 + \pi_2 + \tau_1 + \lambda_2 + \epsilon_{212}$$

$$\text{logit}(P_{A\mid \overline{B}}) = \mu + \gamma_2 + \pi_2 + \tau_1 + \lambda_2 + \epsilon_{2\overline{1}2}$$

where μ =   effect of an overall logit mean;

γ_i =   effect of ith treatment sequence ($i = 1, 2$);

π_j =   effect of jth period ($j = 1, 2$);

τ_k =   direct effect of kth treatment ($k = $ A, B);

λ_r =   carryover effect from treatment r ($r = $ A, B);

ϵ_{i12} = association effect averaged over subjects of sequence i when the period 1 treatment is a success.

$\epsilon_{i\bar{1}2}$ = association effect averaged over subjects of sequence i when the period 1 treatment is a failure.

As stated in Chapter 3, these effects cannot all be separately estimated. Both the sum-to-zero and set-to-zero conventions apply in the categorical case just as they did for the continuous case. Using the sum-to-zero convention, define:

$$\gamma = \gamma_1 = -\gamma_2 \qquad \text{sequence effect}$$
$$\pi = \pi_1 = -\pi_2 \qquad \text{period effect}$$
$$\tau = \tau_1 = -\tau_2 \qquad \text{treatment effect}$$
$$\lambda = \lambda_1 = -\lambda_2 \qquad \text{carryover effect}$$
$$\epsilon_{i12} = -\epsilon_{i\bar{1}2} \qquad \text{association effect for sequence } i$$

With these definitions, the logits can be redefined as:

$$\text{logit}(P_A) = \mu + \gamma + \pi + \tau$$
$$\text{logit}(P_{B|A}) = \mu + \gamma - \pi - \tau + \lambda + \epsilon_{112}$$
$$\text{logit}(P_{B|\overline{A}}) = \mu + \gamma - \pi - \tau + \lambda - \epsilon_{112}$$
$$\text{logit}(P_B) = \mu - \gamma + \pi - \tau$$
$$\text{logit}(P_{A|B}) = \mu - \gamma - \pi + \tau - \lambda + \epsilon_{212}$$
$$\text{logit}(P_{A|\overline{B}}) = \mu - \gamma - \pi + \tau - \lambda - \epsilon_{212}$$

With the exception of the ϵ_{i12}'s, these logits have the same linear interpretation as the cell means for the continuous case of Chapter 3. Although there are six separate logits, compared to only four cell means in the continuous case, the inclusion of ϵ_{112} and ϵ_{212} effectively reduces the system down to four pieces of information. These two association parameters must be in the model; they account for the average correlation between observations on the same subject.

Kenward and Jones (1987) and Jones and Kenward (1989) proposed log-linear models for binary cross-over data that include parameters for all effects stated above, including a form of carryover. Their representation of these models imply that all parameters are estimable. If this were so, then one of two conclusions are possible: either there is more information in the binary case than the continuous case, or their model specification along with the interpretation of model parameters is not correct. Clearly, the first is not true and so, the second conclusion must therefore be.

The logit model proposed in this chapter can be represented as:

$$\text{logit}(P_{ij}) = \ln\left(\frac{P_{ij}}{1 - P_{ij}}\right) = V_{ij}\,\phi$$

where P_{ij} is the probability of success in the ith sequence having defined outcome j ($j = A$, $B|A$ and $B|\overline{A}$ for $i=1$; B, $A|B$ and $A|\overline{B}$ for $i=2$) and V_{ij} is the row of the design matrix V_i ($i = 1, 2$) corresponding to P_{ij} and $\phi = [\mu, \gamma, \pi, \tau, \lambda, \epsilon_{112}, \epsilon_{212}]'$. The logit models for sequences 1 and 2 are, respectively:

$$
\begin{pmatrix}
\text{logit}(P_A) \\
\text{logit}(P_{B|A}) \\
\text{logit}(P_{B|\overline{A}})
\end{pmatrix}
=
\begin{pmatrix}
1 & 1 & 1 & 1 & 0 & 0 & 0 \\
1 & 1 & -1 & -1 & 1 & 1 & 0 \\
1 & 1 & -1 & -1 & 1 & -1 & 0
\end{pmatrix}
\begin{pmatrix}
\mu \\
\gamma \\
\pi \\
\tau \\
\lambda \\
\epsilon_{112} \\
\epsilon_{212}
\end{pmatrix}
$$

and

$$
\begin{pmatrix}
\text{logit}(P_B) \\
\text{logit}(P_{A|B}) \\
\text{logit}(P_{A|\overline{B}})
\end{pmatrix}
=
\begin{pmatrix}
1 & -1 & 1 & -1 & 0 & 0 & 0 \\
1 & -1 & -1 & 1 & -1 & 0 & 1 \\
1 & -1 & -1 & 1 & -1 & 0 & -1
\end{pmatrix}
\begin{pmatrix}
\mu \\
\gamma \\
\pi \\
\tau \\
\lambda \\
\epsilon_{112} \\
\epsilon_{212}
\end{pmatrix}.
$$

As in the continuous case, the model is over-specified, having 7 parameters to estimate, based on 6 logits. Dropping a location parameter from the seven model parameters produces a saturated model and allows for estimation of the remaining six. Dropping either association parameter does no good, because the columns in V_i coding for the association parameters are orthogonal to the columns that code for the location parameters.

The possible model structures that arise by dropping one location parameter were thoroughly discussed in Chapter 3. Because the parameterization discussed here is essentially the same as for a continuous response, we will not repeat the discussion. Instead, we will take the conservative approach and drop the carryover effect parameter λ.

For the most part, the logits defined above are of the same structure as given by Gart (1969), Kenward and Jones (1987), and Jones and Kenward (1989). They differ by the inclusion of a sequence effect and in the case of Gart (1969) by the addition of the carryover effect. Most importantly they differ by the inclusion of the association parameters within the logit. Up to the present, all models for binary cross-over designs have avoided inclusion of these parameters within the logits [Gart (1969), Kenward and Jones (1987), Jones and Kenward (1989), and Le (1984)]. Instead, those authors have either used conditional approaches or have used *ad hoc* procedures to model the association parameters. The reason for this is partly due to the difficulty in modeling the cell probabilities of the contingency tables. The marginal probabilities are mathematically more tractable, but do not allow for the same model specification as the continuous case, because of the association parameters. By basing their models on the marginal distributions, these authors have limited the possible approaches to model development.

The 2-treatment, 2-period, 2-sequence cross-over model with cell probabilities based on the Markov process does present an avenue for using the logits defined above. The cell probabilities are the product of two treatment probabilities as modified by sequence, period, treatment, carryover and association. It is these treatment probabilities with which one should work. They are only equal to the marginal probabilities if the repeated observations made on a subject are independent, that is, the case where no association exists.

Traditional log–linear models were developed for specifying contrasts of cell probabilities. Because the structure of the cell probabilities for categorical cross–over designs can be defined by simple probability arguments as products of conditional probabilities, it makes sense to work directly with this structure and not specifically with the cell probabilities. A log–linear model can be defined to linearize these cell products and the logit definitions can then be applied to the parameters of the log–linear model. A log–linear model for the ith sequence is defined as:

$$\mathbf{F}_i = \exp[\mathbf{X}\beta(\phi)_i] + \delta_i \, ,$$

where $\mathbf{F}_i$ is the 4 x 1 vector of observed frequencies for the ith sequence as cross–classified in the contingency table; $\mathbf{X}$ is a 4 x 7 design matrix of known constants; $\beta(\phi)_i$ is a 7 x 1 parameter vector for the ith sequence that is a function of the elements of the vector ϕ, and δ_i is a 4 x 1 vector of errors associated with the ith sequence such that Plim $\delta_i = 0$. The elements of ϕ are μ, γ, π, τ, λ and ϵ_{i12} $(i = 1, 2)$. For the purposes of discussion, the vector of expected cell means will be defined as $\mu_i = \exp[\mathbf{X}\beta(\phi)_i]$.

Referring back to the table of expected cell frequencies, the parameters of each cell can be separated by taking the natural logarithm. The linearized system is represented in matrix form by pulling off the log transformed parameters and placing them in the vector $\beta(\phi)_i$. The matrix of zeros and ones that remains defines $\mathbf{X}$. Each parameter, along with the complement probabilities, has an associated column in $\mathbf{X}$. A "1" is placed in the column associated with a parameter if the expected frequency of a row contains the parameter, otherwise the element is "0". To better understand this representation, the log–linear models for the AB and BA sequences are, respectively,

$$
\begin{pmatrix} F_{111} \\ F_{112} \\ F_{121} \\ F_{122} \end{pmatrix}
= \exp
\begin{bmatrix}
\begin{pmatrix}
1 & 1 & 0 & 1 & 0 & 0 & 0 \\
1 & 1 & 0 & 0 & 1 & 0 & 0 \\
1 & 0 & 1 & 0 & 0 & 1 & 0 \\
1 & 0 & 1 & 0 & 0 & 0 & 1
\end{pmatrix}
\begin{pmatrix}
\ln(N_1) \\
\ln(P_A) \\
\ln(1-P_A) \\
\ln(P_{B|A}) \\
\ln(1-P_{B|A}) \\
\ln(P_{B|\overline{A}}) \\
\ln(1-P_{B|\overline{A}})
\end{pmatrix}
\end{bmatrix}
+ \delta_1
$$

and

$$
\begin{pmatrix} F_{211} \\ F_{212} \\ F_{221} \\ F_{222} \end{pmatrix} = \exp\left[\begin{pmatrix} 1 & 1 & 0 & 1 & 0 & 0 & 0 \\ 1 & 1 & 0 & 0 & 1 & 0 & 0 \\ 1 & 0 & 1 & 0 & 0 & 1 & 0 \\ 1 & 0 & 1 & 0 & 0 & 0 & 1 \end{pmatrix} \begin{pmatrix} \ln(N_2) \\ \ln(P_B) \\ \ln(1-P_B) \\ \ln(P_{A|B}) \\ \ln(1-P_{A|B}) \\ \ln(P_{A|\overline{B}}) \\ \ln(1-P_{A|\overline{B}}) \end{pmatrix} \right] + \delta_2
$$

Assuming that F_i has a multinomial distribution, the likelihood function can be defined based on the parameterization described above. From this, the first and second derivatives of the likelihood function can be obtained and a Newton–Raphson procedure utilized to compute the maximum likelihood (ML) estimates for ϕ along with the estimated standard errors. This iterative procedure is presented in Appendix 7.A. For the discussion to follow, let $\hat{\phi}$ denote the ML estimator of ϕ. The ML estimates for $\beta(\phi)_i$ and μ_i are denoted $\hat{\beta}(\phi)_i$ and $\hat{\mu}_i$, respectively, and can be computed by evaluating the respective model functions using $\hat{\phi}$.

The algorithm outlined in Appendix 7.A is difficult to implement. An alternative is available to anyone having access to a statistical software package that provides for maximum likelihood logistic regression. Some manipulation of the data will be required so that the procedure will produce the appropriate estimates. The discussion of how to implement such a procedure will be developed in Section 7.3, by example.

If we consider a saturated model where the carryover effect is assumed to be zero, then the remaining parameters of ϕ can be represented as linear functions of the logit probabilities:

$$
\mu = \frac{2\,\text{logit}(P_A) + \text{logit}(P_{A|B}) + \text{logit}(P_{A|\overline{B}}) + 2\,\text{logit}(P_B) + \text{logit}(P_{B|A}) + \text{logit}(P_{B|\overline{A}})}{8}
$$

$$
\gamma = \frac{2\,\text{logit}(P_A) - \text{logit}(P_{A|B}) - \text{logit}(P_{A|\overline{B}}) + 2\,\text{logit}(P_B) + \text{logit}(P_{B|A}) - \text{logit}(P_{B|\overline{A}})}{8}
$$

$$\pi = \frac{2 \ \text{logit}(P_A) - \text{logit}(P_{A|B}) - \text{logit}(P_{A|\overline{B}}) + 2 \ \text{logit}(P_B) - \text{logit}(P_{B|A}) - \text{logit}(P_{B|\overline{A}})}{8}$$

$$\tau = \frac{2 \ \text{logit}(P_A) + \text{logit}(P_{A|B}) + \text{logit}(P_{A|\overline{B}}) - 2 \ \text{logit}(P_B) - \text{logit}(P_{B|A}) - \text{logit}(P_{B|\overline{A}})}{8}$$

$$\epsilon_{112} = \frac{4 \ \text{logit}(P_{B|A}) - 4 \ \text{logit}(P_{B|\overline{A}})}{8}$$

$$\epsilon_{212} = \frac{4 \ \text{logit}(P_{A|B}) - 4 \ \text{logit}(P_{A|\overline{B}})}{8}$$

Focusing attention on τ, one should note that the logits for P_A and P_B are doubly weighted when compared to the logits for $P_{A|B}$, $P_{A|\overline{B}}$, $P_{B|A}$ and $P_{B|\overline{A}}$. This was indicated earlier. For the AB sequence, P_A is estimated from the first period where treatment A was given. The second period information is shared among $P_{B|A}$ and $P_{B|\overline{A}}$. Thus, the logit for P_A is estimated with the same amount of information as the combined logits for $P_{B|A}$ and $P_{B|\overline{A}}$. A similar conclusion can be made concerning the information in the BA sequence.

Cox and Plackett (1980) indicated that one should test for equality of ϵ_{112} and ϵ_{212} before tests of hypothesis for τ should be performed. They reasoned that this would provide some assurance that experimental conditions remained the same for the two sequences. The representation of τ presented above is not affected by differences among ϵ_{112} and ϵ_{212}, so long as the association parameters are represented in the logit model. Only the logits for $P_{A|B}$, $P_{A|\overline{B}}$, $P_{B|A}$ and $P_{B|\overline{A}}$ contain the parameters ϵ_{112} and ϵ_{212}. The logits for $P_{A|B}$ and $P_{A|\overline{B}}$ are identical, except that the logit for $P_{A|B}$ contains ϵ_{212} and the logit for $P_{A|\overline{B}}$ contains $-\epsilon_{212}$. The representation of τ presented above adds the logit of $P_{A|B}$ and $P_{A|\overline{B}}$, and therefore eliminates ϵ_{212} from the representation of τ. By the same logic, ϵ_{112} is also eliminated from the representation of τ because the logits for $P_{B|A}$ and $P_{B|\overline{A}}$ are also added. The statement of Cox and Plackett (1980) that sequences must have equal associations is not true for the proposed model. Unequal association parameters may be an indication of problems in the prosecution of the experiment, but do not invalidate the results. A potentially important indicator of problems is the parameter γ, which determines whether the experimental conditions have remained the same for the two sequences. Unfortunately, a proper test for γ is not possible when subjects, as we have assumed, are random effects. The variance for sequence effects has two components: the first, which is estimable, is the variation due to multinomial sampling; the second, which is not estimable, is the variation between subjects. If subjects were fixed effects, a subject would respond in a constant manner under the same set of conditions, and therefore only the variation due to multinomial sampling would be present.

For the saturated model of Kenward and Jones (1987) the estimator for τ, denoted by $\hat{\tau}_{KJ}$, has the following form: $\hat{\tau}_{KJ} =$

$$\frac{2 \log it(P_A) - 2 \log it(P_B) + \ln \frac{P_{A|\overline{B}}}{1-P_{A|B}} - \ln \frac{P_{B|\overline{A}}}{1-P_{B|A}} + \ln \frac{P_{B|A}}{1-P_{B|\overline{A}}} - \ln \frac{P_{A|B}}{1-P_{A|\overline{B}}}}{4}.$$

When the carryover effect of the Kenward and Jones model (λ^*) is zero, the estimator for τ, denoted by $\hat{\tau}_{KJ}^*$, can be represented in the following form:

$$\hat{\tau}_{KJ}^* = \frac{\log it(P_A) - \log it(P_B) + \ln \frac{P_{A|\overline{B}}}{1-P_{A|B}} - \ln \frac{P_{B|\overline{A}}}{1-P_{B|A}}}{4}.$$

These two representations coincide with the treatment effect presented in this manuscript when the associations for both sequences are zero, that is, under independence. Furthermore, when λ^* is assumed zero the Pearson chi–square statistic for testing τ in the Kenward and Jones log–linear model is identical to the chi–square statistic for Gart's test.

7.3. Hypothesis Testing

In this section, we will assume that the frequencies within each sequence follow the multinomial distribution. If this assumption is correct and we assume that observations between sequences are independent, the joint distribution of all frequencies for a cross–over design will be product multinomial. It is unfortunate that this does not lead to a known small sample distribution for the estimator vector $\hat{\phi}$. However, for reasonably large samples, the asymptotic distribution of $\hat{\phi}$ can be obtained. Applying the multivariate delta method (Agresti, 1990, pp. 420–421), the asymptotic distribution of $\hat{\phi}$ is normally distributed with mean vector ϕ and estimator covariance matrix denoted by $\mathbf{Cov}(\hat{\phi})$. The variance of an estimator, denoted Var($\cdot$), is obtained from the appropriate diagonal element of the estimator covariance matrix. The standard error of an estimator, denoted S.E.($\cdot$), is obtained by taking the square root of the corresponding variance. The properties of the large sample distributions of $\hat{\phi}$ and $\hat{\mu}$ allows for several approaches to hypothesis testing. These approaches include Wald, likelihood ratio and Pearson chi–square tests for simultaneous testing of one or more parameters, and z–tests for a single parameter.

In general, if ϕ is a q x 1 vector of parameters, $\mathbf{H}$ is a r x q matrix of known constants and $\mathbf{h}$ is a r x 1 vector of known constants, then a Wald statistic (Agresti, 1990) can be constructed for testing the hypothesis $\mathbf{H}\phi = \mathbf{h}$ against $\mathbf{H}\phi \neq \mathbf{h}$. The Wald statistic for this test is given by

$$W = (\mathbf{H}\hat{\phi} - \mathbf{h})^{'}(\mathbf{H}\ \mathbf{Cov}(\hat{\phi})\ \mathbf{H}^{'})^{-1}(\mathbf{H}\hat{\phi} - \mathbf{h}),$$

and has asymptotic chi–square distribution with Rank($\mathbf{H}$) degrees of freedom. If the rows of $\mathbf{H}$ are linearly independent then Rank($\mathbf{H}$) = r.

If only a single parameter is to be tested, then a z–test based on the asymptotic distribution of $\hat{\phi}$ for the saturated model can be computed. For example, H_0: $\tau = 0$ vs. H_a: $\tau \neq 0$ can be assessed by

$$z = \frac{\hat{\tau} - 0}{\text{S.E.}\ (\hat{\tau})}.$$

This z statistic has an asymptotic standard normal distribution under the null hypothesis.

Some statistical software packages, such as SAS PROC CATMOD© (SAS Institute Inc., 1990), produce a Wald chi–square statistic for these single parameter tests. The Wald chi–square for this type of test is computed by squaring the above z statistic, and is asymptotically distributed as chi–square with 1 degree of freedom. This same test statistic can be produced by computing W for H having only one row of zero elements, except for a one in the position corresponding to the parameter being tested, and $h = 0$. The rank of H is one and therefore W for this test has an asymptotic chi–square distribution with 1 degree of freedom.

Likelihood ratio and Pearson chi–square statistics (Agresti, 1990) can be computed for testing one or more parameters simultaneously. Let M_1 and M_2 denote two models, where M_1 contains all the parameters of interest and M_2 is a special case of M_1 where the parameters to be tested have been removed. Thus, M_2 is nested within M_1. Let q_1 and q_2 be the number of parameters in models M_1 and M_2, respectively. The ML estimates for the contingency table cell means under models M_1 and M_2 are denoted by $\hat{\mu}(1)$ and $\hat{\mu}(2)$, respectively. The likelihood ratio and Pearson chi–square statistics under model I (I = 1 or 2) are respectively given by:

$$G^2(I) = 2 \sum_i F_i \ln \left(\frac{F_i}{\hat{\mu}(I)_i} \right)$$

$$\chi^2(I) = \sum_i \left(\frac{[F_i - \hat{\mu}(I)_i]^2}{\hat{\mu}(I)_i} \right).$$

The subscript i runs across all possible cells for the cross–over design. The conditional likelihood ratio statistic for testing the effect of the excluded parameters is computed as

$$G^2(M_2 | M_1) = G^2(M_2) - G^2(M_1).$$

When model M_2 is correct, $G^2(M_2 | M_1)$ has asymptotic chi–square distribution with $q_1 - q_2$ degrees of freedom. Similarly, a conditional Pearson type chi–square statistic can be computed in an analogous manner, as

$$\chi^2(M_2 | M_1) = \chi^2(M_2) - \chi^2(M_1).$$

This latter statistic is only an approximation to the equivalent likelihood ratio test statistic.

Agresti (1990) points out that both of the likelihood ratio and Pearson chi–square statistics perform well, even for sparse tables, so long as the number of parameters tested is small. The Wald statistic and z–test should only be used as a guide to further tests. As Hosmer and Lemeshow (1989) point out, these tests tend to be less powerful and behave poorly in some circumstances. The Wald statistic produced from a saturated model may be unrealistically large or small. The likelihood ratio and Pearson statistics tend to behave much better.

7.4. Parameter Estimation using SAS PROC CATMOD©

As previously stated, the algorithms developed in Appendices 7.A and 7.B are difficult to initiate. There is an alternative approach. Most software packages that contain a logistic regression procedure can be used for this analysis, although some rearrangement of the observed data is necessary. To keep in the spirit of this text, SAS PROC CATMOD will be used to analyze categorical data from cross–over designs.

Before PROC CATMOD can be used, the vector of observed frequencies must be redefined. Before discussing the arrangement of the data, consider the table of expected cell probabilities for the 2–treatment, 2–period, 2–sequence cross–over design:

AB Sequence

		Period 2	
		1	2
Period 1	1	$P_A P_{B|A}$	$P_A(1 - P_{B|A})$
	2	$(1 - P_A) P_{B|\overline{A}}$	$(1 - P_A)(1 - P_{B|\overline{A}})$

BA Sequence

Period 2

	1	2
Period 1 1	$P_B P_{A\mid B}$	$P_B(1-P_{A\mid B})$
2	$(1-P_B)P_{A\mid \overline{B}}$	$(1-P_B)(1-P_{A\mid \overline{B}})$

For these two tables, at level 1, the first period marginal probabilities are P_A and P_B for the AB and BA sequences, respectively. The level 2 marginal probabilities are the respective complement probabilities. Therefore, the observed marginal frequencies for the first period can be used to form the logits for P_A and P_B.

Logits for the conditional probabilities, $P_{B\mid A}$, $P_{B\mid \overline{A}}$, $P_{A\mid B}$ and $P_{A\mid \overline{B}}$ are derived from the cross-classified observations and are formed from the ratio of the first and second level expected cell probabilities of the second period. For example, at Period 1, Level 1 of the AB sequence, the logit for $P_{B\mid A}$ is formed from the ratio of cell probabilities for the first level of the second period $P_A P_{B\mid A}$ to the second level of the second period $P_A(1 - P_{B\mid A})$. Therefore, the observed frequencies F_{111} and F_{112} are used to model the logit for $P_{B\mid A}$. The logits for the other conditional probabilities are developed in a similar manner.

As an example, consider the data of Zimmermann and Rahlfs (1978), reproduced here with the permission of Akademie Verlag GmbH:

AB Sequence

Period 2

	1	2
Period 1 1	6	33
2	4	7

BA Sequence

Period 2

		1	2
	1	15	6
Period 1	2	11	18

Using SAS PROC CATMOD, the 2–treatment, 2–period, 2–sequence binary cross–over design would require design variables for the sequence effect (SEQUENCE $= \gamma$), the period effect (PERIOD $= \pi$), the treatment effect (TREAT $= \tau$), and design variables for the association between repeated observations (ASSOC_AB $= \epsilon_{112}$ and ASSOC_BA $= \epsilon_{212}$) within the AB and BA sequences. Also required are the reformulated observed frequencies (COUNT), and the response (Y) which takes on the assigned value of 1 for a success or 2 for a failure. These data are arranged in the following table:

S E Q U E N C E	P E R I O D	T R E A T	A S S - A B	A S S - B A	Y	C O U N T
1	1	1	0	0	1	$39 = F_{111} + F_{112}$
1	1	1	0	0	2	$11 = F_{121} + F_{122}$
1	-1	-1	1	0	1	$6 = F_{111}$
1	-1	-1	1	0	2	$33 = F_{112}$
1	-1	-1	-1	0	1	$4 = F_{121}$
1	-1	-1	-1	0	2	$7 = F_{122}$
-1	1	-1	0	0	1	$21 = F_{211} + F_{212}$
-1	1	-1	0	0	2	$29 = F_{221} + F_{222}$
-1	-1	1	0	1	1	$15 = F_{211}$
-1	-1	1	0	1	2	$6 = F_{212}$
-1	-1	1	0	-1	1	$11 = F_{221}$
-1	-1	1	0	-1	2	$18 = F_{222}$

As with most software packages for logistic regression, PROC CATMOD will automatically insert the intercept that codes for the parameter μ.

If these variables were defined, as in the above table, for the SAS data set "ZIMRAHLF", then the SAS code for the maximum likelihood analysis of Zimmermann and Rahlfs' data is:

```
PROC CATMOD  DATA = ZIMRAHLF;
WEIGHT COUNT;
DIRECT SEQUENCE PERIOD TREAT ASSOC_AB ASSOC_BA;
MODEL Y = SEQUENCE PERIOD TREAT ASSOC_AB ASSOC_BA /
    NOGLS ML;
RUN;
```

For the above example, the SAS output will look as follows (in part):

ANALYSIS OF MAXIMUM LIKELIHOOD ESTIMATES

Effect	Parameter	Estimate	Standard Error	Chi-Square	Prob
INTERCEPT	1	0.0057	0.1660	0.00	0.9728
SEQUENCE	2	0.0611	0.1660	0.14	0.7129
PERIOD	3	0.4658	0.1660	7.87	0.0050
TREAT	4	0.7331	0.1660	19.50	0.0000
ASSOC_AB	5	-0.5726	0.3840	2.22	0.1359
ASSOC_BA	6	0.7044	0.3081	5.23	0.0223

The parameter estimate obtained for treatment can be interpreted in terms of an adjusted odds ratio (Hosmer and Lemeshow, 1989). For the two treatment binary case, the odds of a success for Treatment A relative to a success for Treatment B, while adjusting for the effects of sequence, period and association, is given by $\exp(2\tau)$. An estimate of this value is obtained by substituting the estimate of τ. For the above example, the adjusted odds of a success for Treatment A is 4.333 times as great as the odds of a success for Treatment B.

A P-value of 0.0000 is produced by the Wald chi-square test for the null hypothesis H_o: $\tau = 0$, indicating a significant difference among treatments. Both likelihood ratio and Pearson chi-square statistics can be produced by eliminating the design variable TREAT from the DIRECT and MODEL statements of the SAS code. If this were done, then the likelihood ratio statistic (G^2) would be 20.84 on 1 degree of freedom, producing a P–value of 0.0000. For the Pearson chi–square, hand calculation, based on the predicted cell frequencies, is necessary. The computed value of the Pearson chi–square (χ^2) is 19.33 on 1 degree of freedom, producing a P–value of 0.0000.

If one were to base a test for treatment difference on the first period results only, the usual z–test for two independent proportions applies,

$$z = \frac{0.78 - 0.42}{\left[0.6(1-0.6)\left(\frac{1}{50} + \frac{1}{50}\right)\right]^{1/2}} = 3.67$$

which produces a P–value = 0.000238. For the Kenward and Jones model under the assumption that $\lambda^* \neq 0$, the equivalent Wald test produces $W = 7.779$, with a P–value of 0.00529. All three tests indicate significant differences in treatments, but the test based on the logit model appears to be somewhat more powerful in this example.

Pearson chi–square statistics for testing treatment effect are possible for the logit model presented in this text, the log–linear model of Kenward and Jones (1987), Gart's model (1969) and Prescott's model (1981). For the data of Zimmermann and Rahlfs (1978), the computed chi–square statistics and P–values are:

Estimator	χ^2	P–value
Logit Model	19.329	0.0000110
Log–linear Model ($\lambda^* \neq 0$)	8.383	0.0037875
Gart's Model[1]	16.865	0.0000401
Prescott's Model	23.964	0.0000009

[1] Gart's test and the log–linear model produce identical chi–square statistics when $\lambda^* = 0$.

Although only a single example, these results are consistent with computer simulations conducted by one of us (Evans), which roughly indicated the most powerful approach for detecting treatment effects was given by Prescott's model. This was closely followed by the logit model, then Gart's model and finally the log–linear model of Kenward and Jones when $\lambda^* \neq 0$. The chi–square statistic for Gart's model and the log–linear model of Kenward and Jones are equal when $\lambda^* = 0$. However, the parameters γ and λ^* are confounded, and the test for the null hypothesis $H_0: \lambda^* = 0$ may not be rejected in the presence of a sequence effect. The simulations also indicated that the treatment effect of the log–linear model is confounded with the association parameters. Thus, the log–linear model of Kenward and Jones has some unfortunate operating characteristics.

The simulations indicated that sequence effects, period effects, along with various association patterns had minimal influence on the test for treatments when using either the logit model of this text, Gart's model or Prescott's model.

Before moving on to more complex designs, it is of potential interest to look at one more set of data. The example in point is a contrived data set, very similar to one contrived by Zimmermann and Rahlfs (1978) and is presented in the following table:

AB Sequence

		Period 2	
		1	2
Period 1	1	32	8
	2	8	2

(continued on following page)

BA Sequence

Period 2

		1	2
Period 1	1	2	8
	2	8	32

These data are contrived so as to show how the method proposed by Zimmermann and Rahlfs (1978) will produce highly significant treatment ($\chi^2 =$ 56.25, df $= 1$) and carryover effects ($\chi^2 = 112.5$, df $= 1$) while Gart's test does not detect a difference in treatments ($\chi^2 = 0$, df $= 1$). Although this would appear to be convincing evidence that Gart's method has failed to detect a very large treatment effect, this is not necessarily the case. The "treatment" and "carryover" effects that Zimmermann and Rahlfs believed to be genuine treatment effects can be the result of a large sequence effect. Zimmermann and Rahlfs' model does not contain a sequence component and thus cannot account for this form of variation. Recall from Chapter 3 that several parameterizations are possible for the case of a quantitative response. In particular, a model with a sequence effect, but no carryover, and a model with no sequence effect, but having carryover, are possible. Parameter estimates can be computed using the logit model for both of these parameterizations. The following table presents the estimates for the above data under the two parameterizations:

	Logit Model with $\lambda = 0$			Logit Model with $\gamma = 0$		
Param.	Est.	S.E.	P–value	Est.	S.E.	P–value
μ	0.000	0.2001	1.000	0.000	0.2001	1.000
γ	1.386	0.2001	0.000	———	———	———
π	0.000	0.2001	1.000	0.000	0.2001	1.000
τ	0.000	0.2001	1.000	1.386	0.2500	0.000
λ	———	———	———	2.773	0.4000	0.000
ϵ_{i12}	0.000	0.4419	1.000	0.000	0.4419	1.000

Under both parameterizations, the association effects are estimated to be zero, indicating independence between successive observations. As one can see from the results presented in the table, the choice of model determines which effects are present. Only treatment, carryover and sequence effects may be non–zero, depending upon the model used.

To some extent the most appropriate model for these data can be determined by examining the treatment probabilities. These are as follows:

AB Sequence   BA Sequence

$P_A = 0.8$ $P_B = 0.2$

$P_{B|A} = 0.8$ $P_{A|B} = 0.2$

$P_{B|\overline{A}} = 0.8$ $P_{A|\overline{B}} = 0.2$.

Casually looking over these probabilities might lead one to think that there are strong treatment and carryover effects. This conclusion is drawn from the large probability for Treatment A in the first period of the AB sequence, when compared with the small probability for Treatment B in the first period of the BA sequence. In the AB sequence, the effect of Treatment A appears, at first glance, to negate the effect of Treatment B in the second period. This would appear to indicate a strong carryover effect.

However, one should not jump to premature conclusions. The probabilities for Treatments A and B do not differ within each sequence, only between sequences. There are only two ways in which this can occur. First, if in the AB sequence, Treatment A has a strong carryover that nullified the effect of Treatment B; and in the BA sequence, Treatment B has a strong carryover that nullified the effect of Treatment A, then both treatment and carryover effects are present. This is unreasonable, because Treatment A was dominant in the AB sequence, and therefore should not be nullified in the BA sequence. The second possibility arises from a sequence effect. Those subjects selected for the AB sequence have a natural probability of success of 0.8 and those in the BA sequence have a natural probability of success of 0.2. There would be no treatment or carryover effects. The conclusion of Zimmermann and Rahlfs (1978) that a strong treatment effect and carryover effect exist is quite possibly erroneous and the conclusion of Gart's test is probably not in error.

The 2–treatment, 2–period, 2–sequence design is only a single special case of the broader set of cross–over designs. Because of this, both Prescott's test and

Gart's test are limited in their use. The log–linear model of Kenward and Jones (1987) does allow for the modeling of more general design structures, but lacks the power of the other tests.

7.5. The 2–Treatment, 3–Period, 2–Sequence Binary Cross–over Design

In Chapter 3 the aspects of adding a third period of observation to the 2–treatment, 2–sequence design were discussed for the continuous case. Several different designs were shown to be possible, depending on whether a baseline, wash–out, or treatment were applied in the added period. As indicated, these designs are not equally efficient in the continuous case, so one should not expect these designs to be equally efficient for a categorical response. The 2–treatment, 3–period, 2–sequence design shown to be most efficient in the continuous case has sequences denoted ABB and BAA. Intuitively, the efficiency of this design should carry to the categorical case, although that will not be a topic of discussion.

The observable frequencies from this design can be cross–classified and presented in a pair of 2^3 tables. Let F_{ijkl} denote the joint frequency of response for the ith sequence ($i = 1, 2$), with outcome jkl ($j,k,l = 1, 2$) in the first, second and third periods, respectively. Again, the value "1" will denote a success and the value "2" a failure. With this notation the contingency tables for the joint response frequencies are as follows:

Sequence ABB

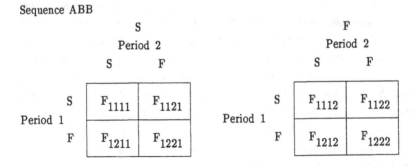

(continued on following page)

Sequence BAA

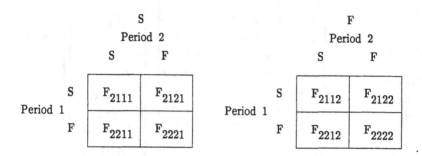

Extending the model of cell probabilities proposed earlier, each cell probability for a three period design will be the product of three conditional probabilities. For the two sequences, these probabilities are as follows:

Sequence ABB	Sequence BAA
P_A	P_B
$P_{B\|A}$	$P_{A\|B}$
$P_{B\|\overline{A}}$	$P_{A\|\overline{B}}$
$P_{B\|AB}$	$P_{A\|BA}$
$P_{B\|A\overline{B}}$	$P_{A\|B\overline{A}}$
$P_{B\|\overline{A}B}$	$P_{A\|\overline{B}A}$
$P_{B\|\overline{AB}}$	$P_{A\|\overline{BA}}$

Only the conditional probabilities from the third period are new, with the first and second period probabilities previously defined. For example, $P_{B\|A\overline{B}}$ is the probability of a success for Treatment B in the third period given success for Treatment A in the first period and failure for Treatment B in the second period. The other seven probabilities for the third period have similar definitions.

The expected cell frequencies under a multinomial assumption are developed by extension of the methods previously described. For example, the expected value of F_{1111} is $N_1 P_A P_{B\|A} P_{B\|AB}$.

A log–linear model for the observable frequencies is given as follows:

$$
\begin{pmatrix} F_{1111} \\ F_{1112} \\ F_{1121} \\ F_{1122} \\ F_{1211} \\ F_{1212} \\ F_{1221} \\ F_{1222} \end{pmatrix} = \exp \left[\begin{pmatrix} 1 & 1 & 0 & 1 & 0 & 0 & 1 & 0 & 0 & 0 & 0 & 0 & 0 & 0 \\ 1 & 1 & 0 & 1 & 0 & 0 & 0 & 1 & 0 & 0 & 0 & 0 & 0 & 0 \\ 1 & 1 & 0 & 0 & 1 & 0 & 0 & 0 & 1 & 0 & 0 & 0 & 0 & 0 \\ 1 & 1 & 0 & 0 & 1 & 0 & 0 & 0 & 0 & 1 & 0 & 0 & 0 & 0 \\ 1 & 0 & 1 & 0 & 0 & 1 & 0 & 0 & 0 & 0 & 1 & 0 & 0 & 0 \\ 1 & 0 & 1 & 0 & 0 & 1 & 0 & 0 & 0 & 0 & 0 & 1 & 0 & 0 \\ 1 & 0 & 1 & 0 & 0 & 0 & 1 & 0 & 0 & 0 & 0 & 0 & 1 & 0 \\ 1 & 0 & 1 & 0 & 0 & 0 & 1 & 0 & 0 & 0 & 0 & 0 & 0 & 1 \end{pmatrix} \begin{pmatrix} \ln(N_1) \\ \ln(P_A) \\ \ln(1-P_A) \\ \ln(P_{B|A}) \\ \ln(1-P_{B|A}) \\ \ln(P_{B|\bar{A}}) \\ \ln(1-P_{B|\bar{A}}) \\ \ln(P_{B|AB}) \\ \ln(1-P_{B|AB}) \\ \ln(P_{B|A\bar{B}}) \\ \ln(1-P_{B|A\bar{B}}) \\ \ln(P_{B|\bar{A}B}) \\ \ln(1-P_{B|\bar{A}B}) \\ \ln(P_{B|\overline{AB}}) \\ \ln(1-P_{B|\overline{AB}}) \end{pmatrix} \right] + \delta_1
$$

$$
\begin{pmatrix} F_{2111} \\ F_{2112} \\ F_{2121} \\ F_{2122} \\ F_{2211} \\ F_{2212} \\ F_{2221} \\ F_{2222} \end{pmatrix} = \exp \left[\begin{pmatrix} 1 & 1 & 0 & 1 & 0 & 0 & 1 & 0 & 0 & 0 & 0 & 0 & 0 & 0 \\ 1 & 1 & 0 & 1 & 0 & 0 & 0 & 1 & 0 & 0 & 0 & 0 & 0 & 0 \\ 1 & 1 & 0 & 0 & 1 & 0 & 0 & 0 & 1 & 0 & 0 & 0 & 0 & 0 \\ 1 & 1 & 0 & 0 & 1 & 0 & 0 & 0 & 0 & 1 & 0 & 0 & 0 & 0 \\ 1 & 0 & 1 & 0 & 0 & 1 & 0 & 0 & 0 & 0 & 1 & 0 & 0 & 0 \\ 1 & 0 & 1 & 0 & 0 & 1 & 0 & 0 & 0 & 0 & 0 & 1 & 0 & 0 \\ 1 & 0 & 1 & 0 & 0 & 0 & 1 & 0 & 0 & 0 & 0 & 0 & 1 & 0 \\ 1 & 0 & 1 & 0 & 0 & 0 & 1 & 0 & 0 & 0 & 0 & 0 & 0 & 1 \end{pmatrix} \begin{pmatrix} \ln(N_2) \\ \ln(P_B) \\ \ln(1-P_B) \\ \ln(P_{A|B}) \\ \ln(1-P_{A|B}) \\ \ln(P_{A|\bar{B}}) \\ \ln(1-P_{A|\bar{B}}) \\ \ln(P_{A|BA}) \\ \ln(1-P_{A|BA}) \\ \ln(P_{A|B\bar{A}}) \\ \ln(1-P_{A|B\bar{A}}) \\ \ln(P_{A|\bar{B}A}) \\ \ln(1-P_{A|\bar{B}A}) \\ \ln(P_{A|\overline{BA}}) \\ \ln(1-P_{A|\overline{BA}}) \end{pmatrix} \right] + \delta_2
$$

The fourteen possible probabilities allow for the development of fourteen logits, indicating a total of fourteen degrees of freedom. Eight of these degrees of freedom must be used to estimate the association parameters. This leaves a total of 6 degrees of freedom available to estimate the location parameters, exactly the same as in the continuous case, where 6 cell means are available. Once again, the same parameter specification, less the association parameters, exists for the binary and continuous cases.

With the addition of a third period, two parameters are necessary to model the period effects, and as in the continuous case, a first order carryover parameter can also be modeled. We will let π_1, π_2 and λ denote these effects.

A third period will necessarily increase the complexity of the association parameters that need to be modeled. For this design, there will be an association parameter for observations made between Periods 1 and 2, between Periods 1 and 3, between Periods 2 and 3, and a combined association between Periods 1, 2 and 3. We will let ϵ_{i12}, ϵ_{i13}, ϵ_{i23} and ϵ_{i123} respectively denote these associations in the ith sequence ($i = 1, 2$). A total of 8 association parameters must be modeled, namely 4 from the first sequence and 4 from the second sequence.

The logits for this model can be represented as linear combinations of the parameters. These logit definitions are as follows:

ABB Sequence

$$\text{logit}(P_A) = \mu + \gamma + \pi_1 + \tau$$
$$\text{logit}(P_{B|A}) = \mu + \gamma + \pi_2 - \tau + \lambda + \epsilon_{112}$$
$$\text{logit}(P_{B|\overline{A}}) = \mu + \gamma + \pi_2 - \tau + \lambda - \epsilon_{112}$$
$$\text{logit}(P_{B|AB}) = \mu + \gamma - \pi_1 - \pi_2 - \tau - \lambda + \epsilon_{113} + \epsilon_{123} + \epsilon_{1123}$$
$$\text{logit}(P_{B|A\overline{B}}) = \mu + \gamma - \pi_1 - \pi_2 - \tau - \lambda + \epsilon_{113} - \epsilon_{123} - \epsilon_{1123}$$
$$\text{logit}(P_{B|\overline{A}B}) = \mu + \gamma - \pi_1 - \pi_2 - \tau - \lambda - \epsilon_{113} + \epsilon_{123} - \epsilon_{1123}$$
$$\text{logit}(P_{B|\overline{A}\,\overline{B}}) = \mu + \gamma - \pi_1 - \pi_2 - \tau - \lambda - \epsilon_{113} - \epsilon_{123} + \epsilon_{1123}$$

BAA Sequence

$$\text{logit}(P_B) = \mu - \gamma + \pi_1 - \tau$$
$$\text{logit}(P_{A|B}) = \mu - \gamma + \pi_2 + \tau - \lambda + \epsilon_{212}$$
$$\text{logit}(P_{A|\overline{B}}) = \mu - \gamma + \pi_2 + \tau - \lambda - \epsilon_{212}$$
$$\text{logit}(P_{A|BA}) = \mu - \gamma - \pi_1 - \pi_2 + \tau + \lambda + \epsilon_{213} + \epsilon_{223} + \epsilon_{2123}$$
$$\text{logit}(P_{A|B\overline{A}}) = \mu - \gamma - \pi_1 - \pi_2 + \tau + \lambda + \epsilon_{213} - \epsilon_{223} - \epsilon_{2123}$$
$$\text{logit}(P_{A|\overline{B}A}) = \mu - \gamma - \pi_1 - \pi_2 + \tau + \lambda - \epsilon_{213} + \epsilon_{223} - \epsilon_{2123}$$
$$\text{logit}(P_{A|\overline{B}\,\overline{A}}) = \mu - \gamma - \pi_1 - \pi_2 + \tau + \lambda - \epsilon_{213} - \epsilon_{223} + \epsilon_{2123}$$

The logit model can be represented in the following matrix notation:

ABB Sequence

$$
\begin{pmatrix}
\text{logit}(P_A) \\
\text{logit}(P_{B|A}) \\
\text{logit}(P_{B|\bar{A}}) \\
\text{logit}(P_{B|AB}) \\
\text{logit}(P_{B|A\bar{B}}) \\
\text{logit}(P_{B|\bar{A}B}) \\
\text{logit}(P_{B|\bar{A}\bar{B}})
\end{pmatrix}
=
\begin{pmatrix}
1 & 1 & 1 & 0 & 1 & 0 & 0 & 0 & 0 & 0 & 0 & 0 & 0 & 0 \\
1 & 1 & 0 & 1 & -1 & 1 & 1 & 0 & 0 & 0 & 0 & 0 & 0 & 0 \\
1 & 1 & 0 & 1 & -1 & 1 & -1 & 0 & 0 & 0 & 0 & 0 & 0 & 0 \\
1 & 1 & -1 & -1 & -1 & -1 & 0 & 1 & 1 & 1 & 0 & 0 & 0 & 0 \\
1 & 1 & -1 & -1 & -1 & -1 & 0 & 1 & -1 & -1 & 0 & 0 & 0 & 0 \\
1 & 1 & -1 & -1 & -1 & -1 & 0 & -1 & 1 & -1 & 0 & 0 & 0 & 0 \\
1 & 1 & -1 & -1 & -1 & -1 & 0 & -1 & -1 & 1 & 0 & 0 & 0 & 0
\end{pmatrix}
\begin{pmatrix}
\mu \\
\gamma \\
\pi_1 \\
\pi_2 \\
\tau \\
\lambda \\
\epsilon_{112} \\
\epsilon_{113} \\
\epsilon_{123} \\
\epsilon_{1123}
\end{pmatrix}
$$

BAA Sequence

$$
\begin{pmatrix}
\text{logit}(P_B) \\
\text{logit}(P_{A|B}) \\
\text{logit}(P_{A|\bar{B}}) \\
\text{logit}(P_{A|BA}) \\
\text{logit}(P_{A|B\bar{A}}) \\
\text{logit}(P_{A|\bar{B}A}) \\
\text{logit}(P_{A|\bar{B}\bar{A}})
\end{pmatrix}
=
\begin{pmatrix}
1 & -1 & 1 & 0 & -1 & 0 & 0 & 0 & 0 & 0 & 0 & 0 & 0 & 0 \\
1 & -1 & 0 & 1 & 1 & -1 & 0 & 0 & 0 & 0 & 1 & 0 & 0 & 0 \\
1 & -1 & 0 & 1 & 1 & -1 & 0 & 0 & 0 & 0 & -1 & 0 & 0 & 0 \\
1 & -1 & -1 & -1 & 1 & 1 & 0 & 0 & 0 & 0 & 0 & 1 & 1 & 1 \\
1 & -1 & -1 & -1 & 1 & 1 & 0 & 0 & 0 & 0 & 0 & 1 & -1 & -1 \\
1 & -1 & -1 & -1 & 1 & 1 & 0 & 0 & 0 & 0 & 0 & -1 & 1 & -1 \\
1 & -1 & -1 & -1 & 1 & 1 & 0 & 0 & 0 & 0 & 0 & -1 & -1 & 1
\end{pmatrix}
\begin{pmatrix}
\mu \\
\gamma \\
\pi_1 \\
\pi_2 \\
\tau \\
\lambda \\
\epsilon_{212} \\
\epsilon_{213} \\
\epsilon_{223} \\
\epsilon_{2123}
\end{pmatrix}
$$

Maximum likelihood estimates can be computed using the Newton–Raphson algorithm of Appendix 7.A, and also by using SAS PROC CATMOD after some manipulation of the observed frequencies. Using PROC CATMOD, the marginal frequencies for the first period are stacked on top of the cross–classified frequencies for the first and second period, which are in turn stacked on top of the cross–classified frequencies for the first, second and third periods. This produces $2 + 4 + 8 = 14$ values for each sequence of the three period binary design. The elements of the design matrix that correspond to periods one and two are the same as in the 2–period design. By considering the structure of the expected cell probabilities for the full cross–classification, the probabilities for the third period that are conditional on the outcomes in first two periods can be determined along with the observed frequencies that are used to estimate these probabilities.

As an example of this method, consider the data set of Ebbutt (1984) that was presented in Chapter 4 (see Table 4.1). These data can be dichotomized by considering a success to be any value greater than the grand median and a failure to be any value less than or equal to the grand median. These data are presented in the following set of 2^3 tables:

Sequence ABB — Period 3

S

		Period 2 S	Period 2 F
Period 1	S	8	0
	F	0	1

F

		Period 2 S	Period 2 F
Period 1	S	2	3
	F	1	7

Sequence BAA — Period 3

S

		Period 2 S	Period 2 F
Period 1	S	7	0
	F	2	2

F

		Period 2 S	Period 2 F
Period 1	S	1	1
	F	4	10

Because some sampling zeros are observed for this data set, it is not possible to estimate parameters for the saturated model, unless a modification is made. A very simple but effective modification is to add a small value to each frequency. One of the authors, Evans, has performed some extensive simulations for log–linear models and found that the addition of 0.25 works well for the binary case with three periods. In general, one might add a value obtained by taking the number of response levels and raising it to the power $1 - P$, where P is the number of periods.

Using SAS PROC CATMOD, the 2–treatment, 3–period, 2–sequence binary cross–over design would require design variables for the sequence effect (SEQUENCE $= \gamma$), the effects for the two periods (PERIOD_1 $= \pi_1$ and PERIOD_2 $= \pi_2$), the treatment effect (TREAT $= \tau$), the first order carryover effect (CARRY $= \lambda$), and design variables for the association between the first and second periods (ASS_112 $= \epsilon_{112}$ and ASS_212 $= \epsilon_{212}$), the first and third periods (ASS_113 $= \epsilon_{113}$ and ASS_213 $= \epsilon_{213}$), the second and third periods (ASS_123 $= \epsilon_{123}$ and ASS_223 $= \epsilon_{223}$), and the three–way association (ASS_1123 $= \epsilon_{1123}$ and ASS_2123 $= \epsilon_{2123}$) within the respective first (ABB) and second (BAA) sequences. Also required are the reorganized observed frequencies (COUNT) and the response (Y) which takes on the assigned value of "1" for a success or "2" for a failure. These data are arranged in the table shown on the following page.

Assuming that these variables have been defined in the SAS data set "EBBUTT", the SAS code for the analysis of this example is as follows:

```
PROC CATMOD DATA = EBBUTT;
WEIGHT COUNT;
DIRECT SEQUENCE PERIOD_1 PERIOD_2 TREAT CARRY ASS_112
    ASS_113 ASS_123 ASS_1123 ASS_212 ASS_213 ASS_223 ASS_2123;
MODEL Y = SEQUENCE PERIOD_1 PERIOD_2 TREAT CARRY ASS_112
    ASS_113 ASS_123 ASS_1123 ASS_212 ASS_213 ASS_223 ASS_2123 /
    NOGLS ML;
RUN;
```

SEQUENCE	PERIOD-1	PERIOD-2	TREAT	CARRY	ASS-112	ASS-113	ASS-123	ASS-1123	ASS-212	ASS-213	ASS-223	ASS-2123	Y	COUNT
1	1	0	1	0	0	0	0	0	0	0	0	0	1	$14.00 = F_{1111}+F_{1112}+F_{1121}+F_{1122}+1$
1	1	0	1	0	0	0	0	0	0	0	0	0	2	$10.00 = F_{1211}+F_{1212}+F_{1221}+F_{1222}+1$
1	0	1	-1	1	1	0	0	0	0	0	0	0	1	$10.50 = F_{1111}+F_{1112}+0.5$
1	0	1	-1	1	1	0	0	0	0	0	0	0	2	$3.50 = F_{1121}+F_{1122}+0.5$
1	0	1	-1	1	-1	0	0	0	0	0	0	0	1	$1.50 = F_{1211}+F_{1212}+0.5$
1	0	1	-1	1	-1	0	0	0	0	0	0	0	2	$8.50 = F_{1221}+F_{1222}+0.5$
1	-1	-1	-1	-1	0	1	1	1	0	0	0	0	1	$8.25 = F_{1111}+0.25$
1	-1	-1	-1	-1	0	1	1	1	0	0	0	0	2	$2.25 = F_{1112}+0.25$
1	-1	-1	-1	-1	0	1	-1	-1	0	0	0	0	1	$0.25 = F_{1121}+0.25$
1	-1	-1	-1	-1	0	1	-1	-1	0	0	0	0	2	$3.25 = F_{1122}+0.25$
1	-1	-1	-1	-1	0	-1	1	-1	0	0	0	0	1	$0.25 = F_{1211}+0.25$
1	-1	-1	-1	-1	0	-1	1	-1	0	0	0	0	2	$1.25 = F_{1212}+0.25$
1	-1	-1	-1	-1	0	-1	-1	1	0	0	0	0	1	$1.25 = F_{1221}+0.25$
1	-1	-1	-1	-1	0	-1	-1	1	0	0	0	0	2	$7.25 = F_{1222}+0.25$
-1	1	0	-1	0	0	0	0	0	0	0	0	0	1	$10.00 = F_{2111}+F_{2112}+F_{2121}+F_{2122}+1$
-1	1	0	-1	0	0	0	0	0	0	0	0	0	2	$19.00 = F_{2211}+F_{2212}+F_{2221}+F_{2222}+1$
-1	0	1	1	-1	0	0	0	0	1	0	0	0	1	$8.50 = F_{2111}+F_{2112}+0.5$
-1	0	1	1	-1	0	0	0	0	1	0	0	0	2	$1.50 = F_{2121}+F_{2122}+0.5$
-1	0	1	1	-1	0	0	0	0	-1	0	0	0	1	$6.50 = F_{2211}+F_{2212}+0.5$
-1	0	1	1	-1	0	0	0	0	-1	0	0	0	2	$12.50 = F_{2221}+F_{2222}+0.5$
-1	-1	-1	1	1	0	0	0	0	0	1	1	1	1	$7.25 = F_{2111}+0.25$
-1	-1	-1	1	1	0	0	0	0	0	1	1	1	2	$1.25 = F_{2112}+0.25$
-1	-1	-1	1	1	0	0	0	0	0	1	-1	-1	1	$0.25 = F_{2121}+0.25$
-1	-1	-1	1	1	0	0	0	0	0	1	-1	-1	2	$1.25 = F_{2122}+0.25$
-1	-1	-1	1	1	0	0	0	0	0	-1	1	-1	1	$2.25 = F_{2211}+0.25$
-1	-1	-1	1	1	0	0	0	0	0	-1	1	-1	2	$4.25 = F_{2212}+0.25$
-1	-1	-1	1	1	0	0	0	0	0	-1	-1	1	1	$2.25 = F_{2221}+0.25$
-1	-1	-1	1	1	0	0	0	0	0	-1	-1	1	2	$10.25 = F_{2222}+0.25$

The SAS output from the preceding code will look as follows (in part):

ANALYSIS OF MAXIMUM LIKELIHOOD ESTIMATES

Effect	Parameter	Estimate	Standard Error	Chi-Square	Prob
INTERCEPT	1	−0.2904	0.2338	1.54	0.2143
SEQUENCE	2	0.0551	0.2143	0.07	0.7970
PERIOD_1	3	0.1377	0.2858	0.23	0.6300
PERIOD_2	4	0.4015	0.3165	1.61	0.2045
TREAT	5	0.4340	0.2143	4.10	0.0429
CARRY	6	−0.0503	0.3205	0.02	0.8754
ASS_112	7	1.4166	0.5397	6.89	0.0087
ASS_113	8	0.5254	0.8144	0.42	0.5188
ASS_123	9	1.0032	0.8144	1.52	0.2180
ASS_1123	10	0.9290	0.8144	1.30	0.2540
ASS_212	11	1.1943	0.5045	5.60	0.0179
ASS_213	12	0.5752	0.6596	0.76	0.3831
ASS_223	13	1.0619	0.6596	2.59	0.1074
ASS_2123	14	0.6217	0.6596	0.89	0.3458

Again, the parameter estimate for treatment effect can be interpreted as an adjusted odds ratio. For the above example, the adjusted odds of a success for Treatment A is $\exp(2 \cdot 0.4340) = 2.382$ times as great as the odds of a success for Treatment B. Because both Treatment A and B were active treatments for the reduction of blood pressure and since a success was defined as a blood pressure greater than the grand median, subjectively, Treatment B is more effective in reducing blood pressure. This conclusion is more rigorously assessed by statistical tests.

Testing for a first order carryover, the Wald chi–square produces a P–value of 0.8754, indicating no evidence of a carryover effect. The Wald chi–square for testing treatments produces a P–value of 0.0429, indicating a marginally significant treatment effect. One can also test these effects with a likelihood ratio test. Using this test, the P–values for the carryover and treatment effects are 0.7054 and 0.0309, respectively. Although the results of these two testing procedures are slightly different, the conclusions are the same. The reader is reminded that the Wald chi–square test is known to behave poorly for small samples and is typically less powerful than the likelihood ratio test (Hosmer and Lemeshow, 1989; Agresti, 1990).

It is possibly of some interest to the reader to review the analysis of Ebbutt's data for a continuous response, found in Chapter 4. For that analysis, the F statistics for carryover and treatment produced P–values of 0.319 and 0.012, respectively. These values are smaller than those of the binary analysis, but not to a great degree. As one might expect, a continuous response contains more information than a binary response.

7.6. The General Model for Cross–over Designs having a Multilevel Categorical Response

The methods described for a binary response naturally extend to a response having $K > 2$ levels. For the sake of simplicity, consider the 2–treatment, 2–period, 2–sequence cross–over design having K response levels. As with the binary response, let F_{ijk} denote the observed frequency for the ith sequence with joint response jk ($j,k = 1, ..., K$) for the first and second periods. For the AB and BA sequences, the frequencies can be cross classified and arranged in two K x K contingency tables.

Again for simplicity's sake, consider the AB and BA sequences when $K = 3$. The observable frequencies for the three level response are neatly presented in the following 3^2 contingency tables:

AB Sequence

		Period 2		
		1	2	3
Period 1	1	F_{111}	F_{112}	F_{113}
	2	F_{121}	F_{122}	F_{123}
	3	F_{131}	F_{132}	F_{133}

(continued on following page)

BA Sequence

Period 2

		1	2	3
Period 1	1	F_{211}	F_{212}	F_{213}
	2	F_{221}	F_{222}	F_{223}
	3	F_{231}	F_{232}	F_{233}

Although Le (1984) extended Gart's model to the general case of K response levels, no simple test statistic was presented for testing treatments. Kenward and Jones (1991) presented a log–linear model for the analysis of cross–over designs having a multilevel response. This model is a direct extension of their earlier work with a binary response. As such, the model for a multilevel response is likely to suffer the same problems as the model for a binary response.

For the AB sequence, define P_{Aj} as the probability of observing category j ($j = 1, 2, ..., K$) for Treatment A in period 1; and $P_{Bk|Aj}$ as the probability of observing category k ($k = 1, 2, ..., K$) for Treatment B in period 2, given category j ($j = 1, 2, ..., K$) was observed for Treatment A in period 1. The BA sequence has probabilities P_{Bj} and $P_{Ak|Bj}$ ($j,k = 1, 2, ..., K$) with similar definitions. Let N_i ($i = 1, 2$) be the number of experimental units sampled for the AB and BA sequences, respectively. Then, the expected cell frequencies for the 2–treatment, 2–period, 2–sequence design having a 3 level response are given in the following tables (see top of following page):

For a K level multinomial response, only K – 1 probabilities, at a given condition of observation, can be independently estimated, because the sum of all K probabilities must equal 1.0. For example, the probabilities for the AB sequence with K = 3 must satisfy the following restrictions: $P_{A1} + P_{A2} + P_{A3}$ = 1.0, and $P_{B1|Aj} + P_{B2|Aj} + P_{B3|Aj}$ = 1.0 given fixed level j ($j = 1, 2,$ or 3) was observed for Treatment A. For K = 3 response levels, each sequence will have twelve probabilities in four sets of three. With this in mind, a baseline response category can be selected, say the Kth, and generalized logits (Agresti,

AB Sequence

Period 2

		1	2	3
	1	$N_1 P_{A1} P_{B1\mid A1}$	$N_1 P_{A1} P_{B2\mid A1}$	$N_1 P_{A1} P_{B3\mid A1}$
Period 1	2	$N_1 P_{A2} P_{B1\mid A2}$	$N_1 P_{A2} P_{B2\mid A2}$	$N_1 P_{A2} P_{B3\mid A2}$
	3	$N_1 P_{A3} P_{B1\mid A3}$	$N_1 P_{A3} P_{B2\mid A3}$	$N_1 P_{A3} P_{B3\mid A3}$

BA Sequence

Period 2

		1	2	3
	1	$N_2 P_{B1} P_{A1\mid B1}$	$N_2 P_{B1} P_{A2\mid B1}$	$N_2 P_{B1} P_{A3\mid B1}$
Period 1	2	$N_2 P_{B2} P_{A1\mid B2}$	$N_2 P_{B2} P_{A2\mid B2}$	$N_2 P_{B2} P_{A3\mid B2}$
	3	$N_2 P_{B3} P_{A1\mid B3}$	$N_2 P_{B3} P_{A2\mid B3}$	$N_2 P_{B3} P_{A3\mid B3}$

1990) constructed of the remaining $K - 1$ probabilities. For the AB sequence with $K = 3$ response levels, four pairs of generalized logits can be constructed. Within each set of two are a logit for the first level of response and one for the second level of response. Because we are assuming a nominal response, there will necessarily be a complete set of location parameters for each response level defined by the logits.

As an example, let's return to the 2–treatment, 2–period, 2–sequence cross–over design with a binary response ($K = 2$). Under the sum to zero convention and assuming no carryover effect, the location parameters of this model are: μ, γ, π and τ. When the response is extended to $K = 3$ levels, we will have two each of these parameters; one set for the first response level and one set for the second. These parameters will be denoted: μ_i, γ_i, π_i and τ_i ($i = 1, 2$).

For the 2-treatment, 2-period, 2-sequence cross-over design with $K \geq 2$, the number of association parameters required per sequence is $(K - 1)^2$. This is the number of degrees of freedom required for testing independence in a $K \times K$ contingency table. In particular, the cross-over design with $K = 3$ response levels will require four association parameters per sequence. These are contrasts among the logits for the conditional or second period probabilities at a particular level of response. The logit for the kth response level ($k = 1, 2$), given a first period response of j ($j = 1, 2$), is contrasted with the logit for the kth level response, given a first period response of $K = 3$. Define for the first sequence (AB), ϵ_{1kjK} as the parameter for the association between observing category k ($k = 1, 2$) in the second period, given category j ($j = 1, 2$) was observed in the first period. Under the sum to zero convention, the association between observing category k ($k = 1, 2$) in the second period, given category $K = 3$ was observed in the first period, is $-\epsilon_{1kjK}$. Thus, the AB sequence will have association parameters: $\epsilon_{1113}, \epsilon_{1123}, \epsilon_{1213}, \epsilon_{1223}$. For the BA sequence the association parameters are defined as: $\epsilon_{2113}, \epsilon_{2123}, \epsilon_{2213}, \epsilon_{2223}$.

Using the last level ($K = 3$) as the baseline response, the 8 possible generalized logits for the AB sequence have the following definition:

$$\ln\left(\frac{P_{A1}}{P_{A3}}\right) = \mu_1 + \gamma_1 + \pi_1 + \tau_1$$

$$\ln\left(\frac{P_{A2}}{P_{A3}}\right) = \mu_2 + \gamma_2 + \pi_2 + \tau_2$$

$$\ln\left(\frac{P_{B1|A1}}{P_{B3|A1}}\right) = \mu_1 + \gamma_1 - \pi_1 - \tau_1 + \epsilon_{1113}$$

$$\ln\left(\frac{P_{B2|A1}}{P_{B3|A1}}\right) = \mu_2 + \gamma_2 - \pi_2 - \tau_2 + \epsilon_{1213}$$

$$\ln\left(\frac{P_{B1|A2}}{P_{B3|A2}}\right) = \mu_1 + \gamma_1 - \pi_1 - \tau_1 + \epsilon_{1123}$$

$$\ln\left(\frac{P_{B2|A2}}{P_{B3|A2}}\right) = \mu_2 + \gamma_2 - \pi_2 - \tau_2 + \epsilon_{1223}$$

$$\ln\left(\frac{P_{B1|A3}}{P_{B3|A3}}\right) = \mu_1 + \gamma_1 - \pi_1 - \tau_1 - \epsilon_{1113} - \epsilon_{1123}$$

$$\ln\left(\frac{P_{B2|A3}}{P_{B3|A3}}\right) = \mu_2 + \gamma_2 - \pi_2 - \tau_2 - \epsilon_{1213} - \epsilon_{1223} \quad .$$

Given these logit definitions, the algorithm of Appendix 7.A can be used for computation of the maximum likelihood estimates. Again, implementation of this algorithm is no easy task. The same trick, used for a binary response, where the first period observations and the cross–classified second period observations are combined, can also be used for the case of a general K level response. SAS PROC CATMOD can then be used to produce the maximum likelihood estimates.

As an example, consider the dysmenorrhea data of Kenward and Jones (1991). These data are extracted from a 6–sequence, 3–treatment cross–over design with a three level response. Treatment A is a placebo and Treatment B is a low dose analgesic. The response has levels: 1 = No Relief; 2 = Moderate Relief; and 3 = Complete Relief. The data from this experiment are presented in the following tables (reproduced with permission of John Wiley and Sons, Ltd):

AB Sequence

		Period 2		
		1	2	3
	1	2	9	2
Period 1	2	0	2	0
	3	0	0	0

(continued on following page)

BA Sequence

Period 2

		1	2	3
	1	1	2	2
Period 1	2	4	1	0
	3	5	0	0

Because some sampling zeros are observed for this data set, it is not possible to estimate all parameters unless a further modification is made. Again a small value is added to each observed frequency. In this case, 1/3 is added.

For these data, we will construct the design variables so that the baseline treatment will be the Placebo (A) and the baseline response level will be No Relief (1). The reason for choosing Treatment A as the baseline is that parameters for the treatments will contrast Treatment B (Low Dose Analgesic) against Treatment A (Placebo). Thus, the parameter will reflect the potential improvement in relief for the active treatments. Choosing the No Relief (1) response level as the baseline for forming the logits will allow for the comparison of the Moderate Relief (2) and Complete Relief (3) levels against the No Relief (1) level.

Using SAS PROC CATMOD, the above design requires design variables for the sequence effect (SEQUENCE codes for γ_i), the period effect (PERIOD codes for π_i), the treatment effect (TREAT codes for τ_i), and design variables for the association between repeated observations (in the AB sequence ASS113 codes for ϵ_{1113} and ϵ_{1213}, ASS123 codes for ϵ_{1123} and ϵ_{1223}, and in the BA sequence ASS213 codes for ϵ_{2113} and ϵ_{2213}, ASS223 codes for ϵ_{2123} and ϵ_{2223}). We will assume there is no carryover effect for these data. Also required are the reorganized observed frequencies (COUNT) and the response (Y) which takes on the assigned value of "1" for No Relief, "2" for Moderate Relief and "3" for Complete Relief. These data are arranged in the following table:

S E Q U E N C E	P E R I O D	T R E A T	A S S 1 1 3	A S S 1 2 3	A S S 2 1 3	A S S 2 2 3	Y	C O U N T
1	1	−1	0	0	0	0	1	$14.000 = F_{11.} + 1$
1	1	−1	0	0	0	0	2	$3.000 = F_{12.} + 1$
1	1	−1	0	0	0	0	3	$1.000 = F_{13.} + 1$
1	−1	1	1	0	0	0	1	$2.333 = F_{111} + 1/3$
1	−1	1	1	0	0	0	2	$9.333 = F_{112} + 1/3$
1	−1	1	1	0	0	0	3	$2.333 = F_{113} + 1/3$
1	−1	1	0	1	0	0	1	$0.333 = F_{121} + 1/3$
1	−1	1	0	1	0	0	2	$2.333 = F_{122} + 1/3$
1	−1	1	0	1	0	0	3	$0.333 = F_{123} + 1/3$
1	−1	1	−1	−1	0	0	1	$0.333 = F_{131} + 1/3$
1	−1	1	−1	−1	0	0	2	$0.333 = F_{132} + 1/3$
1	−1	1	−1	−1	0	0	3	$0.333 = F_{133} + 1/3$
−1	1	1	0	0	0	0	1	$6.000 = F_{21.} + 1$
−1	1	1	0	0	0	0	2	$6.000 = F_{22.} + 1$
−1	1	1	0	0	0	0	3	$6.000 = F_{23.} + 1$
−1	−1	−1	0	0	1	0	1	$1.333 = F_{211} + 1/3$
−1	−1	−1	0	0	1	0	2	$2.333 = F_{212} + 1/3$
−1	−1	−1	0	0	1	0	3	$2.333 = F_{213} + 1/3$
−1	−1	−1	0	0	0	1	1	$4.333 = F_{221} + 1/3$
−1	−1	−1	0	0	0	1	2	$1.333 = F_{222} + 1/3$
−1	−1	−1	0	0	0	1	3	$0.333 = F_{223} + 1/3$
−1	−1	−1	0	0	−1	−1	1	$5.333 = F_{231} + 1/3$
−1	−1	−1	0	0	−1	−1	2	$0.333 = F_{232} + 1/3$
−1	−1	−1	0	0	−1	−1	3	$0.333 = F_{233} + 1/3$

For the above table, the first period marginal frequency for the ith sequence and jth response level are denoted $F_{ij.}$ and are defined as the sum of all observed frequencies for sequence i $(i = 1, 2)$ having level j $(j = 1, 2, 3)$ in the first period.

SAS PROC CATMOD will use the response level with the highest numerical value as the baseline. In order that the first response level is used as the baseline, the data set given above must be ordered in descending order with respect to the response variable Y. Then the ORDER = DATA option in CATMOD can be used to force response level "1" as the baseline.

If these variables were defined, as in the above table, for the SAS data set "DYSMEN2", then the SAS code for the maximum likelihood analysis of the observed data is as follows:

```
PROC SORT DATA = DYSMEN2; BY DESCENDING Y;
PROC CATMOD DATA = DYSMEN2 ORDER = DATA;
WEIGHT COUNT;
DIRECT SEQUENCE PERIOD TREAT ASS113 ASS123 ASS213 ASS223;
MODEL Y = SEQUENCE PERIOD TREAT ASS113 ASS123 ASS213 ASS223 /
    NOGLS ML;
RUN;
```

For the above example, the SAS output will look as follows (in part):

MAXIMUM LIKELIHOOD ANALYSIS OF VARIANCE TABLE

Source	DF	Chi–Square	Prob
INTERCEPT	2	4.92	0.0856
SEQUENCE	2	0.78	0.6779
PERIOD	2	0.99	0.6092
TREAT	2	7.90	0.0192
ASS113	2	0.09	0.9567
ASS123	2	0.50	0.7785
ASS213	2	4.96	0.0837
ASS223	2	0.50	0.7805
LIKELIHOOD RATIO	0	.	.

(continued on following page)

ANALYSIS OF MAXIMUM LIKELIHOOD ESTIMATES

Effect	Parameter	Estimate	Standard Error	Chi-Square	Prob
INTERCEPT	1	−1.0579	0.4794	4.87	0.0273
	2	−0.3901	0.3905	1.00	0.3178
SEQUENCE	3	−0.2616	0.4794	0.30	0.5853
	4	0.1752	0.3905	0.20	0.6536
PERIOD	5	−0.2616	0.4794	0.30	0.5853
	6	−0.3802	0.3905	0.95	0.3302
TREAT	7	1.0579	0.4794	4.87	0.0273
	8	0.9454	0.3905	5.86	0.0155
ASS113	9	−742E−19	1.3093	0.00	1.0000
	10	0.2756	1.1339	0.06	0.8080
ASS123	11	−978E−19	1.8516	0.00	1.0000
	12	0.8352	1.5000	0.31	0.5777
ASS213	13	2.1523	1.1122	3.74	0.0530
	14	1.6902	0.9935	2.89	0.0889
ASS223	15	−0.9723	1.3860	0.49	0.4830
	16	−0.0481	0.9597	0.00	0.9600

As with the binary case, each parameter estimate for treatment can be interpreted as an adjusted odds ratio. For the above example, the adjusted odds of observing Complete Relief (response level 3), relative to No Relief (baseline response level 1), for Treatment B is $\exp(2 \cdot 1.0579) = 8.296$ times as great as the odds for Treatment A. The adjusted odds of observing Moderate Relief (response level 2), relative to No Relief (baseline response level 1), for Treatment B is $\exp(2 \cdot 0.9454) = 6.625$ times as great as the odds for Treatment A. As was discussed for previous analyses, more formal assessment of treatment effects are possible.

The Wald chi–square for treatment effect has a computed value of 7.90 on 2 degrees of freedom, producing a P–value of 0.0192. If the variable coding for treatment effects (TREAT) is removed from both the DIRECT and MODEL statements of the SAS code and the analysis is run again, a likelihood ratio statistic for treatment effect is produced. The computed likelihood ratio statistic for this analysis is 8.33 on 2 degrees of freedom, producing a P–value of 0.0155. Using either test statistic, the treatments give evidence of a difference at $\alpha = 0.05$.

In general, a response having $K \geq 2$ levels will have, under the sum to zero convention, $K - 1$ parameters for the mean response. If there are S sequences, P periods and T treatments, then for the sum to zero convention, there will be $(K - 1) \cdot (S - 1)$ parameters for sequence, $(K - 1) \cdot (P - 1)$ for period, and $(K - 1) \cdot (T - 1)$ for treatment. The number of parameters for carryover will depend on the order of the carryover effect to be modeled. If only first order carryover is considered, then there will be $(K - 1) \cdot (T - 1)$ parameters.

The number of association parameters required for the analysis of a cross–over design depends on the number of sequences, periods and response levels. If there are P periods of observations, then the number of parameters for association within a sequence would be

$$\sum_{i=2}^{P} \binom{P}{i} \cdot (K - 1)^{i} .$$

The number of association parameters for each sequence of the 2–treatment, 2–period, 2–sequence cross–over design having a binary response is one. If the response has $K = 3$ levels, then each sequence requires $(3 - 1)^{2} = 4$ parameters for association between repeated observation. For a $P = 3$ period design with a binary response, each sequence will require $3 \cdot (2 - 1)^{2} + 1 \cdot (2 - 1)^{3} = 4$ association parameters to code for the association between Periods 1 and 2, Periods 1 and 3, Periods 2 and 3, and the three–way association. If the response were of $K = 3$ levels, then modeling the association between the repeated observation of this design would require $3 \cdot (3 - 1)^{2} + 1 \cdot (3 - 1)^{3} = 20$ parameters.

It should be apparent that increasing the number of observation periods or response levels will also increase the maximum number of association parameters. For this reason the examples provided in this text are not extended beyond $K = 3$ response levels.

A final example comes from the dysmenorrhea study presented by Kenward and Jones (1991). The effects of three treatments for the relief of dysmenorrhea was the main interest of this experiment. These treatments are: A = Placebo, B = Low Dose Analgesic, and C = High Doses Analgesic. Six sequences were constructed from all possible permutations of the treatment orderings over three periods. The response has three levels: 1 = No Relief, 2 = Moderate Relief, and 3 = Complete Relief. Let F_{ijkl} denote the observed frequency for the ith sequence with joint response jkl $(j,k,l = 1, 2, 3)$ for the three periods. The 27 observed frequencies for each sequence are presented in the following table.

Outcome	Sequence						Total
	ABC	ACB	BAC	BCA	CAB	CBA	
111	0	2	0	0	3	1	6
112	1	0	0	1	0	0	2
113	1	0	1	0	0	0	2
121	2	0	0	0	0	0	2
122	3	0	1	0	0	0	4
123	4	3	1	0	2	0	10
131	0	0	1	1	0	0	2
132	0	2	0	0	0	0	2
133	2	4	1	0	0	1	8
211	0	1	1	0	0	3	5
212	0	0	2	0	1	1	4
213	0	0	1	0	0	0	1
221	1	0	0	6	1	1	9
222	0	2	1	0	0	0	3
223	1	0	0	0	0	0	1
231	0	0	0	1	0	2	3
232	0	0	0	0	0	0	0
233	0	2	0	0	1	0	3
311	0	0	0	1	0	2	3
312	0	0	2	0	2	1	5
313	0	0	3	0	4	1	8
321	0	0	0	1	0	0	1
322	0	0	0	1	0	0	1
323	0	0	0	0	0	0	0
331	0	0	0	0	0	1	1
332	0	0	0	0	0	0	0
333	0	0	0	0	0	0	0
Total	15	16	15	12	14	14	86

Data from Kenward and Jones (1991), reprinted with permission of John Wiley and Sons, Ltd.

This design allows for the estimation of sequence, period, treatment and first order carryover. The 162 observed frequencies allow for the estimation of at most 156 parameters. Under the sum to zero convention, there will be $(3 - 1) \cdot [1 + (6 - 1) + (3 - 1) + (3 - 1) + (3 - 1)] = 24$ location parameters to be estimated. Each sequence will require $3 \cdot (3 - 1)^2 + 1 \cdot (3 - 1)^3 = 20$ parameters for association. Thus, the total number of parameters required is 144, leaving 12 residual degrees of freedom for this model. Some computer systems cannot handle a problem of this size, but there is an alternative. One can test whether the association parameters are the same across the six sequences. If so, a reduced model coding for the 24 location parameters and the 20 common association parameters can be produced, for a total of 44 parameters modeled.

The common association model can be assessed for appropriate fit by several means. A log–linear model (Bishop *et al.* 1975) can be used to assess equality of the associations across the sequences. Four factors are required for this analysis: SEQUENCE at 6 levels, and PERIOD_1, PERIOD_2 and PERIOD_3, each at 3 levels. Simply, we have 6 tables, each cross–classified by the three periods for a three level response. SAS PROC CATMOD can be used for this analysis. If the SAS data set "COMMON" contained the observed frequencies in the variable COUNT and the dependent classification variables SEQUENCE, PERIOD_1, PERIOD_2 and PERIOD_3, then the SAS code for assessing equality of the associations across sequences is as follows:

```
PROC CATMOD DATA = COMMON;
WEIGHT COUNT;
MODEL SEQUENCE*PERIOD_1*PERIOD_2*PERIOD_3 = _RESPONSE_ /
      ML NOGLS;
LOGLIN SEQUENCE   PERIOD_1    PERIOD_2    PERIOD_3
       SEQUENCE*PERIOD_1    SEQUENCE*PERIOD_2
       SEQUENCE*PERIOD_3    PERIOD_1*PERIOD_2
       PERIOD_1*PERIOD_3    PERIOD_2*PERIOD_3
       PERIOD_1*PERIOD_2*PERIOD_3;
RUN;
```

The likelihood ratio statistic produced by this analysis is $G^2 = 63.39$ on 100 degrees of freedom. This statistic produces a P–value of 0.9984, which indicates that the common association model fits the data very well. Based on this result, the logit model that assumes common associations will be used for the analysis of treatment effects.

An approximate test of the common association model will be obtained from the analysis of this logit model. Under this model there are 112 residual degrees of freedom; 100 of these correspond to the test for common association, and 12 are residual to the full logit model. If the likelihood ratio statistic for this model is not significant, then we would be fairly convinced that the common association model fits the data. This is, however, only an approximate test because both residual fit and common association are simultaneously tested.

Because of the large number of observable cells ($6 \cdot 3 \cdot 3 \cdot 3 = 162$), it is not feasible to present the entire set of design variables for all sequences. Instead, we will produce the design variables for the ABC sequence only. The values of these design variables for the other sequences can be produced by analogy.

If the model having a common association among sequences is used, there is no need to modify the observed frequencies with the addition of a small value. Unfortunately, SAS PROC CATMOD assumes that all sampling zeros in a data set are structural zeros (Agresti, 1990). Therefore, we will analyze the above data with one–ninth (1/9) added to each frequency.

For these data we will construct the design variables so that the baseline treatment will be the Placebo (A) and the baseline response level will be No Relief (1). The reason for choosing Treatment A as the baseline is that parameters for the treatment effect will contrast both Treatment B (Low Dose Analgesic) and C (High Dose Analgesic) against Treatment A. Thus, the parameters will reflect the potential improvement in relief for the two active treatments. Choosing the No Relief (1) response level as the baseline for forming the logits will allow for the comparison of the Moderate Relief (2) and Complete Relief (3) levels against the No Relief (1) level.

Let SEQ_1, SEQ_2, ..., SEQ_5 denote the design variables for the effects of Sequences 1 through 5 contrasted with the 6th Sequence; PERIOD_1 and PERIOD_2 denote the design variables for the effects of Periods 1 and 2 contrasted with the 3rd Period; TREAT_B and TREAT_C denote the design variables for the effect of Treatments B and C contrasted with Treatment A; and CARRY_B and CARRY_C denote the first order carryover effect of Treatments B and C contrasted with Treatment A. The design variables coding for the common associations are: ASS1_12 and ASS2_12 for the 12 association; ASS1_13 and ASS2_13 for the 13 association; ASS1_23 and ASS2_23 for the 23 association; and ASS1_123, ASS2_123, ASS3_123 and ASS4_123 for the 123 association. These design variables along with the response (Y) and the reorganized observed frequencies (COUNT) are presented, for the ABC sequence, in the table appearing on the following page.

Design matrix / effect table (column headers stacked vertically):

```
            P P                             A A A A
            E E T T C C A A A A A A         S S S S
            R R R A A R S S S S S S         S S S S
 S S S S S  I I E E R R S S S S S S         1 2 3 4        C
 E E E E E O O A A R R 1 2 1 2 1 2          ‾ ‾ ‾ ‾        O
 Q Q Q Q Q D D T T Y Y          ‾ ‾ ‾ ‾    1̄ 1̄ 1̄ 1̄       U
                    1̄ 1̄ 1̄ 1̄ 2̄ 2̄  2 2 2 2                 N
 1̄ 2̄ 3̄ 4̄ 5̄ 1̄ 2̄ B̄ C̄ B̄ C̄ 2  2  3  3  3  3  3  3  3  3  Y  T
 ────────────────────────────────────────────────────────────
  1 0 0 0 0  1  0 -1 -1  0  0  0  0  0  0  0  0  0  0  0  0  1   14.000 = F₁₁.. + 1
  1 0 0 0 0  1  0 -1 -1  0  0  0  0  0  0  0  0  0  0  0  0  2    3.000 = F₁₂.. + 1
  1 0 0 0 0  1  0 -1 -1  0  0  0  0  0  0  0  0  0  0  0  0  3    1.000 = F₁₃.. + 1
  1 0 0 0 0  0  1  1  0 -1 -1  1  1  0  0  0  0  0  0  0  0  1    2.333 = F₁₁₁. + 1/3
  1 0 0 0 0  0  1  1  0 -1 -1  1  1  0  0  0  0  0  0  0  0  2    9.333 = F₁₁₂. + 1/3
  1 0 0 0 0  0  1  1  0 -1 -1  1  1  0  0  0  0  0  0  0  0  3    2.333 = F₁₁₃. + 1/3
  1 0 0 0 0  0  1  1  0 -1 -1  0  1  0  0  0  0  0  0  0  0  1    0.333 = F₁₂₁. + 1/3
  1 0 0 0 0  0  1  1  0 -1 -1  0  1  0  0  0  0  0  0  0  0  2    2.333 = F₁₂₂. + 1/3
  1 0 0 0 0  0  1  1  0 -1 -1  0  1  0  0  0  0  0  0  0  0  3    0.333 = F₁₂₃. + 1/3
  1 0 0 0 0  0  1  1  0 -1 -1 -1 -1  0  0  0  0  0  0  0  0  1    0.333 = F₁₃₁. + 1/3
  1 0 0 0 0  0  1  1  0 -1 -1 -1 -1  0  0  0  0  0  0  0  0  2    0.333 = F₁₃₂. + 1/3
  1 0 0 0 0  0  1  1  0 -1 -1 -1 -1  0  0  0  0  0  0  0  0  3    0.333 = F₁₃₃. + 1/3
  1 0 0 0 0 -1 -1  0  1  1  0  0  0  1  0  1  0  1  0  0  0  1    0.111 = F₁₁₁₁ + 1/9
  1 0 0 0 0 -1 -1  0  1  1  0  0  0  1  0  1  0  1  0  0  0  2    0.111 = F₁₁₁₂ + 1/9
  1 0 0 0 0 -1 -1  0  1  1  0  0  0  1  0  1  0  1  0  0  0  3    2.111 = F₁₁₁₃ + 1/9
  1 0 0 0 0 -1 -1  0  1  1  0  0  0  1  0  0  1  0  1  0  0  1    3.111 = F₁₁₂₁ + 1/9
  1 0 0 0 0 -1 -1  0  1  1  0  0  0  1  0  0  1  0  1  0  0  2    4.111 = F₁₁₂₂ + 1/9
  1 0 0 0 0 -1 -1  0  1  1  0  0  0  1  0  0  1  0  1  0  0  3    0.111 = F₁₁₂₃ + 1/9
  1 0 0 0 0 -1 -1  0  1  1  0  0  0  1  0 -1 -1 -1 -1  0  0  1    0.111 = F₁₁₃₁ + 1/9
  1 0 0 0 0 -1 -1  0  1  1  0  0  0  1  0 -1 -1 -1 -1  0  0  2    2.111 = F₁₁₃₂ + 1/9
  1 0 0 0 0 -1 -1  0  1  1  0  0  0  1  0 -1 -1 -1 -1  0  0  3    0.111 = F₁₁₃₃ + 1/9
  1 0 0 0 0 -1 -1  0  1  1  0  0  0  0  1  1  0  0  0  1  0  1    0.111 = F₁₂₁₁ + 1/9
  1 0 0 0 0 -1 -1  0  1  1  0  0  0  0  1  1  0  0  0  1  0  2    0.111 = F₁₂₁₂ + 1/9
  1 0 0 0 0 -1 -1  0  1  1  0  0  0  0  1  1  0  0  0  1  0  3    1.111 = F₁₂₁₃ + 1/9
  1 0 0 0 0 -1 -1  0  1  1  0  0  0  0  1  0  1  0  0  0  1  1    0.111 = F₁₂₂₁ + 1/9
  1 0 0 0 0 -1 -1  0  1  1  0  0  0  0  1  0  1  0  0  0  1  2    1.111 = F₁₂₂₂ + 1/9
  1 0 0 0 0 -1 -1  0  1  1  0  0  0  0  1  0  1  0  0  0  1  3    0.111 = F₁₂₂₃ + 1/9
  1 0 0 0 0 -1 -1  0  1  1  0  0  0  0  1 -1 -1 -1 -1  0  0  1    0.111 = F₁₂₃₁ + 1/9
  1 0 0 0 0 -1 -1  0  1  1  0  0  0  0  1 -1 -1 -1 -1  0  0  2    0.111 = F₁₂₃₂ + 1/9
  1 0 0 0 0 -1 -1  0  1  1  0  0  0  0  1 -1 -1 -1 -1  0  0  3    0.111 = F₁₂₃₃ + 1/9
```

(continued on following page)

```
1 0 0 0  0 -1 -1  0  1  1  0  0  0  1  0 -1 -1 -1  0 -1  0  1  0.111=F₁₃₁₁+1/9
1 0 0 0  0 -1 -1  0  1  1  0  0  0  1  0 -1 -1 -1  0 -1  0  2  0.111=F₁₃₁₂+1/9
1 0 0 0  0 -1 -1  0  1  1  0  0  0  1  0 -1 -1 -1  0 -1  0  3  0.111=F₁₃₁₃+1/9
1 0 0 0  0 -1 -1  0  1  1  0  0  0  0  1  1  0  0 -1  0 -1  1  0.111=F₁₃₂₁+1/9
1 0 0 0  0 -1 -1  0  1  1  0  0  0  0  1  1  0  0 -1  0 -1  2  0.111=F₁₃₂₂+1/9
1 0 0 0  0 -1 -1  0  1  1  0  0  0  0  1  1  0  0 -1  0 -1  3  0.111=F₁₃₂₃+1/9
1 0 0 0  0 -1 -1  0  1  1  0  0  0  0  1  1  0  1  1  1  1  1  0.111=F₁₃₃₁+1/9
1 0 0 0  0 -1 -1  0  1  1  0  0  0  0  1  1  0  1  1  1  1  2  0.111=F₁₃₃₂+1/9
1 0 0 0  0 -1 -1  0  1  1  0  0  0  0  1  1  0  1  1  1  1  3  0.111=F₁₃₃₃+1/9
```

For the above table, the first period marginal frequency for the ith sequence and jth response level are denoted $F_{ij.}$ and are defined as the sum of all observed frequencies for sequence i $(i = 1,...,6)$ having level j $(j = 1, 2, 3)$ in the first period. Similarly, $F_{ijk.}$ will denote the marginal frequency for the ith sequence having the jkth $(j, k = 1, 2, 3)$ joint response for the first two periods of observation.

SAS PROC CATMOD will use the response level with the highest numerical value as the baseline. In order that the first response level is used as the baseline, the data set given above must be ordered in descending order with respect to the response variable Y. Then the ORDER = DATA option in CATMOD can be used to force response level 1 as the baseline.

If these variables were defined, as partly shown in the above table, for the SAS data set "DYSMEN3," then the SAS code for the maximum likelihood analysis of the observed data is as follows:

```
PROC SORT DATA = DYSMEN3; BY DESCENDING Y;
PROC CATMOD DATA = DYSMEN3 ORDER = DATA;
WEIGHT COUNT;
DIRECT  SEQ_1  SEQ_2  SEQ_3  SEQ_4  SEQ_5  PERIOD_1  PERIOD_2
        TREAT_B  TREAT_C  CARRY_B  CARRY_C  ASS1_12  ASS2_12
        ASS1_13  ASS2_13  ASS1_23  ASS2_23  ASS1_123  ASS2_123  ASS3_123
        ASS4_123;
MODEL Y = SEQ_1  SEQ_2  SEQ_3  SEQ_4  SEQ_5  PERIOD_1  PERIOD_2
        TREAT_B  TREAT_C  CARRY_B  CARRY_C  ASS1_12  ASS2_12
        ASS1_13  ASS2_13  ASS1_23  ASS2_23  ASS1_123  ASS2_123  ASS3_123
        ASS4_123 / NOGLS ML;
RUN;
```

For the above example, the SAS output will look as follows (in part):

MAXIMUM LIKELIHOOD ANALYSIS OF VARIANCE TABLE

Source	DF	Chi–Square	Prob
INTERCEPT	2	9.00	0.0111
SEQ_1	2	3.21	0.2007
SEQ_2	2	0.97	0.6146
SEQ_3	2	1.14	0.5662
SEQ_4	2	4.44	0.1087
SEQ_5	2	2.65	0.2652
PERIOD_1	2	1.24	0.5380
PERIOD_2	2	0.13	0.9349
TREAT_B	2	6.14	0.0465
TREAT_C	2	14.95	0.0006
CARRY_B	2	1.40	0.4966
CARRY_C	2	1.03	0.5974
ASS1_12	2	6.04	0.0487
ASS2_12	2	1.75	0.4162
ASS1_13	2	0.64	0.7262
ASS2_13	2	1.85	0.3974
ASS1_23	2	0.98	0.6120
ASS2_23	2	0.47	0.7925
ASS1_123	2	4.65	0.0977
ASS2_123	2	0.87	0.6460
ASS3_123	2	1.25	0.5342
ASS4_123	2	0.73	0.6948
LIKELIHOOD RATIO	112	76.02	0.9963

ANALYSIS OF MAXIMUM LIKELIHOOD ESTIMATES

Effect	Parameter	Estimate	Standard Error	Chi–Square	Prob
INTERCEPT	1	−0.7524	0.2541	8.77	0.0031
	2	−0.2753	0.2061	1.78	0.3178
SEQ_1	3	−0.8438	0.5252	2.58	0.1081
	4	0.0020	0.3951	0.00	0.9959
SEQ_2	5	0.4495	0.4557	0.97	0.3239
	6	0.1876	0.3866	0.24	0.6274
SEQ_3	7	0.3357	0.4157	0.65	0.4193
	8	−0.0621	0.3780	0.03	0.8695
SEQ_4	9	−0.2641	0.5444	0.24	0.6276
	10	0.6528	0.4157	2.47	0.1163
SEQ_5	11	0.3399	0.4344	0.61	0.4339
	12	−0.3516	0.4151	0.72	0.3969

(continued on following page)

PERIOD_1	13	−0.2153	0.3032	0.50	0.4776
	14	0.1367	0.2533	0.29	0.5895
PERIOD_2	15	−0.1121	0.3219	0.12	0.7277
	16	−0.0026	0.2779	0.00	0.9926
TREAT_B	17	0.8064	0.3311	5.93	0.0149
	18	0.3900	0.2688	2.10	0.1469
TREAT_C	19	1.1694	0.3048	14.72	0.0001
	20	0.6533	0.2751	5.64	0.0176
CARRY_B	21	0.5581	0.4719	1.40	0.2369
	22	0.1791	0.3767	0.23	0.6345
CARRY_C	23	−0.3105	0.4475	0.48	0.4877
	24	−0.3733	0.3965	0.89	0.3465
ASS1_12	25	1.1913	0.5054	5.56	0.0184
	26	0.7335	0.4399	2.78	0.0954
ASS2_12	27	0.2019	0.4858	0.17	0.6777
	28	0.5478	0.4174	1.72	0.1895
ASS1_13	29	0.4398	0.5775	0.58	0.4463
	30	0.0390	0.5728	0.00	0.9458
ASS2_13	31	−0.7020	0.5903	1.41	0.2344
	32	−0.6109	0.6486	0.89	0.3462
ASS1_23	33	−0.3720	0.5848	0.40	0.5247
	34	0.2696	0.5399	0.25	0.6175
ASS2_23	35	−0.1186	0.7497	0.03	0.8743
	36	0.3343	0.5851	0.33	0.5678
ASS1_123	37	−1.2786	0.6871	3.46	0.0628
	38	−1.1206	0.6584	2.90	0.0888
ASS2_123	39	0.7440	0.8381	0.79	0.3747
	40	0.0760	0.7237	0.01	0.9164
ASS3_123	41	−0.2274	0.7595	0.09	0.7646
	42	0.6952	0.7405	0.88	0.3478
ASS4_123	43	−0.7475	0.8783	0.72	0.3947
	44	−0.2221	0.7594	0.09	0.7699

The "Maximum likelihood analysis of variance table" provides a likelihood ratio statistic for the common association plus residual model. This statistic has a computed value of $G^2 = 76.02$ on 112 degrees of freedom, producing a P–value of 0.9963. The residual 12 degrees of freedom may adversely affect this statistic; therefore, it can only be considered an approximate test for equal association across sequences. However, the large P–value indicates that the equal association model does fit the data very well, just as the earlier log–linear analysis indicated.

An overall test for first order carryover can be obtained by removing both CARRY_B and CARRY_C from the DIRECT and MODEL statements of the SAS code. If this is done, a likelihood ratio statistic of $G^2 = 78.53$ is produced on 116 degrees of freedom. The conditional likelihood ratio statistic for testing first order carryover is $G^2 = 78.53 - 76.02 = 2.51$ on $116 - 112 = 4$ degrees of freedom. For this statistic a P-value of 0.6428 is produced, implying that there is insufficient evidence of first order carryover. Thus, carryover should not substantially influence the results of any subsequent tests.

An overall test for differences among treatments can be obtained by removing both TREAT_B and TREAT_C from the DIRECT and MODEL statements of the SAS code. If this is done, a likelihood ratio statistic of $G^2 = 115.77$ is produced on 116 degrees of freedom. The conditional likelihood ratio statistic for treatment effects is $G^2 = 115.77 - 76.02 = 39.75$ on $116 - 112 = 4$ degrees of freedom. For this statistic a P-value of 0.0000 is produced, implying a substantial difference among treatments.

Based on the common association model, four parameters are estimated for treatment effects: two for the contrast between the Low Dose Analgesic (B) and Placebo (A), and two for the contrast between the High Dose (C) and Placebo (A). These estimates and their standard errors are, respectively,

Parameter	Standard Error	Wald Chi-Square	P-value
$\tau(AB)_{31} = 0.8064$	0.3311	5.93	0.0149
$\tau(AB)_{21} = 0.3900$	0.2688	2.10	0.1469
$\tau(AC)_{31} = 1.1694$	0.3048	14.72	0.0001
$\tau(AC)_{21} = 0.6533$	0.2751	5.64	0.0176

The subscript 31 corresponds to the contrast of the Complete Relief response level (3) to the baseline or No Relief response level (1). The 21 corresponds to the contrast of the Moderate Relief response level (3) to the baseline or No Relief response level (1). The P-value associated with each estimate provides a single degree of freedom test for comparing the active treatments (B and C) to the Placebo (A), while contrasting the baseline (No Relief) response level to the response levels showing some relief (Moderate and Complete Relief). Only the comparison between the Placebo and Low Dose Analgesic, for the contrast between the baseline (No Relief) and the Moderate Relief response levels, provides no evidence of significance (P-value = 0.1469).

These estimates tend to indicate a trend in both the treatments and response levels. The High Dose treatment produces the greatest probability of response level 3 (Complete Relief). At the other end is the Placebo, which produces the greatest probability of response level 1 (No Relief).

Again, these parameters can be interpreted in terms of odds ratios. The odds of observing Complete Relief (level 3) over No Relief (level 1) for the Low Dose Analgesic treatment (B) is $\exp[0.3900+2\cdot(0.6533)] = 5.455$ times greater than for the Placebo (A). The odds of observing Moderate Relief (level 2) over No Relief (level 1) for the Low Dose Analgesic treatment (B) is $\exp[2\cdot(0.3900)+0.6533] = 4.193$ times greater than for the Placebo (A). The odds of observing Complete Relief (level 3) over No Relief (level 1) for the High Dose Analgesic treatment (C) is $\exp[0.8064+2\cdot(1.1694)] = 23.224$ times greater than for the Placebo (A). The odds of observing Moderate Relief (level 2) over No Relief (level 1) for the High Dose Analgesic treatment (C) is $\exp[2\cdot(0.8064)+1.1694] = 16.155$ times greater than for the Placebo (A). Again, the trend in these odds ratios are exactly as one would expect for an ordinal response and ordinal treatment structure.

7.7. Recommendations about the Techniques and Designs in This Chapter

This chapter outlined a procedure for modeling data from cross–over designs when the response is qualitative. The emphasis has been on the most general procedure. With this in mind, the logit model presented in this chapter is both of a general nature and quite powerful. Unfortunately, this method requires a substantial knowledge before it can be used properly.

For the simplest cross–over design, either Gart's test or Prescott's test can be used. These methods are simple and quite powerful. But beyond this simple design one must employ a method of analysis that is considerably more complex.

One can reduce the complexity of the logit models by limiting the response to the binary case. For example, the full model for the 3–treatment, 3–period, 6–sequence design having 3 response levels requires 120 degrees of freedom for association. If the response were binary, then only 24 degrees of freedom are required to model the association component.

Much as with the number of response levels, the number of observation periods will also greatly affect the complexity of the model structure. Thus, one should carefully weigh how many periods of observation are required to conclude the experiment. One should realize that even a small problem, say 4 periods and 6 sequences with a 4 level response, will require 1296 parameters to fully estimate the association component. This will likely out-strip the limits of nearly all available software.

One area of analysis for cross-over designs that was not covered in this text is the analysis for an ordinal response. The methods described in this chapter can be used for an ordinal response, but the levels for this type of response are treated as nominal. However, a program such as SAS PROC LOGISTIC© that utilizes cumulative logits (Agresti, 1990) can be used for the analysis of a ordinal response data. No change in the design variables over the nominal analysis is necessary. The ordinal response has increased power over the equivalent nominal response because of the narrower parameter space. Each design variable will produce one parameter estimate, whereas the nominal analysis will produce K - 1 parameter estimates per design variable.

Exercises

7.1. In this chapter, the data of Ebbutt (1984) for the ABB and BAA sequences were presented as binary observations (see Section 7.5). Using the first two periods of observation, a realization for the cross-over design having 2-treatments, 2-periods, and 2-sequences is produced. Collapsing these data over the third period produces the tables:

AB Sequence

Period 2

		1	2
	1	10	3
Period 1	2	1	8

and

BA Sequence

Period 2

	1	2
Period 1 1	8	1
2	6	12

(a) Use the logit model of this chapter to test for equality of treatments.

(b) Compare this result with those of Gart's test and Prescott's test.

7.2. The first two periods of the ACB and CAB sequences of the Kenward and Jones (1991) data having three response levels are presented below (reproduced with permission of John Wiley and Sons, Ltd).

AC Sequence

Period 2

Period 1	1	2	3
1	2	3	6
2	1	2	2
3	0	0	0

CA Sequence

Period 2

Period 1	1	2	3
1	3	2	0
2	1	1	1
3	6	0	0

(a) Compute the maximum likelihood estimates.

(b) Produce the odds ratios using the first response level (No Relief) as the baseline. Do these values make sense when compared with the same analysis for the AB and BA sequences?

(c) Using the logit model, test whether the treatments are equal.

7.3. By collapsing the Moderate and Complete Relief response levels of the Kenward and Jones (1991) data, a realization of the 3–treatment, 3–period, 6–sequence design with a binary response is produced. The data from this experiment are presented in the following table (reproduced with permission of John Wiley and Sons, Ltd):

Outcome	ABC	ACB	Sequence BAC	BCA	CAB	CBA	Total
111	0	2	0	0	3	1	6
112	2	0	1	1	0	0	4
121	2	0	1	1	0	0	4
122	9	9	3	0	2	1	24
211	0	1	1	1	0	5	8
212	0	0	8	0	7	3	18
221	1	0	0	8	1	4	14
222	1	4	1	1	1	0	8
Total	15	16	15	12	14	14	86

(a) Compute the maximum likelihood estimates for these data. Include in the model parameters for sequences, treatments, periods and first order carryover. Also include the 24 parameters coding for association. The design variables should be formed so as to contrast the active treatments (B and C) with the Placebo (A). The No Relief response level (1) should be used as the baseline for the logits.

(b) Test for a significant treatment effect. Based on the parameter estimates, determine the relationships among the treatments.

(c) Compute the odds ratios for this experiment and interpret their meaning.

Appendix 7.A. Methods for Computing Maximum Likelihood Estimates

Maximum likelihood estimates for the logit model parameter vector can be computed using a modified Newton–Raphson algorithm. The algorithm has the following form:

$$\phi_{n+1} = \phi_n - P_n^{-1} G_n$$

where ϕ_n is the nth iterated estimate of the parameter vector ϕ. The terms G_n and P_n represent the first and second partial derivatives of the multinomial likelihood function evaluated at ϕ_n. For the cross–over design having I sequences, the vector of first partial derivatives and the matrix of second partial derivatives are, respectively,

$$G_n = \sum_{i=1}^{I} V_i' W_{i_n}' Y_{i_n}' X' (F_i - \mu_{i_n})$$

$$P_n = -\sum_{i=1}^{I} V_i' W_{i_n}' Y_{i_n}' X' D_{\mu_{in}} X Y_{i_n} W_{i_n} V_i .$$

The matrices $\mathbf{X}$ and $\mathbf{V_i}$ ($i = 1, 2,..., I$) have already been defined for log–linear and logit models, respectively. The vector μ_{1_n} is the nth iterated estimate of $\mu_i = \exp(\mathbf{X}\beta_i)$ for the ith sequence. $\mathbf{F_i}$ is the vector of observable frequencies for the ith sequence. The matrix $\mathbf{D}_{\mu_{1n}}$ is a diagonal matrix with μ_{1n} down the main diagonal. The matrices $\mathbf{Y_{i_n}}$ and $\mathbf{W_{i_n}}$ have more complicated structures.

The matrix $\mathbf{Y_{i_n}}$ is block diagonal with k x k-1 blocks, $\mathbf{Y_{ij_n}}$, down the diagonal, except that the first row of $\mathbf{Y_{i_n}}$ has all zero elements. To visualize this, we define $\mathbf{Y_{i_n}}$ as follows:

$$
Y_{i_n} =
\begin{bmatrix}
0 & 0 & 0 & . & . & . & 0 \\
Y_{i1_n} & 0 & 0 & . & . & . & 0 \\
0 & Y_{i2_n} & 0 & . & . & . & 0 \\
. & . & . & . & . & . & . \\
. & . & . & . & . & . & . \\
. & . & . & . & . & . & . \\
0 & 0 & 0 & . & . & . & Y_{iJ_n}
\end{bmatrix} ,
$$

where the block matrices Y_{ij_n} have the following structure,

$$
Y_{ij_n} =
\begin{bmatrix}
\dfrac{1}{P_{ij1_n}} & 0 & 0 & 0 & . & . & . & 0 \\
0 & \dfrac{1}{P_{ij2_n}} & 0 & 0 & . & . & . & 0 \\
0 & 0 & \dfrac{1}{P_{ij3_n}} & 0 & . & . & . & 0 \\
. & . & . & & & & . & 0 \\
. & . & . & & & & . & 0 \\
0 & 0 & 0 & 0 & . & . & . & \dfrac{1}{P_{ij(k-1)_n}} \\
\dfrac{-1}{P_{ijk_n}} & \dfrac{-1}{P_{ijk_n}} & \dfrac{-1}{P_{ijk_n}} & \dfrac{-1}{P_{ijk_n}} & . & . & . & \dfrac{-1}{P_{ijk_n}}
\end{bmatrix}
$$

For this matrix, we let P_{ijk_n} denote the nth iterated estimate of a probability P_{ijk} having definition j at the kth response level in the ith sequence.

The matrix W_{i_n} is also block diagonal in nature and can be represented as:

$$
\mathbf{W}_{i_n} =
\begin{bmatrix}
\mathbf{W}_{i1_n} & 0 & 0 & \cdots & 0 \\
0 & \mathbf{W}_{i2_n} & 0 & \cdots & 0 \\
0 & 0 & \mathbf{W}_{i3_n} & \cdots & 0 \\
\cdot & \cdot & \cdot & \cdot \cdot & \cdot \\
\cdot & \cdot & \cdot & \cdot \cdot \cdot & \cdot \\
\cdot & \cdot & \cdot & \cdot \cdot \cdot & \cdot \\
0 & 0 & 0 & \cdots & \mathbf{W}_{iJ_n}
\end{bmatrix} ,
$$

where the block matrices $\mathbf{W}_{ij_n}$ have the following structure:

$$
\mathbf{W}_{i_n} =
\begin{bmatrix}
P_{ij1_n}(1 - P_{ij1_n}) & -P_{ij1_n}P_{ij2_n} & \cdots & -P_{ij1_n}P_{ijK_n} \\
-P_{ij2_n}P_{ij1_n} & P_{ij2_n}(1 - P_{ij2_n}) & \cdots & -P_{ij2_n}P_{ijK_n} \\
\cdot & \cdot & \cdot \cdot & \cdot \\
\cdot & \cdot & \cdot \cdot \cdot & \cdot \\
\cdot & \cdot & \cdot \cdot \cdot & \cdot \\
-P_{ijK_n}P_{ij1_n} & -P_{ijK_n}P_{ij2_n} & \cdots & P_{ijK_n}(1 - P_{ijK_n})
\end{bmatrix} .
$$

As an example, consider the AB and BA sequences of a 2-treatment, 2-period, 2-sequence design with a binary response to be the $i = 1$ and $i = 2$ sequence, respectively. The probabilities from the $i = 1$, or AB sequence, can be denoted by $P_{111_n} = P_A$, $P_{112_n} = (1 - P_A)$, $P_{121_n} = P_{B|A}$, $P_{122_n} = (1 - P_{B|A})$, $P_{131_n} = P_{B|\overline{A}}$ and $P_{132_n} = (1 - P_{B|\overline{A}})$. Therefore, the matrix $\mathbf{Y}_{1_n}$ for this example is given by the following:

$$Y_{1_n} = \begin{bmatrix} 0 & 0 & 0 \\ P_A & 0 & 0 \\ \frac{-1}{(1-P_A)} & 0 & 0 \\ 0 & P_{B|A} & 0 \\ 0 & \frac{-1}{(1-P_{B|A})} & 0 \\ 0 & 0 & P_{B|\overline{A}} \\ 0 & 0 & \frac{-1}{(1-P_{B|\overline{A}})} \end{bmatrix},$$

and the matrix $\mathbf{W}_{1_n}$ is

$$\begin{bmatrix} P_A(1-P_A) & -P_A(1-P_A) & 0 & 0 & 0 & 0 \\ -(1-P_B)P_A & (1-P_A)P_A & 0 & 0 & 0 & 0 \\ 0 & 0 & P_{B|A}(1-P_{B|A}) & -P_{B|A}(1-P_{B|A}) & 0 & 0 \\ 0 & 0 & -(1-P_{B|A})P_{B|A} & (1-P_{B|A})P_{B|A} & 0 & 0 \\ 0 & 0 & 0 & 0 & P_{B|\overline{A}}(1-P_{B|\overline{A}}) & -P_{B|\overline{A}}(1-P_{B|\overline{A}}) \\ 0 & 0 & 0 & 0 & -(1-P_{B|\overline{A}})P_{B|\overline{A}} & (1-P_{B|\overline{A}})P_{B|\overline{A}} \end{bmatrix}.$$

Appendix 7.B. Covariance Matrix Estimation

The nonlinear nature of the estimation problem for categorical data precludes an analytic solution to the small sample covariance matrix for $\hat{\phi}$. It is, however, possible to develop a large sample estimator of this covariance matrix, $\text{Cov}(\hat{\phi})$. We will not give a theoretical development for this estimator, but will only present its form. Anyone desiring a more substantive discussion is referred to Bishop *et al.* (1975) and Agresti (1990). The asymptotic covariance matrix for $\hat{\phi}$ is given by

$$\text{Cov}(\hat{\phi}) = P^{-1}$$

where P^{-1} is the value of P_n^{-1} from the final iteration of the Newton–Raphson algorithm.

Estimated variances for a particular estimated logit parameter can be obtained from the appropriate diagonal element of $\text{Cov}(\hat{\phi})$. Estimated standard errors are obtained by taking the square root of the appropriate variance.

8

Ordinary Least Squares Estimation Versus Other Criteria of Estimation: Justification for Using the Methodology Presented in This Book

The methodology associated with the statistical analysis of cross-over designs presented in Section 1.1 rests wholly upon the assumption that least squares estimation, also called ordinary least squares and abbreviated OLS, is valid. In this chapter we examine the assumptions that have to be made to justify the use of this widely used criterion of estimation in regression analysis and the analysis of variance of designed experiments. Cross-over designs differ from many other designs in use in applied science by virtue of the fact that the experimental units (the subjects) are used more than once, thus making it unlikely that the measurements at different time periods for the same subject are independent of each other. It is the form in which the dependence occurs which determines whether or not the OLS estimator will be appropriate. Thus, we examine several possible error structures and see how closely actual data sets meet the assumptions needed to justify the use of OLS.

8.1. The Assumptions Underlying OLS

It is sometimes stated that the use of ordinary least squares rests upon the assumption of independent errors. Although the independent error case is a sufficient condition for the validity of OLS, provided of course that these errors are also identically normally distributed, it is not a necessary condition. It is also sometimes stated that a more general condition of correlated errors, known as "uniform covariance" or "compound symmetry", is required, but this too is a sufficient, not a necessary, condition to justify OLS. The most general condition, which is both sufficient and necessary as a justification for the use of OLS estimation, is that the errors obey a covariance structure called the Type H structure studied by Huynh and Feldt (1970). In this section, we will describe these various assumptions, and in Section 8.2 test them using real data sets.

Consider a cross–over design that has p periods, T treatments, s sequences, and n experimental units that are randomly assigned to the s sequences. Power considerations dictate that each sequence be assigned the same number of units, but this is not always possible, and deaths and dropouts during the trial may result in unequal numbers in the sequences. The statistical model for this design is the same as used in Chapter 2, when there is more than one experimental unit per sequence,

$$Y_{ijk} = \mu + \gamma_i + \pi_j + \tau_t + \delta_j \alpha_r + \xi_{i(k)} + \epsilon_{ijk} \tag{8.1}$$

where Y_{ijk} is the observed response in sequence i and period j on sampling unit k, μ is an overall mean, γ_i is the effect of sequence i, $i=1,2,...,s$, $\xi_{i(k)}$ is a random effect due to sampling unit k, $k=1,2,...,n_i$, n_i being the number of subjects in sequence i, π_j is the effect of period j, $j=1,2,...,p$, τ_t is the effect of treatment t, $t=1,2,...,T$, α_r is the carryover effect due to treatment r having been applied in the preceding period, δ_j is an indicator variable whose value is zero in the first period and unity in subsequent periods, and ϵ_{ijk} represents random experimental error corresponding to sequence i, period j, and sampling unit k. This model is often referred to as a "mixed model", because it contains a second random component (in addition to the experimental error ϵ_{ijk}), which represents the variation due to the kth sampling unit. It is also referred to as a "split–plot" model, since the time periods at which the measurements are made can be considered to be "sub–plots" nested under the "main plots", the individual subjects.

Since n_i is the number of subjects in sequence i, the total number of subjects n in the experiment is given by

$$n = \sum_{i=1}^{s} n_i .$$

Measurements on individual subjects are assumed to be uncorrelated with measurements on other subjects, but the measurments over the p time periods on any one subject are correlated. Making the testable assumption that each subject, irrespective of the sequence that that subject is in, has the same underlying covariance structure, the covariance matrix for the np responses may be written as follows:

$$\begin{pmatrix} \Sigma & 0 & \cdots & 0 \\ 0 & \Sigma & \cdots & 0 \\ \vdots & \vdots & \cdots & \vdots \\ 0 & 0 & \cdots & \Sigma \end{pmatrix},$$

where Σ is the p x p covariance matrix for the observations on each individual. In this chapter, we examine the consequences of different assumed or observed properties for Σ.

Following Chinchilli and Elswick (1989), we write Equation (8.1) as

$$Y_{mj} = \mu + \gamma_i + \pi_j + \tau_t + \delta_j \alpha_r + \xi_m + \epsilon_{mj} \tag{8.2}$$

where the index m has combined the sequence subscript i and the experimental unit k of the ith sequence. Thus, m ranges from 1 to n. Here, ξ_m is a random effect for the ith subject, and $\xi_1, \xi_2, ..., \xi_n$ are independently and identically distributed with a normal distribution with mean zero and variance ν^2, commonly denoted $N(0, \nu^2)$, whereas ϵ_{mj} is a random effect representing the measurement error during the jth period on the mth subject, and is assumed to be independent and identically distributed according to $N(0, \sigma^2)$. The fixed effects of the model, that is, the grand mean, sequence, period, treatment and carryover effects, can now be combined into a parameter vector β of length r, and Equation (8.2) can be written as

$$Y_{mj} = X'_{mj} \beta + \xi_m + \epsilon_{mj}. \tag{8.3}$$

where X_{mj} represents a vector of length r containing the design coefficients for the fixed effects. The expectation of this mixed model contains only the fixed effects, viz.

$$E(Y_{mj}) = X'_{mj} \beta,$$

since the terms involving the random components are zero. The covariance matrix of the response in Equation (8.3) has no contribution from the fixed effect components, but has a contribution from each of the random effects resulting in the following:

$$\text{Cov}(Y_{mj}, Y_{lk}) = \begin{cases} \nu^2 + \sigma^2, & \text{if } m=l \text{ and } j=k \\ \nu^2, & \text{if } m=l \text{ and } j \neq k \\ 0, & \text{if } m \neq l . \end{cases}$$

The above conditions tell us that the covariance is zero if we are dealing with different subjects, but if it is the same subject in question, there is an error component ν^2 which each subject contributes in each period, plus a component σ^2 of measurement error if the period is the same. The above type of covariance structure is known variously as compound symmetry or uniform covariance. The symmetry or uniformity is apparent if we write down the variance–covariance matrix of the p responses for each individual, denoted Σ:

$$\Sigma = \sigma^2 I_p + \nu^2 J_{pxp} = \begin{pmatrix} \sigma^2 + \nu^2 & \nu^2 & \cdots & \nu^2 \\ \nu^2 & \sigma^2 + \nu^2 & \cdots & \nu^2 \\ \vdots & \vdots & \ddots & \vdots \\ \nu^2 & \nu^2 & \cdots & \sigma^2 + \nu^2 \end{pmatrix} .$$

Here, the matrix I_p is a pxp identity matrix, and J_{pxp} is a pxp matrix of ones. The uniformity is apparent since all the diagonal elements (the variances) are the same, and all the off–diagonal elements (the covariances) are the same. The special case of independent errors occurs when ν^2 is zero, resulting in a matrix that can be written as $\sigma^2 I_p$.

A more general covariance structure that is both necessary and sufficient for allowing OLS estimation is the Type H structure of Huynh and Feldt (1970) which can be written, if j_p is a p-vector of ones, and $\gamma = (\gamma_1, \gamma_2, ..., \gamma_p)'$ a vector of parameters, as

$$\Sigma = \lambda I_p + \eta j_p' + j_p \gamma' = \begin{pmatrix} \lambda + 2\gamma_1 & \gamma_1 + \gamma_2 & \cdots & \gamma_1 + \gamma_p \\ \gamma_1 + \gamma_2 & \lambda + 2\gamma_2 & \cdots & \gamma_2 + \gamma_p \\ \cdots & \cdots & \cdots & \cdots \\ \gamma_1 + \gamma_p & \gamma_2 + \gamma_p & \cdots & \lambda + 2\gamma_p \end{pmatrix}.$$

The model underlying the Type H structure is that the differences between observations on the same subject at different time periods are equally variable. That is, the condition that the variances of all possible differences $Y_{mj} - Y_{mk}$, where $j \neq k$, are equal leads directly to the above Type H structure. It is readily observed that if $\gamma_1 = \gamma_2 = \cdots = \gamma_p$, the Type H structure reduces to the compound symmetry condition, and that if $\gamma_i = 0$ for all i, the independent error case is obtained.

8.2. Examining the Various Covariance Structures

We will now use two sets of data, previously considered in this book, to obtain some evidence about what form or forms of covariance structure are actually observed in real data sets from cross-over designs. The first data set is the one from Patel (1983), given in Table 3.2. In that data set, there are two sequences, with 8 subjects in the first sequence and 9 subjects in the second sequence. There are four periods, the first being a run-in period, then a treatment period, then a wash-out period, and finally a second treatment period. In Sequence 1, the two treatments were given in the order AB, and in Sequence 2, the order was BA.

It will also be assumed that the variances and covariances at corresponding time periods are the same in each sequence, so that they can be pooled. A further assumption is that the data are normally distributed, so that corrected sums of squares become distributed as multiples of the chi-square distribution, enabling significance testing to be carried out. The fact that the different time periods have different treatments and carryover effects is of no consequence, as these are fixed effects and variances and covariances are invariant to location changes in the data.

Below are presented the corrected sums of squares and cross-products for the data in Table 3.2.

Period

	1	2	3	4
Sequence 1:	1.65915	1.82715	1.51483	1.93260
	1.82715	2.28855	1.80758	2.57600
	1.51483	1.80758	1.91209	2.54960
	1.93260	2.57600	2.54960	3.74180

Period

	1	2	3	4
Sequence 2:	4.22182	3.03664	3.18277	3.12032
	3.03664	4.22568	2.75223	2.85064
	3.18277	2.75223	3.55340	3.66227
	3.12032	2.85064	3.66227	4.17362

The entries for Sequence 2 appear to be somewhat higher than those for Sequence 1, even allowing for the fact that there are 9 subjects in Sequence 2 compared with 8 subjects in Sequence 1. A formal test of whether two or more covariance matrices are equal is presented in Morrison (1976, Section 7.4). Under the null hypothesis that k covariance matrices, each having p periods, are equal, one calculates a test statistic

$$M = [\sum_{i=1}^{k} (n_i-1)] \ln |S| - \sum_{i=1}^{k} (n_i-1) \ln |S_i|,$$

where S_i is the ith sample covariance matrix, S is the pooled sample covariance matrix, and n_i is the sample size of the ith matrix. The pooled matrix S is a weighted average of the individual covariance matrices obtained from the following formula:

$$S = \sum_{i=1}^{k} (n_i-1) S_i / \sum_{i=1}^{k} (n_i-1).$$

Box (1949) introduced a scale factor

$$1/C = 1 - \frac{2p^2 + 3p - 1}{6(p + 1)(k - 1)} [\sum_{i=1}^{k} 1/(n_i-1) - 1/\sum_{i=1}^{k} (n_i-1)]$$

such that M/C is approximately distributed as chi–square with $(k-1)p(p+1)/2$ degrees of freedom. The chi–square approximation is reasonably good for k and p not in excess of 4 to 5 and n_i of 20 or more. If we try the test on the data of Patel (1983), we note that the n_i are much smaller ($n_1 = 9$ and $n_2 = 8$) than the above requirement, so the test may lack power. Dividing the Sequence 1 SSCP matrix by $n_1-1 = 8$ results in a covariance matrix whose determinant is $|S_1| = 0.0000042851$, and dividing the Sequence 2 SSCP matrix by $n_2-1 = 7$ results in $|S_2| = 0.0014194639$, a rather higher figure. The pooled matrix S has a determinant of 0.0003287876, and

$$M = (8 + 7)\log_e(0.0003287876)-8\log_e(0.0000042851)-7\log_e(0.0014194639)$$
$$= 24.4837.$$

$$1/C = 1 - 43(0.267857-0.066667)/30 = 0.711627.$$

$M/C = 17.42$ is approximately distributed as chi–square with $1(4)(4+1)/2 = 10$ degrees of freedom. The P–value corresponding to this statistic is 0.0656, so there is marginal evidence of a difference between covariance matrices, but hardly enough to invalidate procedures that are based on a pooled matrix.

The pooled covariance matrix for these data, with 15 (= 9 + 8 − 2) degrees of freedom, is as follows:

$$\hat{\Sigma} = \begin{pmatrix} 0.3921 & 0.3243 & 0.3132 & 0.3369 \\ 0.3243 & 0.4343 & 0.3040 & 0.3618 \\ 0.3132 & 0.3040 & 0.3644 & 0.4141 \\ 0.3369 & 0.3618 & 0.4141 & 0.5277 \end{pmatrix}.$$

It is clear that the above pooled covariance matrix is far from being consistent with the special case of independent error. The correlations between measurements in the different time periods are rather high, reflecting the fact that subjects who have high scores in one period tend to also have high scores in the other periods, and that subjects having low scores in one period tend to have low scores in other periods as well. Since the Type H structure is necessary and sufficient to justify use of the mixed model approach of Chapter 1, we will now try to assess whether the above matrix is consistent with that structure. One could begin by testing whether the above matrix satisfies the conditions of uniform covariance, on the grounds that that structure is simpler than that of the Type H matrix. The test, due to Box (1950), is presented in Appendix 8.A with a worked example. However, the model for the covariance structure is secondary to that of estimating and testing whether the means of treatments, carryover, etc. are significantly different. Since the Type H structure is necessary as well as sufficient to justify OLS, we proceed directly to examine whether the sample pooled covariance matrix conforms to that structure.

There are five parameters in the Type H matrix for the four–period data set of Patel (1983), λ, γ_1, γ_2, γ_3 and γ_4. There are 10 distinct variances or covariances in the above data matrix. We can equate the observed elements to their expectations, to obtain the following 10 equations in five unknowns:

$$
\begin{aligned}
0.3921 &= \lambda + 2\gamma_1 & 0.3040 &= \gamma_2 + \gamma_3 \\
0.3243 &= \gamma_1 + \gamma_2 & 0.3618 &= \gamma_2 + \gamma_4 \\
0.3132 &= \gamma_1 + \gamma_3 & 0.3644 &= \lambda + 2\gamma_3 \\
0.3369 &= \gamma_1 + \gamma_4 & 0.4141 &= \gamma_3 + \gamma_4 \\
0.4343 &= \lambda + 2\gamma_2 & 0.5277 &= \lambda + 2\gamma_4 .
\end{aligned}
$$

The parameter estimators can now be evaluated by least squares, although other methods of estimation may be used. Writing the above equations in matrix form yields

$$
\begin{pmatrix}
1 & 2 & 0 & 0 & 0 \\
0 & 1 & 1 & 0 & 0 \\
0 & 1 & 0 & 1 & 0 \\
0 & 1 & 0 & 0 & 1 \\
1 & 0 & 2 & 0 & 0 \\
0 & 0 & 1 & 1 & 0 \\
0 & 0 & 1 & 0 & 1 \\
1 & 0 & 0 & 2 & 0 \\
0 & 0 & 0 & 1 & 1 \\
1 & 0 & 0 & 0 & 2
\end{pmatrix}
\begin{pmatrix}
\lambda \\
\gamma_1 \\
\gamma_2 \\
\gamma_3 \\
\gamma_4
\end{pmatrix}
=
\begin{pmatrix}
0.3921 \\
0.3243 \\
0.3132 \\
0.3369 \\
0.4343 \\
0.3040 \\
0.3618 \\
0.3644 \\
0.4141 \\
0.5277
\end{pmatrix} ,
$$

for which the solution is as follows:

$$
\begin{pmatrix} \hat{\lambda} \\ \hat{\gamma}_1 \\ \hat{\gamma}_2 \\ \hat{\gamma}_3 \\ \hat{\gamma}_4 \end{pmatrix} = \begin{pmatrix} 0.08723 \\ 0.14991 \\ 0.16658 \\ 0.15016 \\ 0.21816 \end{pmatrix} .
$$

Using these parameter estimates in place of the unknown parameters, the Type H matrix of the form

$$
\Sigma = \lambda I + \gamma j' + j \gamma' = \begin{pmatrix} \lambda + 2\gamma_1 & \gamma_1 + \gamma_2 & \gamma_1 + \gamma_3 & \gamma_1 + \gamma_4 \\ \gamma_1 + \gamma_2 & \lambda + 2\gamma_2 & \gamma_2 + \gamma_3 & \gamma_2 + \gamma_4 \\ \gamma_1 + \gamma_3 & \gamma_2 + \gamma_3 & \lambda + 2\gamma_3 & \gamma_3 + \gamma_4 \\ \gamma_1 + \gamma_4 & \gamma_2 + \gamma_4 & \gamma_3 + \gamma_4 & \lambda + 2\gamma_4 \end{pmatrix}
$$

is estimated to be

$$
\hat{\Sigma} = \begin{pmatrix} 0.3870 & 0.3165 & 0.3000 & 0.3681 \\ 0.3165 & 0.4204 & 0.3167 & 0.3847 \\ 0.3000 & 0.3167 & 0.3875 & 0.3683 \\ 0.3681 & 0.3847 & 0.3683 & 0.5235 \end{pmatrix}
$$

compared with the variances–covariances from the data themselves,

$$
\hat{\Sigma} = \begin{pmatrix} 0.3921 & 0.3243 & 0.3132 & 0.3369 \\ 0.3243 & 0.4343 & 0.3040 & 0.3618 \\ 0.3132 & 0.3040 & 0.3644 & 0.4141 \\ 0.3369 & 0.3618 & 0.4141 & 0.5277 \end{pmatrix} .
$$

In the absence of a formal statistical test, it appears that the agreement between the two matrices is rather good. When a formal test is made (see Appendix 8.B) of whether the Type H structure fits the above sample pooled covariance matrix and the individual matrices from each sequence, it is found that whereas the matrix from the second sequence closely fits the Type H structure ($P = 0.6176$), that from the first sequence has to be rejected (P=0.0065). This suggests that the use of the usual least squares procedure described in Chapter 1 may not be a valid way of analyzing this data set, producing biased results. We return to a discussion of this point in the next section.

We now estimate a uniform covariance structure for Σ using the results of the analysis of the data from SAS PROC GLM. This will give one a better insight as to the meaning of this covariance structure. The analysis of variance of Patel's data was shown in Table 3.7. That analysis considered both first- and second-order carryover effects, so that the model was "saturated". That is, ignoring the individual subjects and concentrating on the sequence–by–period cell means, there are 8 parameters (a grand mean, one sequence, three period, one treatment and two carryover parameters) to be estimated from the 8 cell means.

The residual mean square was 0.087239, and is our best estimate of σ^2. From the between–subjects stratum, the mean square for subjects within sequences, 1.456690, is obtained. The expectation of this mean square, obtained from the PROC GLM output, is $\sigma^2 + 4\ \nu^2$, the multiplying factor 4 appearing because there were four periods for that set of data, with baseline and wash–out readings included. Equating the observed mean square to its expectation, in a "method of moments" solution employing 0.087239 as the estimate of σ^2, one obtains

$$\nu^2 = (1.456690 - 0.087239)/\ 4 = 0.342363 \quad .$$

Having estimates now for both ν^2 and σ^2, it is easy to obtain the uniform covariance matrix for this data set, with diagonal elements equal to $\hat{\nu}^2 + \hat{\sigma}^2$ and off–diagonal elements equal to $\hat{\nu}^2$. This results in the following matrix, rounded to four decimal places,

$$\hat{\Sigma} = \begin{pmatrix} 0.4296 & 0.3424 & 0.3424 & 0.3424 \\ 0.3424 & 0.4296 & 0.3424 & 0.3424 \\ 0.3424 & 0.3424 & 0.4296 & 0.3424 \\ 0.3424 & 0.3424 & 0.3424 & 0.4296 \end{pmatrix} \quad ,$$

which is not as close to the observed data matrix

$$\hat{\Sigma} = \begin{pmatrix} 0.3921 & 0.3243 & 0.3132 & 0.3369 \\ 0.3243 & 0.4343 & 0.3040 & 0.3618 \\ 0.3132 & 0.3040 & 0.3644 & 0.4141 \\ 0.3369 & 0.3618 & 0.4141 & 0.5277 \end{pmatrix} ,$$

as was the estimated Type H matrix. If one averages the four diagonal elements of the data matrix, 0.3921, 0.4343, 0.3644, and 0.5277, one obtains 0.4296, which is exactly equal to $\hat{\sigma}^2 + \hat{\nu}^2$, and if one averages the six unique off–diagonal elements of the data matrix, 0.3243, 0.3132, 0.3369, 0.3040, 0.3618, and 0.4141, one obtains 0.3424, which is exactly equal to $\hat{\nu}^2$. This is no accident; it is a consequence of the model being saturated, with 8 parameters estimated from 8 cell means. Whenever a model is saturated, the diagonal elements of the uniform covariance matrix will be the average of the diagonal elements of the data matrix, and the covariances will be the average of the off–diagonal elements of the data matrix.

Now let us consider a second set of data, those in Table 4.1 obtained by Ebbutt (1984). This data set corresponds to the 2–treatment, 2–sequence, 3–period model given in Design 4.1.1, viz.

		Period		
		1	2	3
Sequence	1	A	B	B
	2	B	A	A

Using the data themselves from the 22 subjects in Sequence 1 and the 27 subjects in Sequence 2, the following corrected sums of squares and cross-products matrices may be obtained:

	Period		
	1	2	3

Sequence 1:

$$\begin{pmatrix} 7279.82 & 4132.27 & 4210.18 \\ 4132.27 & 8493.09 & 4059.73 \\ 4210.18 & 4059.73 & 6817.82 \end{pmatrix}$$

	Period		
	1	2	3

Sequence 2:

$$\begin{pmatrix} 6550.67 & 3297.33 & 2131.67 \\ 3297.33 & 6420.67 & 4563.33 \\ 2131.67 & 4563.33 & 11120.67 \end{pmatrix} .$$

The test of homogeneity of the two covariance matrices obtained by dividing the above Sequence 1 matrix by 21 and the above Sequence 2 matrix by 26 leads to $M/C = 5.690$ with six degrees of freedom ($P = 0.4588$). Hence, the matrices may be considered homogeneous and combined to give the following pooled covariance matrix with 47 degrees of freedom:

$$\hat{\Sigma} = \begin{pmatrix} 294.3 & 158.1 & 134.9 \\ 158.1 & 317.3 & 183.5 \\ 134.9 & 183.5 & 381.7 \end{pmatrix} .$$

As was the case with the other data set examined in this section, the above matrix is not consistent with the independent error case, although the correlations between readings in successive periods are not as high as in that data set. We will now estimate the parameters λ, γ_1, γ_2 and γ_3 of the Type H matrix for this three–period problem. There are six distinct variances or covariances in the above data matrix. Equating the observed elements to their expectations, one obtains, in matrix form

$$\begin{pmatrix} 1 & 2 & 0 & 0 \\ 0 & 1 & 1 & 0 \\ 0 & 1 & 0 & 1 \\ 1 & 0 & 2 & 0 \\ 0 & 0 & 1 & 1 \\ 1 & 0 & 0 & 2 \end{pmatrix} \begin{pmatrix} \lambda \\ \gamma_1 \\ \gamma_2 \\ \gamma_3 \end{pmatrix} = \begin{pmatrix} 294.3 \\ 158.1 \\ 134.9 \\ 317.3 \\ 183.5 \\ 381.7 \end{pmatrix},$$

for which the following solution results:

$$
\begin{pmatrix} \hat{\lambda} \\ \hat{\gamma}_1 \\ \hat{\gamma}_2 \\ \hat{\gamma}_3 \end{pmatrix} = \begin{pmatrix} 172.267 \\ 59.763 \\ 78.683 \\ 99.803 \end{pmatrix} .
$$

Making use of these parameter estimates, the Type H matrix of the form

$$
\Sigma = \lambda I_3 + \gamma j_3' + j_3 \gamma' = \begin{pmatrix} \lambda+2\gamma_1 & \gamma_1+\gamma_2 & \gamma_1+\gamma_3 \\ \gamma_1+\gamma_2 & \lambda+2\gamma_2 & \gamma_2+\gamma_3 \\ \gamma_1+\gamma_3 & \gamma_2+\gamma_3 & \lambda+2\gamma_3 \end{pmatrix}
$$

is estimated to be

$$
\hat{\Sigma} = \begin{pmatrix} 291.8 & 138.4 & 159.6 \\ 138.4 & 329.6 & 178.5 \\ 159.6 & 178.5 & 371.9 \end{pmatrix} ,
$$

compared with the variances and covariances from the data themselves,

$$
\hat{\Sigma} = \begin{pmatrix} 294.3 & 158.1 & 134.9 \\ 158.1 & 317.3 & 183.5 \\ 134.9 & 183.5 & 381.7 \end{pmatrix} .
$$

As was the case with the other data set, it can be said, even in the absence of a formal statistical test, that the agreement between the Huynh–Feldt matrix and the data matrix is quite good. Using the formal "sphericity" test in Appendix 8.B, the result is that the data matrix conforms closely (P=0.4324) to the Type H structure. This suggests that the results obtained using OLS on this data set should be valid.

We now estimate a uniform covariance structure for Σ using results from the analysis of the data by means of SAS PROC GLM. Six cell means are used to estimate 6 parameters, so the model is saturated. The analysis of variance of the above data set is shown in Table 4.2. The residual mean square of 172.257 is the estimate of σ^2. The subjects within sequences mean square of 648.736 estimates $\sigma^2 + 3\ \nu^2$, the multiplying factor of 3 being present because there are three periods in this design. Equating the mean square to its expectation and employing 172.257 for σ^2, one obtains

$$\nu^2 = (648.736 - 172.257)/\ 3 = 158.8\ .$$

With these estimates of ν^2 and σ^2, one can obtain the uniform covariance matrix for this data set as

$$\hat{\Sigma} = \begin{pmatrix} 331.1 & 158.8 & 158.8 \\ 158.8 & 331.1 & 158.8 \\ 158.8 & 158.8 & 331.1 \end{pmatrix}$$

which is not as close to the observed data matrix

$$\hat{\Sigma} = \begin{pmatrix} 294.3 & 158.1 & 134.9 \\ 158.1 & 317.3 & 183.5 \\ 134.9 & 183.5 & 381.7 \end{pmatrix}$$

as was the estimated Type H matrix. Nevertheless, the formal test (see Appendix 8.A) shows close agreement (P = 0.6525) between the data and the uniform covariance assumption. It is easy to show that the average of the three diagonal elements of the data matrix is exactly the same as $\hat{\sigma}^2 + \hat{\nu}^2 = 331.1$ and that the average of the off–diagonal elements is 158.8, the same as $\hat{\sigma}^2$. This is due to the fact that a saturated model was used.

8.3. Consequences of Using Different Covariance Structures

We will now examine the effects of employing different covariance structures on the estimates of the parameters and their standard errors. If the covariance structure were known, then use of a generalized least squares (GLS) approach is possible. Following Chinchilli and Elswick (1989), we can write, for the mth experimental unit, $m=1, 2,..., n$, the observation vector for the p periods and the design matrix for the p periods as

$$Y_m = \begin{pmatrix} Y_{m1} \\ Y_{m2} \\ . \\ . \\ Y_{mp} \end{pmatrix} \qquad \text{and} \qquad X_m = \begin{pmatrix} x'_{m1} \\ x'_{m2} \\ . \\ . \\ x'_{mp} \end{pmatrix}$$

respectively. The dimensions of X_m are $p \times r$, where r is the number of fixed–effect parameters to be estimated. Under the assumption that Σ is <u>known</u>, the GLS estimator of the parameter vector β is

$$\hat{\beta} = (\sum_{m=1}^{n} X'_m \Sigma^{-1} X_m)^{-1} (\sum_{m=1}^{n} X'_m \Sigma^{-1} Y_m) \qquad (8.4)$$

and its variance–covariance matrix, of dimension $r \times r$, is

$$\Omega = (\sum_{m=1}^{n} X'_m \Sigma^{-1} X_m)^{-1} . \qquad (8.5)$$

The GLS estimator is a "BLUE", a best linear unbiased estimator (see Judge *et al.*, 1982) for the model under consideration here. The problem with the GLS estimator is that the $p \times p$ covariance matrix Σ of the data is not known. We will now use the two sets of data that were studied in Section 8.2, to see what effect different estimates of Σ have on the estimates of β and its standard error.

The first data set is that of Patel (1983), given in Table 3.2. In Section 8.2, we derived the uniform covariance matrix for this data set, to four decimal places. Here it is given to five decimal places:

$$\hat{\Sigma} = \begin{pmatrix} 0.42960 & 0.34236 & 0.34236 & 0.34236 \\ 0.34236 & 0.42960 & 0.34236 & 0.34236 \\ 0.34236 & 0.34236 & 0.42960 & 0.34236 \\ 0.34236 & 0.34236 & 0.34236 & 0.42960 \end{pmatrix} .$$

We must realize that this is an estimate of Σ, as the true covariance matrix is not known. Nevertheless, one can use this matrix in place of Σ in Equation (8.4) to obtain *empirical* or *estimated* GLS estimators. The resulting estimates of the treatment parameter τ, the first–order carryover parameter λ, and the second–order carryover parameter θ, are

$$(\hat{\tau}, \hat{\lambda}, \hat{\theta})' = (-0.1502778, 0.0563194, -0.0440278)',$$

and the portion of the covariance matrix for these parameters from Equation (8.5) is

$$\Omega = \begin{pmatrix} 0.010299167 & 0.005149583 & 0.015448750 \\ 0.005149583 & 0.010299167 & 0.010299167 \\ 0.015448750 & 0.010299167 & 0.030897500 \end{pmatrix}.$$

This leads to the following parameter estimates and their standard errors,

τ: -0.1502778 ± 0.1014848
λ: 0.0563194 ± 0.1014848
θ: -0.0440278 ± 0.1757768

These results are <u>identical</u> to the estimates and standard errors produced when PROC GLM was used on these data, provided that those quantities are divided by 2, reflecting a difference in the conventions used. Because the model is saturated, with 8 parameters estimated from 8 cell means, the same parameter estimates will be produced, whatever the numerical values in Σ. However, since Ω depends upon Σ, the standard errors will vary as the numerical values in Σ vary. The fact that the standard errors are identical to those from PROC GLM demonstrates that the standard OLS solution for a saturated model is in effect a procedure which uses average values of the variances and covariances of the data matrix just as a uniform covariance matrix does. The averaging process makes all of the diagonal elements of the uniform covariance matrix equal to the average of the diagonal elements of the data matrix, and makes all of the off–diagonal elements equal to the average of the off–diagonal elements of the data matrix.

Now let us use the Type H form of Σ that was derived in Section 8.2. That matrix was

$$\hat{\Sigma} = \begin{pmatrix} 0.3870 & 0.3165 & 0.3000 & 0.3681 \\ 0.3165 & 0.4204 & 0.3167 & 0.3847 \\ 0.3000 & 0.3167 & 0.3875 & 0.3683 \\ 0.3681 & 0.3847 & 0.3683 & 0.5235 \end{pmatrix} .$$

When this matrix is used in Equation (8.4), exactly the same parameter estimates are obtained as when the uniform covariance matrix is used, since the model is saturated. However, different variances and covariances for the treatment and carryover parameters result:

$$\Omega = \begin{pmatrix} 0.010294444 & 0.005147222 & 0.015435764 \\ 0.005147222 & 0.010300347 & 0.010294444 \\ 0.015435764 & 0.010294444 & 0.030865625 \end{pmatrix} .$$

These values are similar to, but not identical to, the values obtained using the uniform covariance matrix. The estimates and standard errors of the treatment and carryover parameters are as follows,

τ: -0.1502778 ± 0.1014615
λ: 0.0563194 ± 0.1014906
θ: -0.0440278 ± 0.1756862 .

Now let's see what happens when one employs the covariance matrix obtained from the data themselves. In Section 8.2, that was shown to be

$$\hat{\Sigma} = \begin{pmatrix} 0.3921 & 0.3243 & 0.3132 & 0.3369 \\ 0.3243 & 0.4343 & 0.3040 & 0.3618 \\ 0.3132 & 0.3040 & 0.3644 & 0.4141 \\ 0.3369 & 0.3618 & 0.4141 & 0.5277 \end{pmatrix} .$$

Employing this Σ matrix in Equation (8.4), the resulting empirical GLS estimates of the parameters are identical to what they were with the other Σ matrices, as the model is saturated, but the variances and covariances corresponding to τ, λ and θ are different:

$$\Omega = \begin{pmatrix} 0.01049514 & 0.00345903 & 0.01596701 \\ 0.00345903 & 0.00767951 & 0.01267326 \\ 0.01596701 & 0.01267326 & 0.03595972 \end{pmatrix} .$$

This leads to the following estimates and standard errors:

τ: -0.1502778 ± 0.1021458
λ: 0.0563194 ± 0.0876328
θ: -0.0440278 ± 0.1896305 .

These standard errors differ from the uniform covariance and Type H solutions much more than those two solutions differed from each other. This may be attributable to the fact that the variability present in the data matrix has not been "smoothed" in the way in which the other two matrices have been.

We now turn our attention to the unsaturated case. Since the second-order carryover parameter θ was non-significant, its estimated standard error being much larger than the estimate irrespective of the assumed Σ matrix, we now consider a model that lacks this parameter. This model will have only seven parameters (a grand mean, one sequence, three period, one treatment, and one first-order carryover). We consider the three different estimates of Σ, viz. the uniform covariance, the Type H, and the covariance matrix obtained from the data. The results from use of these three different approaches are presented below in compact form, giving just the estimates for the treatment and first-order carryover parameters and their standard errors.

From the uniform covariance estimate of Σ:

τ: -0.1282639 ± 0.0507424
λ: 0.0709954 ± 0.0828620 .

From the Type H estimate of Σ:

τ: -0.1282597 ± 0.0507453
λ: 0.0710038 ± 0.0828667 .

From the estimate of Σ from the data:

τ: −0.1307284 ± 0.0583557
λ: 0.0718361 ± 0.0566841 .

These results demonstrate that, unlike the saturated model case, different parameter estimates are obtained from the different Σ matrices. Like the saturated case, there is better agreement between the standard errors which result from using the uniform covariance and Type H matrices than between either of those and the matrix obtained directly from the data.

We now consider a higher degree of unsaturation, by eliminating the first–order carryover parameter, since the above results indicate that that effect is not significant. The results using the three different estimates of Σ are given below.

From the uniform covariance estimate of Σ:

τ: −0.1282639 ± 0.0507424 .

From the Type H estimate of Σ:

τ: −0.1282495 ± 0.0507453 .

From the estimate of Σ from the data:

τ: −0.0822527 ± 0.0440710 .

The estimates and standard errors resulting from the use of the uniform covariance matrix are identical to the answers obtained when first–order carryover is in the model, and those obtained using the Type H matrix are almost identical to what they were before. However, the estimates obtained using the Σ matrix obtained from the data are quite different from what they were when carryover parameters were in the model.

The second data set to be considered here is that of Ebbutt (1984), which was presented in Table 4.1. Since there are six cell means and six parameters to estimate (a grand mean, one sequence, two period, one treatment, and one carryover), the model is saturated. In Section 8.2, we estimated the uniform covariance matrix, and that was given by

$$\hat{\Sigma} = \begin{pmatrix} 331.1 & 158.8 & 158.8 \\ 158.8 & 331.1 & 158.8 \\ 158.8 & 158.8 & 331.1 \end{pmatrix} .$$

Substituting this estimate of Σ into Equation (8.4) yields the following estimates of the treatment and carryover parameters and their standard errors.

τ: 2.960859 ± 1.154194

λ: 1.335859 ± 1.332749 .

These are identical to the estimates and standard errors produced by PROC GLM, provided the latter are divided by two. Any Σ matrix will produce exactly the same parameter estimates for this saturated model, but the fact that the above standard errors agree with the OLS solution using SAS is due to the fact that the latter procedure is in effect using a uniform covariance matrix constructed from the residual mean square and the subjects within sequence mean square.

The estimated Type H matrix was shown in Section 8.2 to be

$$\hat{\Sigma} = \begin{pmatrix} 291.8 & 138.4 & 159.6 \\ 138.4 & 329.6 & 178.5 \\ 159.6 & 178.5 & 371.9 \end{pmatrix} .$$

Substituting this matrix into Equations (8.4) and (8.5) results in the following parameter estimates and their standard errors,

τ: 2.960859 ± 1.154282

λ: 1.335859 ± 1.332722 .

The parameter estimates are identical to before, as expected, and the standard errors are very close to the previous values. Using the estimate of Σ obtained from the data themselves,

$$\hat{\Sigma} = \begin{pmatrix} 294.3 & 158.1 & 134.9 \\ 158.1 & 317.3 & 183.5 \\ 134.9 & 183.5 & 381.7 \end{pmatrix}.$$

results in the following estimates from use of Equations (8.4) and (8.5),

τ: 2.960859 ± 1.175033
λ: 1.335859 ± 1.308320 .

As was observed for the first data set, the standard errors using the data matrix are not as close to the other two solutions as those two are to each other.

We now look at the unsaturated case, where the non–significant carryover parameter λ is not present in the model. Thus, there are five parameters to be estimated (a grand mean, a sequence, two periods, a treatment). The three different estimates of Σ are used, giving the following results for the parameter estimates and their standard errors:

From the uniform covariance estimate of Σ:

τ: 2.960859 ± 1.154338 .

From the Type H estimate of Σ:

τ: 2.961053 ± 1.154282 .

From the estimate of Σ from the data themselves:

τ: 2.737947 ± 1.154574 .

As was observed previously, the estimates resulting from use of the covariance matrix obtained from the data themselves tend to be more deviant than the estimates resulting from use of either of the more specific covariance structures.

The information obtained in this section of this chapter can be summarized by stating that the parameter estimates and their standard errors will depend upon the choice of the covariance matrix Σ. If Σ were known, then a BLUE estimate of β is obtained directly from Equation (8.4) with an estimate of the variances and covariances of the parameters of β from Equation (8.5). However, it will be a rare circumstance when Σ is known. A good estimate of Σ might be obtained from previous experiments on the same or similar experimental units. More usual is the circumstance that Σ will be unknown, and has to be estimated from the data. Little can be said in general about the properties of the variances of empirical GLS estimators. They may or may not be more efficient than OLS estimators. The best situation is when there is little difference between the estimates and their standard errors produced by both methods. This is more likely to be the case when the model is saturated or close to saturation.

In the second of the two data sets used as illustrations in this chapter, the Type H structure was closely approximated by the variance–covariance matrix of the data observations obtained by pooling information on variation and covariation among subjects in sequences. Since the validity of the OLS solution for cross–over design data depends upon the data satisfying the Type H structure, the OLS solution is probably quite valid for that data set. The best situation is where the model is saturated, that is, where there are as many parameters to estimate as there are cell means. This is often the case in cross–over designs, where interactions between treatments and periods and other interactions are often of interest. Often, however, these parameters of interest cannot be included in the model for lack of sufficient cell means.

With a saturated model, one at least knows that the parameter estimates will be independent of Σ, but their standard errors will depend, as can be seen from Equation (8.5), upon Σ. As one moves further and further away from saturation, by deleting parameters and estimating only a subset of the parameters in the saturated case, the discrepancies between results obtained for different estimates of Σ become greater and greater. Unless one is confident that the Type H structure holds, it may not be possible to decide which method of estimation is more accurate. In the next section of this chapter, we will examine two other estimation procedures that may be considered when one is not confident that the Type H structure is sufficiently closely approximated by the data to justify the use of ordinary least squares.

8.4. Other Methods of Analysis

If the variances and covariances of the data do not satisfy the Type H structure, then OLS is not appropriate for the analysis of cross–over designs. In this section, we explore other approaches to the analysis of such data. Patel (1986) advanced a multivariate model specifically to deal with the cross–over design with changing covariates, which includes the fact that different treatments and carryover effects appear in different sequences in a given period. A limitation of this approach, as noted by Chinchilli and Elswick (1989) for both this design and the multiple design multivariate model of McDonald (1975), where a different design matrix occurs for each of the p periods, is that p sets of parameters are needed to model the expected responses. They concluded that it would be ideal to model the responses with only one set of parameters, but to allow for more general covariance structures.

One can express the linear model for the cross–over experiment as

$$\mathbf{Y}_m = \mathbf{X}_m \beta + \epsilon_m \tag{8.6}$$

for $m=1, 2,...,n$, where $\mathbf{X}_m$ and $\mathbf{Y}_m$ are as defined in Section 8.3, and ϵ_m is a p–vector of random errors such that $\Sigma = \text{Var}(\epsilon_m)$ is a $p \times p$ positive definite matrix.

Under the mixed model analysis of variance approach, Σ was restricted to being of the Type H structure to justify use of ordinary least squares. Diggle (1988) proposed a structure for repeated measures designs that can be adapted to cross–over design experiments. The structure is given by

$$\Sigma = \sigma^2 \mathbf{I}_p + \nu^2 \mathbf{J}_{p \times p} + \zeta^2 \mathbf{R}(t),$$

where t denotes the p–vector of times in the experiment and $\mathbf{R}(t)$ is a $p \times p$ matrix whose (j,j')th element, $j,j'=1, 2,...,p$, is given by

$$\exp(-\alpha |t_j - t_{j'}|^c),$$

with c chosen to be either 1 or 2, justifying c=1 as a continuous–time analog of a first–order autoregressive process, and c=2 as an intrinsically smoother process. The model is an extension of the uniform covariance structure defined by the first two terms of the model, $\sigma^2 I_p + \nu^2 J_{pxp}$, the magnitude of α determining how much departure there is from the compound symmetry structure. Since the Type H structure is more general than that of uniform covariance, it seems more plausible to add Diggle's extension to that structure rather than as above. This leads to

$$\Sigma = \lambda I_p + \eta j_p' + j_p \gamma' + \zeta^2 R(t) \tag{8.7}$$

where the diagonal element of the pth row or column is given by

$$\lambda + 2\gamma_p + \zeta^2$$

and the off–diagonal element of the jth row and j'th column is given by

$$\gamma_j + \gamma_{j'} + \zeta^2 \exp(-\alpha|t_j - t_{j'}|^c).$$

The maximum likelihood (ML) estimators of β and of ζ^2 are

$$\hat{\beta} = (\sum_{m=1}^{n} X_m' \Sigma_0^{-1} X_m)^{-1} (\sum_{m=1}^{n} X_m' \Sigma_0^{-1} Y_m) \tag{8.8}$$

$$\hat{\zeta}^2 = n^{-1} (\sum_{m=1}^{n} (Y_m - X_m \hat{\beta})' \Sigma_0^{-1} (Y_m - X_m \hat{\beta}), \tag{8.9}$$

where Σ_0 is the empirical $p \times p$ covariance matrix which minimizes the function

$$\log |\Sigma_0| + n \log [n^{-1} (\sum_{m=1}^{n} (Y_m - X_m \hat{\beta})' \Sigma_0^{-1} (Y_m - X_m \hat{\beta})].$$

When the optimum Σ_0 is found, the estimated covariance matrix of $\hat{\beta}$ is given by

$$\Omega = (\sum_{m=1}^{n} X_m' \Sigma_0^{-1} X_m)^{-1} .$$

Diggle (1988) also proposed a restricted maximum likelihood (REML) solution to the above problem. The REML estimator of β has the same formula as before,

$$\hat{\beta} = (\sum_{m=1}^{n} X_m' \Sigma_0^{-1} X_m)^{-1} (\sum_{m=1}^{n} X_m' \Sigma_0^{-1} Y_m),$$

with

$$\hat{\zeta}^2 = (n-p)^{-1} (\sum_{m=1}^{n} (Y_m - X_m \hat{\beta})' \Sigma_0^{-1} (Y_m - X_m \hat{\beta}), \qquad (8.10)$$

which is likely to be less biased than the corresponding expression from the maximum likelihood solution, with Σ_0 being optimized to minimize the function

$$\log |\Sigma_0| + \log \left| \sum_{m=1}^{n} X_m' \Sigma_0^{-1} X_m) \right| +$$

$$(n-p) \log [(n-p)^{-1} (\sum_{m=1}^{n} (Y_m - X_m \hat{\beta})' \Sigma_0^{-1} (Y_m - X_m \hat{\beta})].$$

When we attempted to fit the modified Diggle model to the two data sets considered in this section, that is, to estimate the parameters λ, $\gamma_1, \gamma_2, ..., \gamma_p$, ζ^2, and α, with $c=1$, convergence was not achieved. Instead, the model behaved in an unstable manner, which may be attributed to the fact that the two data sets used in this section are not far from satisfying a Type H structure. Thus, the term $\zeta^2 R(t)$ in Equation (8.7) should be close to zero. However, this is impossible as ζ^2, given by Equation (8.9) or (8.10), depending upon whether the ML or REML solutions are used, is non-zero, and $R(t)$ is an identity matrix at one extreme (when α is large), or a matrix of ones (when α is zero). Thus, at the first extreme, Equation (8.7) collapses to

$$\Sigma = (\lambda + \zeta^2)I_p + \eta'_p + j_p\gamma'$$

while at the other extreme, the equation collapses to

$$\Sigma = \sigma^2 I_p + (\nu^2 + \zeta^2)J_{p\times p} \, ,$$

a uniform covariance structure. Since λ cannot be estimated separately from ζ^2, and since ν^2 cannot be estimated separately from ζ^2, there will be difficulty of estimation whenever the structure is close to one of those extremes. The method may work when the underlying structure is far from a Type H structure, but it is fairly complicated, involving minimization of a function of at least $p + 2$ parameters (fixing c at either 1 or 2). Diggle (1988) used the Nelder–Mead (1965) simplex method for this purpose, as did we.

An alternative approach to the above is one due to Ware (1985), who considered various structures for Σ or no structure at all. Ware, like Diggle, considered more general repeated measures applications with complete and balanced data. For the latter case of complete, balanced data and no structure for Σ, the maximum likelihood equations for β and Σ are

$$\hat{\beta} = (\sum_{m=1}^{n} X'_m \Sigma_0^{-1} X_m)^{-1} (\sum_{m=1}^{n} X'_m \Sigma_0^{-1} Y_m)$$

and

$$\Sigma_0 = n^{-1}(\sum_{m=1}^{n} (Y_m - X_m\hat{\beta})(Y_m - X_m\hat{\beta})' . \tag{8.11}$$

The solution is obtained by iteration between the two equations until convergence of $\hat{\beta}$ and Σ_0 occurs. Intuitively, it seems better to replace the divisor n in

Equation (8.11) by $n\text{-}s$, where s is the number of sequences, in the same manner as was used in obtaining the pooled covariance matrix directly from the raw data. Hence, in place of Equation (8.11), we use

$$\Sigma_0 = (n\text{-}s)^{-1}(\sum_{m=1}^{n} (\mathrm{Y}_m - \mathrm{X}_m\hat{\beta})(\mathrm{Y}_m - \mathrm{X}_m\hat{\beta})'. \tag{8.12}$$

We tested this approach on the two data sets in this section. For the Patel (1983) data set, one can start the iteration process using either the estimated Type H structure or the pooled covariance matrix from the raw data. In either case, rapid convergence occurs. We used an unsaturated model which has only seven parameters. A program written in the matrix language GAUSS (1988) to perform these calculations is given below, where the matrix labelled *hf* contains the estimated Type H structure of Huynh and Feldt (1970):

```
output file=d:\gauss\patel.out;output on;
let x1={1 1 1 0 0 0 0,1 1 0 1 0 1 0,1 1 0 0 1 0 1,1 1 -1 -1 -1 -1 0};
let x2={1 -1 1 0 0 0 0,1 -1 0 1 0 -1 0,1 -1 0 0 1 0 -1,1 -1 -1 -1 -1 -1 1 0};
let y1={1.09 1.28 1.24 1.33,1.38 1.60 1.90 2.21,2.27 2.46 2.19 2.43,
        1.34 1.41 1.47 1.81,1.31 1.40 0.85 0.85,0.96 1.12 1.12 1.20,
        0.66 0.90 0.78 0.90,1.69 2.41 1.90 2.79};
let y2={1.74 3.06 1.54 1.38,2.41 2.68 2.13 2.10,3.05 2.60 2.18 2.32,
        1.20 1.48 1.41 1.30,1.70 2.08 2.21 2.34,1.89 2.72 2.05 2.48,
        0.89 1.94 0.72 1.11,2.41 3.35 2.83 3.23, 0.96 1.16 1.01 1.25};
let hf={0.3870 0.3165 0.3000 0.3681, 0.3165 0.4204 0.3167 0.3847,
        0.3000 0.3167 0.3875 0.3683,0.3681 0.3847 0.3683 0.5235};
count=1; beta2=zeros(7,1);crit=1.0e-22;s=hf;
do while count <=10;
sinv=inv(s);x3=x1'sinv*x1;x4=x2'sinv*x2;
x5=inv(8.*x3+9.*x4);
y3=(x1'sinv*y1')';y4=sumc(y3);
y5=(x2'sinv*y2')';y6=sumc(y5);
y7=y4+y6;beta=x5*y7;beta;
x1b=x1*beta;x2b=x2*beta;x1m=ones(1,8);x2m=ones(1,9);
x1r=(x1b*x1m)';x2r=(x2b*x2m)';xb=x1r|x2r;
y8=y1|y2;s=(y8-xb)'(y8-xb)/15;s;
sumsq=(beta-beta2)'(beta-beta2);
if sumsq<crit;stop;endif;
beta2=beta;
count=count+1;
endo;
```

The above program produced the following converged estimates of $\hat{\beta}$ and Σ_0, with only the estimates relating to the treatment parameter τ_1 and carryover parameter λ_1 being shown:

$$\begin{pmatrix} \hat{\tau}_1 \\ \hat{\lambda}_1 \end{pmatrix} = \begin{pmatrix} -0.1307 \\ 0.0718 \end{pmatrix},$$

and

$$\hat{\Sigma}_0 = \begin{pmatrix} 0.39215 & 0.32415 & 0.31311 & 0.33670 \\ 0.32415 & 0.43441 & 0.30407 & 0.36196 \\ 0.31311 & 0.30407 & 0.36442 & 0.41424 \\ 0.33670 & 0.36196 & 0.41424 & 0.52797 \end{pmatrix}.$$

This is rather close to the matrix obtained from the data themselves, which is perhaps not too surprising, even though the model that was fitted had seven parameters only, one short of saturation.

We now consider the second set of data, that of Ebbutt (1984), shown in Table 4.1. In Section (8.2), we found that the estimated uniform covariance matrix for this data set was

$$\hat{\Sigma} = \begin{pmatrix} 331.1 & 158.8 & 158.8 \\ 158.8 & 331.1 & 158.8 \\ 158.8 & 158.8 & 331.1 \end{pmatrix}$$

This matrix was used as the starting point in an iterative method employing the following code using SAS/IML©, in which the carryover parameter is not included in the model so as to avoid having a saturated model:

```
proc iml;
x1={1 1 1 0 1 ,1 1 0 1 -1 ,1 1 -1 -1 -1 };
x2={1 -1 1 0 -1 ,1 -1 0 1 1 ,1 -1 -1 -1 1 };
y1={159 140 137,153 172 155,160 156 140,160 200 132,170 170 160,174 132 130,
    175 155 155,154 138 150,160 170 168,160 160 170,145 140 140,148 154 138,
    170 170 150,125 130 130,140 112 95,125 140 125,150 150 145,136 130 140,
    150 140 160,150 140 150,202 181 170,190 150 170};
```

(continued on next page)

```
y2={165 154 173,160 165 140,140 150 180,140 125 130,158 160 180,180 165 160,
    170 160 160,140 158 148,126 170 200,130 125 150,144 140 120,140 160 140,
    120 145 120,145 150 150,155 130 140, 168 168 168,150 160 180,120 120 140,
    150 150 160,150 140 130,175 180 160,140 170 150,150 160 130,150 130 125,
    140 150 160,140 140 130,126 140 138};
s={331.1 158.8 158.8,158.8 331.1 158.8,158.8 158.8 331.1};
count=1;beta2=j(5,1,0);crit=1.0e-20;
do while (count<=10);
sinv=inv(s);x3=t(x1)*sinv*x1;x4=t(x2)*sinv*x2;x5=inv(22*x3+27*x4);
y3=t((t(x1)*sinv*t(y1)));y4=y3[+,];y5=t((t(x2)*sinv*t(y2)));
y6=y5[+,];y7=t(y4+y6);beta=x5*y7;print beta;
x1b=x1*beta;x2b=x2*beta;x1m=j(1,22,1);x2m=j(1,27,1);
x1r=t((x1b*x1m));x2r=t((x2b*x2m));xb=x1r//x2r;
y8=y1//y2;s=t((y8-xb))*(y8-xb)/47;print s;
sumsq=t((beta-beta2))*(beta-beta2);
if sumsq>crit then do;
  beta2=beta;
  count=count+1;
  end;
else do;
  print x5;se=sqrt(diag(x5));print se;
  stop;
end;
end;
```

The above program produced the following converged estimates and standard errors for $\hat{\beta}$ (with only the estimate for τ_1 being shown):

$$\hat{\tau}_1 = 2.73829 \pm 1.15499 ,$$

which is close to the empirical GLS solution using the data matrix as the estimate of Σ. The estimated covariance matrix at convergence was

$$\hat{\Sigma} = \begin{pmatrix} 294.3 & 158.3 & 134.6 \\ 158.3 & 318.5 & 181.7 \\ 134.6 & 181.7 & 384.3 \end{pmatrix}$$

which is very close to the pooled covariance matrix calculated from the raw data:

$$\hat{\Sigma} = \begin{pmatrix} 294.3 & 158.1 & 134.9 \\ 158.1 & 317.3 & 183.5 \\ 134.9 & 183.5 & 381.7 \end{pmatrix} .$$

Thus, the method of Ware (1985), which is easy to implement, produces an estimate of the covariance matrix which is not far from that derived from the data. Had the model been saturated, the converged covariance matrix would have been identical to the data matrix.

8.5. A Time Series Approach to the Correlation Structure

Another approach to the examination of the question of the efficiency of ordinary least squares was undertaken by Matthews (1990). He considered the data obtained in a typical cross–over trial to be a large number (one for each subject) of very short time series (of length p, the number of periods).

Matthews (1990) considered both autoregressive and moving average models of order one for these short time series. Thus, for the first–order stationary autoregressive case when there are three time periods, the covariance structure would have the following form,

$$\hat{\Sigma} = c\,(1-\rho^2)^{-1} \begin{pmatrix} 1 & \rho & \rho^2 \\ \rho & 1 & \rho \\ \rho^2 & \rho & 1 \end{pmatrix}$$

where c is some constant and ρ is a correlation coefficient. This matrix is incompatible with the matrices of the Type H form that were considered in Sections 8.2–8.4, except for the special case of $\rho=0$, which is the independence case.

For the first–order stationary moving average process when there are three time periods, the covariance structure would have the following form:

$$\hat{\Sigma} = c \begin{pmatrix} 1 & \rho & 0 \\ \rho & 1 & \rho \\ 0 & \rho & 1 \end{pmatrix}.$$

This is rather an unrealistic form for most cross–over trials, since the correlation between measurements on the same subject is likely to be high even when the periods are far apart. Matthews (1990) obtained results for a wide range of assumed true values for the correlation coefficient ρ, by comparing results for the variances obtained from GLS estimation, from the true OLS estimator if the covariance matrix were known, and the usual estimate of the covariance obtained by fitting the data to the model and using the unbiased estimator of σ^2 in place of the unknown experimental error variance. He considered two quantities, R_1 and R_2, defined as

$$1 + R_1 = \frac{\text{estimated standard error using OLS}}{\text{true standard error under OLS}}$$

and

$$R_2 = \frac{\text{true standard error using GLS}}{\text{true standard error using OLS}}$$

Matthews (1990) calculated the values of R_1 and R_2 using PROC MATRIX© of SAS. He presented results only for various cross–over designs with two treatments, including many of those covered in Chapter 4 of this book. Since the moving average case is less likely to occur, we present some of the results obtained by Matthews in Table 8.1, but only for the autoregressive case and only for those designs that were shown in Chapter 4 to have good estimation properties.

This shows that there is a rather wide range of values of ρ for which R_1 is less than 0.10 in absolute magnitude, and that there is also a wide range of values for which R_2 exceeds 0.90. The efficiency of OLS is generally less affected by large positive values of ρ rather than by large negative values of ρ. This is fortunate, since large positive correlations are more likely in practice than large negative correlations. Looking at the two data sets considered in this chapter, we convert the covariance matrix from the data set of Patel (1983) to correlation form to yield

$$\hat{\Sigma} = \begin{pmatrix} 1.0 & 0.7859 & 0.8286 & 0.7406 \\ 0.7859 & 1.0 & 0.7642 & 0.7558 \\ 0.8286 & 0.7642 & 1.0 & 0.9443 \\ 0.7406 & 0.7558 & 0.9443 & 1.0 \end{pmatrix},$$

and the data set of Ebbutt (1984) to yield

$$
\hat{\Sigma} = \begin{pmatrix} 1.0 & 0.517 & 0.403 \\ 0.517 & 1.0 & 0.527 \\ 0.403 & 0.527 & 1.0 \end{pmatrix}
$$

The data set of Patel (1983), in particular, contains some rather high correlations, which raises the question of whether OLS is justified in view of the work of Matthews (1990). We must remember that the first–order autoregressive stationary model is incompatible with a Huynh–Feldt structure except for $\rho = 0$. Thus, the information contained in Table 8.1 is really a comparison of models having non–zero ρ against the special case of independence. A more appropriate comparison would be against the estimated Type H matrices derived in Section 8.2 for these data sets.

Table 8.1. Ranges of the correlation ρ for various designs with three or four periods, equal allocation of subjects to the two, four or six sequences being assumed (after Matthews, 1990, with the permission of The Biometric Society).

	Intervals of ρ for which			
	$|R_1|$ is less than		R_2 exceeds	
Design	0.05	0.10	0.95	0.90
ABB/BAA	$(-0.25, 0.35)$	$(-0.45, 0.90)$	All ρ	
ABB/BAA/ AAB/BBA	$(-0.20, 0.30)$	$(-0.35, 0.75)$	$(-0.45, 0.70)$	$(-0.60, 0.95)$
ABB/BAA/ ABA/BAB	$(-0.25, 0.35)$	$(-0.40, 0.90)$	$(-0.65, 0.95)$	$(-0.75, 0.95)$
ABB/BAA/ AAB/BAA/ ABA/BAB	$(-0.30, 0.45)$	$(-0.55, 0.95)$	$(-0.45, 0.60)$	$(-0.55, 0.95)$
ABBA/BAAB ABBB/BAAA AABB/BBAA	$(-0.30, 0.45)$ $(-0.75, 0.80)$ $(-0.10, 0.10)$	$(-0.40, 0.65)$ $(-0.95, 0.95)$ $(-0.20, 0.30)$	$(-0.85, 0.95)$ $(-0.45, 0.95)$ $(-0.65, 0.50)$	$(-0.90, 0.95)$ $(-0.55, 0.95)$ $(-0.80, 0.75)$
ABBA/BABB/ AABB/BBAA	$(-0.15, 0.25)$	$(-0.25, 0.95)$	$(-0.55, 0.50)$	$(-0.85, 0.75)$

We now consider a data set that appeared in Gill (1978, p. 248) [reproduced here with the permission of The Iowa State University Press], which has a pooled covariance matrix from the data that appears to be consistent with a first order autoregressive structure. This enables us to see the consequences of using OLS estimation when a covariance structure different from a Type H structure appears to apply. The data set involves a dairy feeding experiment which compared two diets [A = whey supplemented ration, B = control ration] using a "switchback" design, with one group of 12 Guernsey cows having the treatment sequence ABAB and a second group of 12 Guernseys having the sequence BABA. [Note: as described in Chapter 4, such a design in which the two treatments alternate in successive periods is a poor one in the presence of carryover effects, since carryover from A is always into B, and carryover from B is always into A. This design should be used only in situations where carryover effects are deemed to be impossible, a rare occurrence.]

ABAB	Period				BABA	Period			
Cow	1	2	3	4	Cow	1	2	3	4
1	30.8	31.0	29.6	21.4	13	31.2	31.3	29.2	28.1
2	35.4	33.3	29.2	30.2	14	35.3	36.7	34.4	31.9
3	35.5	35.5	37 0	29.9	15	28.2	26.8	24.2	22.0
4	31.3	28.0	29.0	26.3	16	26.4	26.7	27.0	25.8
5	30.2	27.3	26.2	24.9	17	21.7	21.4	20.3	19.1
6	30.2	32.2	30.4	29.5	18	25.6	27.1	25.8	25.1
7	22.7	22.6	20.4	19.1	19	31.3	30.2	26.8	25.5
8	31.7	30.3	27.3	24.5	20	29.7	27.2	25.6	22.2
9	34.1	29.5	27.4	25.7	21	30.0	28.6	26.8	23.5
10	32.4	29.5	28.9	25.8	22	37.6	36.4	33.5	29.2
11	36.3	33.8	30.0	26.7	23	32.6	27.9	26.5	22.7
12	25.6	23.9	20.1	19.2	24	25.0	25.5	25.7	24.1

The estimated corrected sum of squares and cross–products matrix with 11 degrees of freedom for the results of Sequence 1 (ABAB) is as follows,

$$
SSCP_1 = \begin{pmatrix} 177.750 & 152.415 & 160.705 & 133.640 \\ 152.415 & 164.469 & 174.741 & 133.507 \\ 160.705 & 174.741 & 220.809 & 151.703 \\ 133.640 & 133.507 & 151.703 & 157.827 \end{pmatrix},
$$

and that for Sequence 2 (BABA), also with 11 degrees of freedom, is

$$SSCP_2 = \begin{pmatrix} 222.850 & 199.695 & 163.360 & 122.910 \\ 199.695 & 209.409 & 177.435 & 152.607 \\ 163.360 & 177.435 & 159.730 & 139.550 \\ 122.910 & 152.607 & 139.550 & 135.707 \end{pmatrix}.$$

Although the variances in Sequence 2 show a decreasing trend with time, no similar trend applies in Sequence 1. Using the test for equal covariances described in Section 8.2, one finds that the determinant of the covariance matrix $SSCP_1/11 = 630.698$, compared to a much smaller value of 16.040 for the determinant of the covariance matrix $SSCP_2/11$. The statistic $M/C = 21.33$, resulting in a P-value of 0.019. Hence there is a marginally significant difference between the two covariance matrices. Since the evidence of a difference is not overwhelming, we proceed as though the matrices were homogeneous.

The pooled within-sequence covariance matrix is

$$\hat{\Sigma} = \begin{pmatrix} 18.2091 & 16.0050 & 14.7302 & 11.6614 \\ 16.0050 & 16.9945 & 16.0080 & 13.0052 \\ 14.7302 & 16.0080 & 17.2972 & 13.2388 \\ 11.6614 & 13.0052 & 13.2388 & 13.3424 \end{pmatrix},$$

which, if converted into correlation form, gives

$$\hat{\rho} = \begin{pmatrix} 1.0 & 0.90982 & 0.83000 & 0.74815 \\ 0.90982 & 1.0 & 0.93367 & 0.86366 \\ 0.83000 & 0.93367 & 1.0 & 0.87145 \\ 0.74815 & 0.86366 & 0.87145 & 1.0 \end{pmatrix},$$

which has all the appearance of satisfying a first-order stationary autoregressive time series process whose general form for $p = 4$ is as follows, where c is a constant,

$$\hat{\Sigma} = c(1-\rho^2)^{-1} \begin{pmatrix} 1 & \rho & \rho^2 & \rho^3 \\ \rho & 1 & \rho & \rho^2 \\ \rho^2 & \rho & 1 & \rho \\ \rho^3 & \rho^2 & \rho & 1 \end{pmatrix}.$$

Adopting nonlinear least squares as a practical means of finding the value of ρ that minimizes the function

$$(\rho-0.90982)^2+(\rho-0.93367)^2+(\rho-0.87145)^2+(\rho^2-0.83000)^2+$$
$$(\rho^2-0.86366)^2+(\rho^3-0.74815)^2$$

results in $\hat{\rho} = 0.9125$ as the best estimate of the autocorrelation. An estimate of the average variance may be obtained by averaging the four pooled variances of the data matrix, as follows:

$$(18.2091 + 16.9945 + 17.2972 + 13.3424)/4 = 16.461.$$

Multiplying this average variance by $\hat{\rho}$, $\hat{\rho}^2$ and $\hat{\rho}^3$, respectively, results in the following estimated first–order autoregressive matrix:

$$\hat{\Sigma} = \begin{pmatrix} 16.461 & 15.021 & 13.706 & 12.507 \\ 15.021 & 16.461 & 15.021 & 13.706 \\ 13.706 & 15.021 & 16.461 & 15.021 \\ 12.507 & 13.706 & 15.021 & 16.491 \end{pmatrix}.$$

To use this matrix in Equation (8.4), it is necessary to write down a model to be estimated, as the model determines the parameters in X. Since the data set contains two sequences of four periods each, there are eight cell means, so a saturated model will have eight parameters. In addition to the overall mean, there are six standard additional parameters (one sequence, three period, one treatment and one carryover) leaving one short of a saturated model. One of the choices for the remaining parameter is the treatment–by–carryover interaction. This interaction has been used by Jones and Kenward (1989), but is generally not recommended in this book because such an interaction may be difficult to interpret. To code this interaction, one simply multiplies the elements of the sixth column (treatment) by the elements of the seventh column (carryover). We will examine whether the model is estimable with these eight parameters using the ECHELON function of SAS/IML software (see Section 1.5).

The matrix consists of eight rows and eight columns, the rows being in the order of periods within sequences and the columns being in the order μ, γ_1, π_1,

π_2, π_3, τ_1, λ_1 and $(\tau\lambda)_{11}$, the last term being the treatment–by–carryover interaction.

$$\mathbf{X} = \begin{pmatrix} 1 & 1 & 1 & 0 & 0 & 1 & 0 & 0 \\ 1 & 1 & 0 & 1 & 0 & -1 & 1 & -1 \\ 1 & 1 & 0 & 0 & 1 & 1 & -1 & -1 \\ 1 & 1 & -1 & -1 & -1 & -1 & 1 & -1 \\ 1 & -1 & 1 & 0 & 0 & -1 & 0 & 0 \\ 1 & -1 & 0 & 1 & 0 & 1 & -1 & -1 \\ 1 & -1 & 0 & 0 & 1 & -1 & 1 & -1 \\ 1 & -1 & -1 & -1 & -1 & 1 & -1 & -1 \end{pmatrix} .$$

The echelon form of this matrix is

$$\begin{pmatrix} 1 & 0 & 0 & 0 & 0 & 0 & 0 & -1 \\ 0 & 1 & 0 & 0 & 0 & 0 & 0 & 0 \\ 0 & 0 & 1 & 0 & 0 & 0 & 0 & 1 \\ 0 & 0 & 0 & 1 & 0 & 0 & 0 & 0 \\ 0 & 0 & 0 & 0 & 1 & 0 & 0 & 0 \\ 0 & 0 & 0 & 0 & 0 & 1 & 0 & 0 \\ 0 & 0 & 0 & 0 & 0 & 0 & 1 & 0 \end{pmatrix} ,$$

which is not an identity matrix. The interaction parameter $(\tau\lambda)_{11}$ is confounded with the grand mean and with the first period parameter.

If, instead of the treatment–by–carryover interaction, one uses the sequence–by–treatment interaction, then the eighth column of the design matrix would be made up of the products of the elements of the second (sequence) and seventh (carryover) columns of $\mathbf{X}$ to give $(0,1,-1,1,0,1,-1,1)'$. This also results in an echelon form which is not an identity matrix. Similarly, the attempt to use a sequence–by–carryover interaction also fails to result in an identity matrix. Thus, none of the "natural" interactions leads to an estimable solution. It is possible, however, to find an eighth column vector which, together with the other seven columns, constitutes a saturated model. Consider the following design matrix which contains the same first seven columns (and therefore corresponding parameters) as above, but with an eighth column vector that leads to an identity matrix when put into echelon form:

$$X = \begin{pmatrix} 1 & 1 & 1 & 0 & 0 & 1 & 0 & 1 \\ 1 & 1 & 0 & 1 & 0 & -1 & 1 & 0 \\ 1 & 1 & 0 & 0 & 1 & 1 & -1 & 0 \\ 1 & 1 & -1 & -1 & -1 & -1 & 1 & -1 \\ 1 & -1 & 1 & 0 & 0 & -1 & 0 & -1 \\ 1 & -1 & 0 & 1 & 0 & 1 & -1 & 0 \\ 1 & -1 & 0 & 0 & 1 & -1 & 1 & 0 \\ 1 & -1 & -1 & -1 & -1 & 1 & -1 & 1 \end{pmatrix}.$$

The eighth column vector is the product of the elements of the sequence (γ_1) and first period (π_1) vectors.

If the above saturated model is used in conjunction with the estimated covariance matrix having a first–order autoregressive structure, the latter being substituted for Σ in Equation (8.4), the following empirical GLS estimates of the treatment and carryover parameters and their standard errors based on asymptotic theory, are obtained for the saturated model case:

$$\hat{\tau}_1 = 0.1645833 \pm 0.5288352$$

$$\hat{\lambda}_1 = 0.1958333 \pm 0.4996667$$

and

$$(\widehat{\tau\pi})_{11} = 0.3 \pm 0.4791486 .$$

If we now drop the parameter $(\tau\pi)_{11}$, as it is clearly non–significant, to give an *unsaturated* model having the seven parameters μ, γ_1, π_1, π_2, π_3, τ_1, λ_1, the empirical GLS estimates and standard errors, again using the autoregressive covariance structure, are

$$\hat{\tau}_1 = 0.4027049 \pm 0.3674566$$

and

$$\hat{\lambda}_1 = 0.3589549 \pm 0.4263683 .$$

One can remove the carryover parameter λ_1 from the model, as it is not significant, and estimate only six parameters, viz. μ, γ_1, π_1, π_2, π_3, τ_1, to yield the following estimate of τ_1:

$$\hat{\tau}_1 = 0.1052862 \pm 0.1016367 .$$

Note that the estimates of τ_1 vary greatly depending upon which other parameters are in the model. This is a consequence of the fact that the design ABAB/BABA is a poor one for separating the effect of treatment from the effect of carryover and the interaction between these two effects. There is a high variance inflation factor when the interaction terms are in the model. This is shown by the fact that the standard errors progressively decrease as parameters are dropped from the model. Since the treatment parameter is not significant, there does not appear to be any effect of treatment for this data set.

Now we examine the consequences of using other covariance matrices in Equation (8.4) to give an empirical GLS solution. We confine attention only to the unsaturated case where there are treatment and carryover parameters in the model, but not their interaction. That is, there are seven parameters, μ, γ_1, π_1, π_2, π_3, τ_1, λ_1. If the estimated uniform covariance structure is used, the estimates and standard errors of the two main parameters of interest are

$$\hat{\tau}_1 = 0.5395833 \pm 0.3671611$$

and

$$\hat{\lambda}_1 = 0.4958333 \pm 0.4428130.$$

These estimates are larger than those obtained using the estimated autoregressive structure, but the standard errors are about the same. The inference remains unchanged; one would not reject the null hypothesis of no difference for either treatment or carryover with either set of results.

If one now uses the estimated Type H covariance matrix of Huynh and Feldt (1970) in place of Σ, the results are

$$\hat{\tau}_1 = 0.5395833 \pm 0.3671391$$

and

$$\hat{\lambda}_1 = 0.4958333 \pm 0.4427864.$$

Thus, the estimates are identical to those obtained using the uniform covariance structure, and the standard errors are almost the same. Altogether, there is very little difference in the results despite the use of different forms of covariance structure.

If one now uses the method of Ware (1985) [see Section (8.4)] on this data set, that is, uses Equation (8.12) iteratively with the equation for β until convergence occurs, starting with the covariance matrix obtained from the data, the solution for the treatment and carryover parameters in the seven parameter unsaturated case is as follows:

$$\hat{\tau}_1 = 0.3977149 \pm 0.3997828$$

and

$$\hat{\lambda}_1 = 0.3866643 \pm 0.4775345.$$

Although these differ somewhat from the estimates obtained making other assumptions, the inference is the same. There does not seem to be a substantial difference between the various solutions, and this may be attributable to the fact that although the data set fits an autoregressive structure better than any other structure, the differences between them were not striking. This is not to say that other data sets that fit one covariance structure much better than any other will not show bigger differences in their parameter estimates and standard errors than what was observed with the data set examined here. It does appear, however, that when it is difficult to decide which covariance structure is the best, there will be non–substantial differences between the inference obtained. Any of the proposed methods should lead to similar inferences, whether it be the mixed-model OLS solution, the empirical GLS solution using an assumed or estimated covariance structure, or Ware's iterative method starting with any convenient covariance structure (as the converged solution will be the same irrespective of the starting point).

Exercises

8.1. Consider the data set of Gill (1978), p. 248, presented in Section 8.5.

(a) Fit this data set using SAS PROC GLM or some other package for the standard mixed model which contains sequence effects, cows within sequences, periods, treatments, and carryover. Note that the model will not be saturated.

(b) From the pooled covariance matrix determined from the raw data, construct a uniform covariance matrix. Using the test given in Appendix 8.A, test whether the uniform covariance matrix is consistent with the data.

(c) Consider the Type H covariance structure of Huynh and Feldt (1970). Starting with the pooled covariance matrix determined from the raw data, find, by use of least–squares, the best fitting values of λ, γ_1, γ_2, γ_3 and γ_4.

(d) Using the test in Appendix 8.B, test whether the fitted Type H matrix provides a good representation of the pooled covariance matrix.

Appendix 8.A. Formal Test for Uniform Covariance Structure

This test for whether an observed covariance matrix conforms to a uniform covariance structure is due to Box (1950). It is necessary to calculate the determinant of the $p \times p$ data matrix and also the determinant of the uniform covariance matrix obtained by replacing all diagonal elements by the average of the p variances, and replacing all the off–diagonal elements by the average of the $p(p-1)/2$ covariances located above the main diagonal. Thus, for the Ebbutt (1984) data set, the determinant of the data matrix

$$\hat{\Sigma} = \begin{pmatrix} 294.3 & 158.1 & 134.9 \\ 158.1 & 317.3 & 183.5 \\ 134.9 & 183.5 & 381.7 \end{pmatrix},$$

is calculated to be 18,246,143 while the determinant of the uniform covariance matrix

$$\hat{\Sigma} = \begin{pmatrix} 331.1 & 158.8 & 158.8 \\ 158.8 & 331.1 & 158.8 \\ 158.8 & 158.8 & 331.1 \end{pmatrix}$$

is calculated to be 19,258,145. The test involves calculation of the ratio of the determinant of the data matrix to the determinant of the uniform covariance matrix. This yields

$$\Delta = \frac{18246143}{19258145} = 0.94745 \ .$$

The next step is to calculate

$$M = \nu \log_e \Delta^{-1},$$

where ν is the number of degrees of freedom of the sample covariance matrix. For the Ebbutt data, this is 47, and so M becomes

$$M = 47 \log_e(0.94745)^{-1} = 2.5371.$$

Next one calculates $f_1 = (p^2 + p - 4)/2$

and

$$A_1 = \frac{\{p(p + 1)^2(2p - 3)\}}{\{6\nu(p - 1)(p^2 + p - 4)\}}$$

and refers $(1 - A_1)M$ to a table of the chi–square distribution with f_1 degrees of freedom.

For the Ebbutt data, $f_1 = 4$, as is the case for any data set of three periods, and $A_1 = 0.031915$, obtained by substituting $p = 3$ and $\nu = 47$ into the above equations. This results in

$$(1 - A_1) \, M = (1 - 0.031915) \, 2.5371 = 2.456$$

with $f_1 = 4$ degrees of freedom. The P–value corresponding to the above statistic is 0.6525. Clearly, the data produce a covariance matrix which is consistent with the uniform covariance assumption.

Appendix 8.B. Formal Test for Huynh–Feldt Type H Structure

The formal test for whether a pooled within–sequence covariance matrix, obtained from a cross–over trial data set, satisfies the Type H structure which is necessary and sufficient to ensure the validity of the F–test in the usual split–plot type of analysis advocated in Section 1.1, is given by Theorem 4 of Huynh and Feldt (1970). The test is based on the sphericity criterion of Mauchly (1940). One proceeds by choosing a $(p - 1)$ x p matrix C which has orthogonal rows which are also orthogonal to a row matrix of 1's. For $p = 3$, such a matrix can be chosen to be

$$C \; = \; \begin{pmatrix} 1/\sqrt{2} & -1/\sqrt{2} & 0 \\ 1/\sqrt{6} & 1/\sqrt{6} & -2/\sqrt{6} \end{pmatrix},$$

while for $p = 4$, the matrix can be written as

$$C = \begin{pmatrix} 3/\sqrt{12} & -1/\sqrt{12} & -1/\sqrt{12} & -1/\sqrt{12} \\ 0 & 2/\sqrt{6} & -1/\sqrt{6} & -1/\sqrt{6} \\ 0 & 0 & 1/\sqrt{2} & -1/\sqrt{2} \end{pmatrix}.$$

The sample covariance matrix $\hat{\Sigma}$ is premultiplied by the C matrix and postmultiplied by the transpose of C. Then the determinant of this matrix product is taken. Thus, for the Ebbutt (1984) data,

$$\hat{\Sigma} = \begin{pmatrix} 294.3 & 158.1 & 134.9 \\ 158.1 & 317.3 & 183.5 \\ 134.9 & 183.5 & 381.7 \end{pmatrix},$$

and one obtains

$$|C\hat{\Sigma}C'| \; = \; 28613.5 \; .$$

The trace of that same matrix is found to be

$$\text{tr } C\hat{\Sigma}C' \; = \; 344.533$$

and Mauchly's W is

$$W = |C\hat{\Sigma}C'|/[\text{tr } C\hat{\Sigma}C'/q]^q,$$

where $q = p - 1$, one less than the number of periods.

For the Ebbutt data set, this yields

$$W = 28613.5/29675.8 = 0.9642.$$

The closer W is to 1.0, the more closely the covariance matrix conforms to a Type H matrix. The statistic

$$-\nu \, d \log_e W \, ,$$

where $d = 1 - (2q^2 + q + 2)/6q\nu$

and ν is the degrees of freedom of the covariance matrix, is distributed as chi–square with

$$f = pq/2 - 1$$

degrees of freedom. For the Ebbutt data,

$$\nu = 47, \ d = 46/47 = 0.97872, \text{ and}$$

$$-\nu \, d \log_e W = -47 \, (0.97872) \log_e (0.9642) = 1.677$$

with $f = 2$ degrees of freedom. The P–value corresponding to this is 0.432, and clearly the Ebbutt data set has a covariance matrix which conforms well with the Type H structure.

9
Other Topics in Cross-over Designs

9.1. The Bayesian Analysis of Cross-over Designs

The methods expounded in the preceding chapters can be classified as falling under the umbrella of the "frequentist", "traditional", "classical" or "sampling theory" school of statistical inference. Interestingly, practitioners of that brand of inference rarely apply these labels to themselves, but the terms are often used by people who consider themselves to be "Bayesians", after the Rev. Thomas Bayes, whose treatises on inference were published two years after his death in April, 1761. Although Bayes' Theorem, which is taught in introductory courses in probability theory, is of unquestioned importance, Bayes himself is believed to have had grave reservations about the inferential method that now bears his name. In consequence, he is said to have held back two papers on the subject, rather than seek their publication. They were forwarded for publication in the Philosophical Transactions of the Royal Society of London after Bayes' death by his friend Richard Price, who felt that the papers were of sufficient importance to to merit publication. Readers interested in more details should see the accounts by Press (1989) and Stigler (1986).

Bayesian inference differs from classical inference essentially in that the likelihood function (that is, the likelihood of the observed set of data as a function of the unknown parameters) is multiplied by another function representing one's prior belief in the values of the parameters under consideration. The product of these two functions, referred to as the posterior distribution, can then be described in various ways. For example, a 95% (or any other percentage) highest posterior density (HPD) region may be calculated, which becomes the equivalent, in the "classical" approach, to a 95% confidence interval. A large part of the controversy about the Bayesian school of inference has centered around the difficulty, and subjectivity, of defining prior probabilities. Differing individuals may have different "degrees of belief" in these priors, resulting in potentially different inferences. Attempts have been made to make the choice of prior probabilities less subjective by introducing such concepts as "uninformative" or "vague" priors, and "natural conjugate priors" which have convenient mathematical properties (and are therefore also called "convenience" priors). Since the stated aim of Bayesian inference is to incorporate prior information into the analysis, the manner in which such information is incorporated becomes a question of vital importance for that system of inference. Part of the reason why the Bayesian outlook has not been widely adopted by practicing statisticians may be largely due to the failure of Bayesian protagonists to provide examples from real data situations in which results from previous studies are used to calculate prior probabilities which are then employed as priors in a future study. Instead, much of the debate and rhetoric about Bayesian inference seems to rest on theoretical arguments, such as about the virtue of various theoretical priors. The Bayesian cause is also not helped by its insistence on such concepts as the "likelihood principle" (not to be confused with Fisher's maximum likelihood) and "coherence", principles whose adoption immediately excludes the classical school of inference on an *a priori* basis.

In the field of cross–over designs, there has been a series of papers on the use of Bayesian inference by A. P. Grieve and co–workers [Racine *et al.*, 1986, Grieve (1985), Grieve (1987a and b)]. One of these (Racine *et al.*, 1986), read before a meeting of the Royal Statistical Society, with discussion, provoked the discussants to raise many questions regarding the use of Bayesian methodology in practice. Regrettably, the only cross–over trial that was illustrated in that and other papers is the 2–sequence, 2–treatment, 2–period trial, which we have discussed in detail in Chapter 3 for the case of a continuous response variable and in Chapter 7 for the case of a binary response variable. For both kinds of data, we have shown that there is insufficient information for estimation of an overall mean together with sequence, period, treatment and carryover parameters.

Therefore, the model used by Racine *et al.* (1986) did not include sequence parameters, an omission pointed out by P. R. Freeman in the discussion to that paper. Instead, in the fashion that has become typical of those whose think that a carryover effect and a sequence effect are synonymous, the Bayesian analysis of Racine *et al.* (1986) proceeds without a sequence parameter. It can be of no surprise, therefore, that the marginal posterior distribution for the treatment parameter in the presence of a carryover parameter is very different from the marginal posterior distribution for the treatment parameter when the carryover parameter is not in the model (see Fig. 3 of that paper), which might lead to vastly different conclusions about the treatment effect. We emphasize that this problem has nothing to do with differences between classical and Bayesian viewpoints, as both systems of inference are unable to cope with unreasonable models. In the case of the 2 x 2 cross-over design, the impossibility of estimating five parameters from four cell means should cause practitioners to abandon that design, rather than to attempt to salvage information by making the untestable assumption that there will be no sequence differences. The fact that the treatment and carryover parameters are poorly separable, as described in Chapter 3 using the measure presented in Section 1.2, is the real source of the large discrepancy between the two marginal posterior distributions. This result is directly analogous to the behavior observed when this design was examined by traditional methods in Chapter 3. The explanation was seen there to be closely tied to the related concepts of "variance inflation" and "efficiency". J.N.S. Matthews, in his discussion of Racine *et al.* (1986), noted that the advantage of the 2 x 2 trial over parallel groups trials of comparing treatments using only within–subject variation "is vitiated if a carryover effect is present", and recommended alternative two–treatment designs such as are found in Kershner and Federer (1981). Some of these designs have been discussed in Chapter 4, where we have strongly recommended the 2–sequence, 3–period design ABB/BAA as having exceptionally good properties.

There is no doubt that Bayesian methodology will play an increasingly important role in the analysis of cross–over trials in the future, especially as software for carrying out the sometimes complex analyses becomes more readily available. We await the publication of Bayesian analyses of results from better cross–over designs than the 2 x 2 design, and their comparison with conclusions drawn from the classical system of inference such as have been presented in this book. It will be of interest to see what indeed are the differences in conclusions drawn by the two approaches.

9.2. Repeated Measurements Within a Cross–over Design

As pointed out in the Introduction to this book, a cross–over trial is a form of repeated measurements experiment, since each experimental unit is used on more than one occasion. A second level of repeated measurements may occur if more than one measurement, over time, is taken after each treatment period. That is, considering the milk yield example of Section 1.1, suppose that, instead of a single cumulative milk yield for the entire period after the application of the specific experimental diet to each animal, the milk yields over a series of successive three–day periods were recorded, with the idea that there may have been a peak or trough in the milk production curve. That would be an example where the response would consist of a series of measurements over time, rather than a single response for each treatment period. Now consider the experiment of Ciminera and Wolfe (1953), whose aim was to study the effect of two mixtures of a new preparation of insulin, designated as NPH insulin, on a patient's blood sugar, in the management of diabetes. One of the mixtures, labeled B, contained 5% less protamine than the mixture labeled A. The experimental subjects consisted of 22 female rabbits, randomly allocated to two sequence groups, ABAB and BABA, and injected with the mixtures at weekly intervals. The scheme may be summarized as follows.

	Period (Week)			
Seq.	1	2	3	4
↓				
1	A	B	A	B
2	B	A	B	A

It should immediately be remarked that this is a very poor design in terms of the separability of treatment and carryover effects, since Treatment A is always followed by Treatment B, and vice versa, except at the end of the sequence. A contingency table relating treatment and carryover can be written down as follows:

<pre>
                          CARRYOVER
                        0   A   B

      TREATMENT   A   1   0   3   4
                   B   1   3   0   4

                        2   3   3   8 .
</pre>

The Pearson chi–square value for this design is 6.0. When converted into Cramèr's V (see Section 1.2),

$$V = [(6/8)/\min(1,2)]^{1/2} = 0.866.$$

It follows that the "efficiency" of separation of treatment and carryover effects is a very low 13.4%. It appears that designs of this type, where two treatments alternate in successive periods, has a long history. For example, such trials were described by Brandt (1938) under the names "reversal" or "switchback" trials. It must be said that in such experiments, no thought was given to the possibility of carryover effects. Once an experimenter realizes that the possibility of a carryover effect must be taken into account in the choice of experimental design, a switchback trial then is not a suitable choice.

Nevertheless, we will persevere with the analysis of the data from the experiment of Ciminera and Wolfe (1953), since the *method* of analysis, as is the case for all other designs in this book, does not depend upon the efficiency of the design. In that experiment, after each weekly injection of the appropriate insulin mixture, blood sugar levels were determined from each rabbit at 0 (initial level), 1.5, 3.0, 4.5 and 6.0 hours after injection. The data from this experiment are given in Table II of Ciminera and Wolfe (1953), and will not be given in full here. As pointed out by those authors, "the physiological relation between time and blood sugar levels, following insulin injection, is quite complex and is not adequately described by any simple function". What is obvious from a graph of blood sugar level versus time is that the level falls to a minimum, and then increases again. One way of analyzing repeated measurements data in time is to fit some model to the time series curve, which results in certain parameters that describe the curve. Then each parameter can be analyzed in the univariate sense. For the present data, since each animal experiences a minimum sugar level, a suitable variable for analysis seems to be that minimum level if it can be found. One approach could be to fit a model, such as a quadratic, which is capable of describing a minimum in a curve. However, in view of the physiologically complex nature of the response, a quadratic is not a suitably accurate model. For simplicity, we will use the lowest value observed in any of the five time periods as the variable for analysis. Although this is not the true minimum response, this observed minimum almost always occurred either at 1.5 hours or 3.0 hours after injection. There were few cases where the minimum occurred at 4.5 hours, and it never occurred either at 0 hours or 6.0 hours, except for Rabbit 8 in Period 1 where the minimum observed after 3.0 hours was repeated at 6.0 hours. Table 9.1 lists the data of Ciminera and Wolfe (1953), modified so that only the minimum observed blood sugar level is shown.

One can now analyze these data in the same fashion as in Section 1.1. The code using SAS might look like this (see top of following page):

Table 9.1.   Minimum (observed) blood sugar levels in two groups of rabbits injected with one of two insulin mixtures in Sequence 1 (ABAB) or Sequence 2 (BABA).   [Adapted from Ciminera and Wolfe (1953), reproduced with the permission of the Biometric Society.]

Sequence 1		Period 1	Period 2	Period 3	Period 4
Treatment		A	B	A	B
Rabbit No.	1	35	47	35	60
	2	43	52	26	60
	3	39	30	39	60
	4	22	47	26	26
	5	52	60	18	47
	6	22	30	56	73
	7	47	26	18	43
	8	22	8	22	18
	9	47	39	26	47
	10	12	26	56	22
	11	56	56	47	52
Sequence 2		Period 1	Period 2	Period 3	Period 4
Treatment		B	A	B	A
Rabbit No.	12	22	30	26	52
	13	56	43	22	52
	14	8	30	12	47
	15	35	39	12	52
	16	35	47	35	52
	17	47	35	18	43
	18	35	30	56	60
	19	12	12	8	12
	20	22	26	22	64
	21	33	39	26	35
	22	35	26	22	35

```
DATA CIMINERA;
INPUT Y RABBIT PERIOD TREAT $ CARRY $ SEQUENCE @@;
IF CARRY = '0' THEN CARRY='B';
CARDS;
35  1 1 A 0 1     47  1 2 B A 1     35  1 3 A B 1     60  1 4 B A 1
43  2 1 A 0 1     52  2 2 B A 1     26  2 3 A B 1     60  2 4 B A 1
39  3 1 A 0 1     30  3 2 B A 1     39  3 3 A B 1     60  3 4 B A 1
22  4 1 A 0 1     47  4 2 B A 1     26  4 3 A B 1     26  4 4 B A 1
52  5 1 A 0 1     60  5 2 B A 1     18  5 3 A B 1     47  5 4 B A 1
22  6 1 A 0 1     30  6 2 B A 1     56  6 3 A B 1     73  6 4 B A 1
47  7 1 A 0 1     26  7 2 B A 1     18  7 3 A B 1     43  7 4 B A 1
22  8 1 A 0 1      8  8 2 B A 1     22  8 3 A B 1     18  8 4 B A 1
47  9 1 A 0 1     39  9 2 B A 1     26  9 3 A B 1     47  9 4 B A 1
12 10 1 A 0 1     26 10 2 B A 1     56 10 3 A B 1     22 10 4 B A 1
56 11 1 A 0 1     56 11 2 B A 1     47 11 3 A B 1     52 11 4 B A 1
22 12 1 B 0 2     30 12 2 A B 2     26 12 3 B A 2     52 12 4 A B 2
56 13 1 B 0 2     43 13 2 A B 2     22 13 3 B A 2     52 13 4 A B 2
 8 14 1 B 0 2     30 14 2 A B 2     12 14 3 B A 2     47 14 4 A B 2
35 15 1 B 0 2     39 15 2 A B 2     12 15 3 B A 2     52 15 4 A B 2
35 16 1 B 0 2     47 16 2 A B 2     35 16 3 B A 2     52 16 4 A B 2
47 17 1 B 0 2     35 17 2 A B 2     18 17 3 B A 2     43 17 4 A B 2
35 18 1 B 0 2     30 18 2 A B 2     56 18 3 B A 2     60 18 4 A B 2
12 19 1 B 0 2     12 19 2 A B 2      8 19 3 B A 2     12 19 4 A B 2
22 20 1 B 0 2     26 20 2 A B 2     22 20 3 B A 2     64 20 4 A B 2
33 21 1 B 0 2     39 21 2 A B 2     26 21 3 B A 2     35 21 4 A B 2
35 22 1 B 0 2     26 22 2 A B 2     22 22 3 B A 2     35 22 4 A B 2

PROC GLM;
CLASS RABBIT PERIOD TREAT CARRY;
MODEL Y=RABBIT PERIOD TREAT CARRY/SOLUTION SS1 SS2;
RANDOM RABBIT;
LSMEANS TREAT CARRY/PDIFF;
RUN;
```

Below is shown an edited portion of the output from the above code:

SOURCE	DF	SUM OF SQUARES	MEAN SQUARE	F VALUE
MODEL	26	12057.57955	463.75306	3.18587
ERROR	61	8879.50000	145.56557	
TOTAL	87	20937.07955		

Source	df	Type II SS	Mean Square	F value	Pr > F
TREAT	1	7.43802	7.43802	0.05110	0.8219
CARRY	1	63.84091	63.84091	0.43857	0.5103
Error	61	8879.50000	145.56557		

From the above output, it is clear that neither carry effects nor treatment effects appear to be significant. It is possible, however, that the low F–values (and high P–values) are a manifestation of the "collinearity" effect described in Section 1.3, resulting from the design that was used being a poor one for separating treatment and carryover effects.

Hence, we carry out a second run, in which the carryover term is eliminated from the model. In the absense of carryover, there is no need for the statement

IF CARRY = '0' THEN CARRY='B';,

and this should be removed from the data step. Eliminating the "carry" term from the model statement, so that the latter reads

MODEL Y=RABBIT PERIOD TREAT/SOLUTION SS1 SS2;,

results in an orthogonal analysis, in which the Type I and Type II sums of squares are the same. The resulting analysis of variance table is as follows:

Source	df	Sum of Squares	Mean Square	F value	Pr > F
RABBIT	21	8315.32955	395.9681	2.74506	0.0011
PERIOD	3	3567.03409	1189.0114	8.24286	0.0001
TREAT	1	111.37500	111.3750	0.77211	0.3830
Error	62	8943.34091	144.24743		

Treatment effects remain non–significant, but the sum of squares of 111.375 is considerably higher than the value obtained (7.438) when the carryover parameters were in the model. This is a further illustration of the phenomenon described in Section 1.3, where a large reduction in the treatment sum of squares is experienced when a carryover term is added to the model in a poor design for separating treatment and carryover effects. This is one persuasive reason why such a design should not be used for a cross–over trial, where the possibility of a significant carryover effect is an ever–present one.

The conclusion obtained for the above example is that the two treatments do not differ in their ability to control a patient's blood sugar content over a relatively long period of time. However, the above table indicates a very strong period effect. The period means are given in the following table. Since the model in the absence of carryover effects is orthogonal, these means are the raw means as well as the "least squares" means.

Period	Average Blood Sugar Levels		
	Seq. 1	Seq. 2	Overall Average
1	36.09	30.91	33.5
2	38.27	32.45	35.4
3	33.55	23.55	28.5
4	46.18	45.82	46.0

From these means, it is clear that there is a marked increase in blood sugar level in the fourth period after a decrease in the third period. Since this occurs in both sequences, it cannot be ascribed to differences between the two treatments. Ciminera and Wolfe (1953) attribute the change in blood sugar level "to the rabbits becoming less sensitive to insulin in the last two periods of the experiment". Although this may explain the results in the fourth period, it does not explain the low blood sugar results in the third period, suggesting that the rabbits were *more* sensitive to the insulin at that stage of the experiment.

In this section, we have only illustrated one approach to dealing with repeated measurements within a cross–over trial. In the example, we reduced the set of five measurements during each period to a single measurement, the observed minimum blood sugar content. This seems reasonable, since the purpose of the experiment was to test the comparative ability of two insulin muixtures to lower blood sugar level. Suppose, however, that the measure of efficacy included assessing the speed at which the blood sugar was lowered. Assuming that the insulin mixture would be considered ineffective if the blood sugar level did not respond within a specified period of time, say 3 hours, then the response variable of interest might be the reading at 3 hours, rather than the minimum reading. In other applications, it may also be appropriate to reduce a set of repeated measurements to one or a few summary statistics. For example, the slope of a response curve at time zero might be a more suitable measure of the rate at which a process is occurring than the set of readings taken over time. The slope would then be analyzed in the univariate sense, as has been illustrated in the above example. Other, more complex methods of analysis of repeated measurements data are outside the scope of this book.

9.3. Optimality in Cross-over Experimental Designs

In Chapters 2–4, we tried to select the best designs among classes of designs, such as Latin squares, or designs having a fixed number of treatments and periods, by calculating formulae for the variances of the estimators of the treatment and carryover parameters. If the variances of the treatment and carryover parameters in a particular design were less than the corresponding values for some other design, we concluded that the first design was the better one. This method of calculation can be tedious, involving the setting up of matrix equations and their inversion. Various authors (for example, Hedayat and Afsarinejad, 1978; Laska, Meisner and Kushner, 1983; Cheng and Wu, 1980; Dey, Gupta and Singh, 1983; Afsarinejad, 1985; Laska and Meisner, 1985; Kershner, 1986; Gill and Shukla, 1987; Matthews, 1987) have sought methods of finding designs that can be considered to be "optimal", in some sense. Various criteria for optimality exist in the statistical literature, such as D–, A– and E–optimality, which respectively involve the determinant, the trace, and the maximum eigenvalue of the covariance matrix of the parameters or contrasts of interest in the design. Kiefer (1975) introduced the concept of "universal optimality", which subsumed D–, A– and E–optimality; that is, a design that is universally optimal also satisfies the other optimality conditions.

Finding the universally optimal cross–over design for any given class (for example, of t–treatment, p–period designs of any given number of sequences) is complicated by the fact that the dependence between the responses in different periods on the same individual may be of difference forms, and the results are, in general, dependent upon a particular structure. The mixed–model type of analysis of variance that we have presented in Section 1.1 and which we have used extensively throughout this book assumes that, as has been detailed in Chapter 8, the variances and covariances between the observations on individuals satisfy the Type H structure of Huynh and Feldt (1970). Special cases of that necessary and sufficient condition are the sufficient (but not necessary) conditions of uniform covariance (compound symmetry) and independence. [Note: a common mistake in the statistical literature, for example Gill and Shukla (1987),

Matthews (1987), and Williams (1987), is to state that the standard mixed model assumes uncorrelated errors between successive measurements on individuals. It assumes, rather, a particular form of correlated error structure, namely the Type H form.] Sometimes an autocorrelated error structure is assumed, generally a first–order autoregressive (Markov) structure. For example, Gill and Shukla (1987) did not include a carryover term in their linear model, attempting to account for carryover by an autoregressive term on the previous response. This leads to certain universally optimal designs that would not, in our opinion, be good designs to use in cross–over trials. For example, Gill and Shukla (1987) obtained the following optimal design for three treatments and six periods:

	Period					
Seq. ↓	1	2	3	4	5	6
1	A	B	C	A	B	C
2	C	A	B	C	A	B
3	B	C	A	B	C	A

It is difficult to reconcile the supposed optimality of this design with the principles enunciated in Section 1.2, where "separability" of treatment and carryover effects was emphasized. In the above design, Treatment A is always followed by Treatment B (except when A appears in the last period), Treatment B is always followed by Treatment C, etc. This results in the following contingency table,

		CARRYOVER				
		0	A	B	C	
	A	1	0	0	5	6
TREATMENT	B	1	5	0	0	6
	C	1	0	5	0	6
		3	5	5	5	18

which has a Pearson chi–square of 30.0. When converted into Cramèr's V (see Section 1.2),

$$V = [(30/18)/\min(2,3)]^{1/2} = 0.913.$$

It follows that the "efficiency" of separation of treatment and carryover effects is

$$S = 100(1 - V) = 8.71\%,$$

an extremely low efficiency of separation of direct and residual treatment effects in this design.

Another example of a universally optimum design given by Gill and Shukla, obtained using their first-order autoregressive model, is as follows:

	Period				
Seq. ↓	1	2	3	4	5
1	A	B	C	E	D
2	B	C	E	D	A
3	C	E	D	A	B
4	E	D	A	B	C
5	D	A	B	C	E
6	A	C	D	B	E
7	C	D	B	E	A
8	D	B	E	A	C
9	B	E	A	C	D
10	E	A	C	D	B

This results in the following contingency table,

				CARRYOVER				
		0	A	B	C	D	E	
	A	2	0	0	0	4	4	10
	B	2	4	0	0	4	0	10
TREATMENT	C	2	4	4	0	0	0	10
	D	2	0	0	4	0	4	10
	E	2	0	4	4	0	0	10
		10	8	8	8	8	8	50 .

which has a Pearson chi-square of 60.0. When converted into Cramèr's V, one obtains

$$V = [(60/50)/\min(4,5)]^{1/2} = 0.548,$$

and it follows that the separation efficiency of treatment and carryover effects is

$$S = 100(1 - V) = 45.2\%.$$

As was the case with the previous example, the source of the poor behavior is the fact that each treatment is not equally preceded and followed by each other treatment. Although the above separability is not as low an value as in the previous example, one would not wish to use this design when others are available which have higher efficiencies.

The above examples show that an optimally efficient cross-over design in the sense of Kiefer (1975) is not necessarily a "good" design for practical use. For example, Laska and Meisner (1985) using a random-subjects mixed model such as we have employed throughout this book, in concert with a uniform covariance structure, show that the optimum design having two treatments and two periods is the AB/BA/AA/BB design that we have examined in Section 4.2. This is one of the two-period designs of Balaam (1968) and was previously labelled as Design 4.2 in Section 4.2, viz.

		Period	
		1	2
	1	A	B
Sequence	2	B	A
	3	A	A
	4	B	B

In that section, we showed that the inverse of the portion of the $\mathbf{X}'\mathbf{X}$ matrix pertaining to the treatment parameter τ_1 and the carryover parameter λ_1 was

$$(\mathbf{X}'\mathbf{X})^{-1} = \begin{pmatrix} 0.5 & 0.5 \\ 0.5 & 1.0 \end{pmatrix},$$

so that the correlation coefficient between these two parameters is a rather high

0.707. Examination of the variance inflation factors (VIF's) in the designs in the Appendices of Chapters 5 and 6 show that two period designs are always inefficient, a fact pointed out by Patterson and Lucas (1962). Thus, it is of little solace to a user of cross–over designs to know that a design may satisfy certain optimality criteria and still not be considered a "good" design.

For two treatment designs with more than two periods, Laska and Meisner (1985) found that, for $p=3$, the optimal design is ABB/BAA, that is, Design 4.1.1 of Chapter 4; for $p=4$, the optimal design is ABBA/BAAB/AABB/BBAA, that is, Design 4.4.11 of Chapter 4; for $p=5$, any one of three choices, viz. ABBAA/BAABB or AABBA/BBAAB or these first two choices combined, is optimal; for $p=6$, any one of three choices, viz. ABBAAB/BAABBA/AABBBA/ BBAAAB or ABBBAA/BAAABB/ABBAAB/BAABBA or these first two choices combined, is optimal. In all of the above cases, equal numbers of subjects are allocated to each sequence. The optimality does not depend upon the magnitude of the correlation between observations in successive times, provided a uniform covariance structure is assumed. Laska and Meisner (1985) also examined optimality using the same random–subjects mixed model with a carryover parameter but combining it with a first–order autoregressive process. This does not lead to any optimal solutions except for $p=2$, where the Balaam design AB/BA/AA/BB with one–quarter of the subjects allocated to each sequence proved, as before, to be optimal. For $p>2$, optimality depended upon the magnitude ρ, the Markov parameter.

Laska and Meisner (1985) found that, for $p=3$, the optimum design utilizes only the sequences AAB/BBA/ABB/BAA, but not in equal numbers. Depending upon ρ, the proportions of each sequence vary, but AAB and its dual BBA always had more than 90% of the subjects. This contrasts with the results when a uniform covariance, rather than first–order autoregressive structure, is assumed. In the former case, only the pair ABB/BAA was used. For $p=4$, Laska and Meisner (1985) found that, as before, the only sequences in the optimal design were AABB and ABBA and their duals BBAA and BAAB. For $\rho > 0.42$, only ABBA/BAAB were used in the optimal design, with AABB/BBAA appearing only for smaller correlations. Matthews (1987) studied a similar model to Laska and Meisner, with the sole difference that he considered subject effects to be fixed, rather than random. The question of whether subject effects should be considered to be fixed or random is always a vexed one. In this book we have taken the view that it is the purpose of an experiment to obtain information on the underlying population, rather than on the specific sample chosen. We recognize, however, that a truly random sample from that population is seldom available, with the experimental subjects often serving as a convenience sample. For example, in experiments involving animals, the ones chosen as subjects may be among a small set of those animals actually available; in clinical trials, the subjects may be the first n people who have a certain condition who have come in for treatment and have been asked to take part in a scientific experiment. For $p=4$, the results obtained by Matthews (1987) agree closely with those of Laska and Meisner (1985), but for $p=3$, there were some discrepancies. For $\rho > -0.2$, the use of ABB and its dual BAA tends to result in high efficiency, but for large negative ρ, it was better to use an unequal mixture of the four sequences ABB/BAA/AAB/BBA. In practice, a large negative correlation is not likely. In many experimental situations, a large positive correlation coefficient is caused by the fact that a subject who scores high in one period of the trial also tends to score high in the other periods. Although an antagonistic effect is possible, where a high performance in Period 1 results in a low performance in Period 2, followed by a high score in Period 3, and so on, such an outcome is unlikely and may not be tolerated for ethical reasons.

It is a rare circumstance when an experimenter knows in advance what is the correct form of the data's covariance matrix. We have seen for the two data sets used as examples in Chapter 8 and for the data in Exercise 8.1 that the Type H structure of Huynh and Feldt (1970) provides a reasonably good approximation to the covariance structure, and its special case of uniform covariance is also a reasonable approximation. For one data set a stationary first-order autoregressive process with a large positive estimate of ρ provided a good fit. Since there is a wide measure of agreement as to the optimal design for these covariance structures, the designer of cross-over trials probably will not be far wrong by choosing solutions that are optimal when a uniform covariance structure applies. This means that for two treatments, the designs recommended in Chapter 4 are likely to be good ones to use.

9.4. Missing Values in Cross-over Designs

The general method for coding treatment and carryover effects described in Section 1.1 may be employed without modification for the case where there are some missing values in the data set. This is no surprise, as the basis for the statistical analysis is a *regression* model rather than an *analysis of variance* model. Since orthogonality of the effects is not a requirement in regression models, missing data are easily handled. To illustrate the procedure, consider the milk yield data set of Table 1.1. This is reproduced below as Table 9.1, modified such that the reading in Period 2 for Cow 2 is now assumed to be missing, for the sake of illustration. To designate this, the corresponding milk yield is replaced by a dash to indicate a missing value.

Table 9.1. Pair of Balanced Latin Squares for Milk Yield Example,
Cochran and Cox (1957), with one reading assumed missing

	Square 1				Square 2		
Period→	1	2	3	Period→	1	2	3
Cow↓				Cow↓			
1	A 38	B 25	C 15	4	A 86	C 76	B 46
2	B 109	C –	A 39	5	B 75	A 35	C 34
3	C 124	A 72	B 27	6	C 101	B 63	A 1

The coding using SAS is exactly the same as in Section 1.1, with one or two exceptions. The first is that the symbol '.' is used to represent the missing reading for Period 2 of Cow 2 in the DATA step. A second modification may occur to the character variable CARRY (representing carryover) in the *following* period, depending upon the circumstances which caused the missing value. Had the prescribed diet 'C' been given to Cow 2 as planned, but some event occurred later that resulted in the milk yield being incorrectly or inaccurately recorded, or unavailable (for example, suppose that the cow had kicked over the milk pail during one of the milking sessions, an ever–present hazard in the era before the invention of milking machines!), there would be no modification to the coded value of 'CARRY' in Period 3. That is, one would still use a 'C' for the carryover variable, since Treatment C was indeed the treatment that was actually given in Period 2. A different situation would occur if, for some reason, that treatment period had been omitted. This is more likely to occur in clinical trials with humans, where a subject may indeed completely miss a treatment period. Both of these cases will be illustrated.

For the case where the correct diet was given, but the reading of milk yield for Cow 2, Period 2, was inaccurate or unavailable, the code would look as follows:

```
DATA MILKYLD;
INPUT Y SQUARE COW PERIOD TREAT $ CARRY $ @@;
IF CARRY='0' THEN CARRY='C';
CARDS;
    38  1  1  1  A  0      25  1  1  2  B  A      15  1  1  3  C  B
   109  1  2  1  B  0       .  1  2  2  C  B      39  1  2  3  A  C
   124  1  3  1  C  0      72  1  3  2  A  C      27  1  3  3  B  A
    86  2  4  1  A  0      76  2  4  2  C  A      46  2  4  3  B  C
    75  2  5  1  B  0      35  2  5  2  A  B      34  2  5  3  C  A
   101  2  6  1  C  0      63  2  6  2  B  C       1  2  6  3  A  B
RUN;

PROC GLM;
CLASS COW PERIOD SQUARE TREAT CARRY;
MODEL Y=COW PERIOD(SQUARE) TREAT CARRY/SOLUTION
    SS1 SS2 E1 E2;
RANDOM COW;
LSMEANS TREAT CARRY/PDIFF;
RUN;
```

Note that the statement

IF CARRY='0' THEN CARRY='C';

is necessary if one is to obtain estimable least squares means for treatment and carryover effects, because the confounding between carryover into Period 1 and the Period 1 parameter, which is a characteristic of cross–over designs, still prevails. The output (edited) from the above commands looks as follows:

SOURCE	DF	SUM OF SQUARES	MEAN SQUARE	F VALUE
MODEL	13	19471.72059	1497.82466	51.80
ERROR	3	86.75000	28.91667	
TOTAL	16	19558.47059		

The missing observation results in the loss of a degree of freedom in the residual or "error" term. The part of the output that relates to the test of whether there are significant treatment or carryover effects is as follows:

Source	df	Type II SS	Mean Square	F value	Pr >F
TREAT	2	2705.0	1352.5	46.77	0.00548
CARRY	2	547.2	273.6	9.46	0.05062 .

One can see that only very minor changes have occurred, by comparing the above results to the output in Section 1.1, as a result of assuming that one of the values of the response variable was lost, contaminated, or otherwise uncertain, making it advisable to assume it to be missing. Treatment effects are still highly significant, and there is still some marginal evidence of carryover effects.

Now let us consider a second assumed case where the diet that was supposed to have been given to Cow 2 in Period 2, namely Treatment C, was not actually given, but that the cow was given a standard diet that was not one of the treatment diets. Although this exercise is a hypothetical one, it is not unusual in agricultural experiments to use a "washout" period that consists of returning the experimental animals to a standard diet, which may not be one of the treatment diets. Thus, let us assume that a mistake was made, and that Cow 2, instead of being given the Treatment C diet in Period 2, was left on the standard diet. The consequence of this, assuming that carryover effects never exceed first order, is that there is no carryover effect into Period 3. Thus, the carryover is similar to that which occurs from the "zero" period into Period 1, and can also be coded as '0'. Thus, the code using SAS would be as follows,

```
DATA MILKYLD;
INPUT Y SQUARE COW PERIOD TREAT $ CARRY $ @@;
CARDS;
    38  1  1  1  A  0     25  1  1  2  B  A     15  1  1  3  C  B
   109  1  2  1  B  0      .  1  2  2  C  B     39  1  2  3  A  0
   124  1  3  1  C  0     72  1  3  2  A  C     27  1  3  3  B  A
    86  2  4  1  A  0     76  2  4  2  C  A     46  2  4  3  B  C
    75  2  5  1  B  0     35  2  5  2  A  B     34  2  5  3  C  A
   101  2  6  1  C  0     63  2  6  2  B  C      1  2  6  3  A  B
RUN;

PROC GLM;
CLASS COW PERIOD SQUARE TREAT CARRY;
MODEL Y=COW PERIOD(SQUARE) TREAT CARRY/SOLUTION
    SS1 SS2 E1 E2;
RANDOM COW;
LSMEANS TREAT CARRY/PDIFF;
RUN;
```

the differences being that carryover for Cow 2, Period 3, is coded as '0' instead of as 'C', and the fact that the statement,

IF CARRY='0' THEN CARRY='C'

is not present. The reason for this is that the carryover value coded '0' for Cow 2 in Period 3 means that '0' carryover values no longer *always* occur into Period 1, breaking the previous confounding between period and carryover. The output resulting from the above coding is as follows:

SOURCE	DF	SUM OF SQUARES	MEAN SQUARE	F VALUE
MODEL	14	19475.87443	1391.13389	33.69
ERROR	2	82.59615	41.29808	
TOTAL	16	19558.47059		

with the portion of the analysis of variance table that is appropriate for testing the treatment and carryover effects being

Source	df	Type II SS	Mean Square	F value	Pr >F
TREAT	2	2576.7	1288.4	31.20	0.03106
CARRY	3	551.4	183.8	4.45	0.18892 .

Thus, it should be clear that the presence of missing values poses no problems for the analysis of cross-over data. Incomplete data, from subjects who do not complete the trial, or who have missed one or more of intermediate periods in the trial, can thus be utilized. Even subjects who enter the trial late, having missed the first period, can still be accommodated, the modification to the coding for the carryover parameter being obvious (see Exercise 9.1).

9.5. Summary and Conclusions

In this book, we have attempted to present a consistent approach to the analysis of cross-over designs. We have shown that, for a continuous response variable, there is a general method given in Section 1.1 of coding data sets from *any* cross–over design which enables treatment and carryover effects to be estimated. Because the approach is model–based, it is free of *ad hoc* decisions. Few assumptions are made; these are, that the response variable is normally distributed, that the covariance matrix underlying the observations is homogeneous for each subject, and that its form, in its most general sense, satisfies the Type H structure of Huynh and Feldt (see Chapter 8). When these conditions are fulfilled, the analysis of variance described in Chapter 1 is a correct analysis.

In this book, we have not confined our attention exclusively to the analysis of cross–over data, but also to the design of appropriate cross–over experiments. We have examined the 2–treatment, 2–period, 2–sequence cross–over design in great detail in Chapters 3 and 7 and have shown that it is a very flawed one, since it cannot be used to estimate true carryover. Its use can only be contemplated when there is unambiguously no possibility of the effect of a treatment persisting into the next treatment period, but it is difficult to identify the circumstances when one can be absolutely sure of that. We recommend that the best policy is to abandon use of the 2 x 2 design, and to seek alternatives. An attractive alternative when two treatments are being tested is the 2–sequence, 3–period design, Design 4.1.1, described in Section 4.1,

Design 4.1.1

	Period		
Seq. ↓	1	2	3
1	A	B	B
2	B	A	A

This design has the desirable property of orthogonality of the treatment and carryover effects. Thus, in terms of the concept of "separability" of these two kinds of effects, an omnibus test for which was given in Section 1.2, these two effects are completely separable. A consequence of this is that the estimates of the treatment parameters and its contribution to the explained sum of squares is the same whether or not the carryover parameters are in the model. This design is optimal (see Section 9.3) among designs with two treatments and three periods.

For more than two treatments, the digram–balanced Latin square designs (see Chapter 3) are very attractive, as they have a high degree of separability of treatment and carryover effects. The addition of an extra period to these designs makes the two kinds of effects orthogonal, maximizing their efficiency. The only barrier to *always* using extra–period designs is the fact that there are other factors in experimentation which dictate trying to make the number of treatment periods as few as possible. These include the increased tendency for subjects to "drop out" of the trial if it is prolonged, and ethical considerations which would apply if it is clear that a particular treatment is having a detrimental effect. In Chapter 5 and 6, a wide variety of designs are catalogued, of varying degrees of efficiency. For example, it is seen that *all* two period designs are inefficient. Therefore, the dictates of experimentation that suggest a small number of periods is countered by the fact that efficiency tends to increase with increasing number of periods, with two period designs being particularly inefficient.

Although balanced designs tend to be more powerful than ones having missing values, the method of coding and the approach to the analysis of data recommended in Section 1.1 can handle any degree of imbalance, as described in Section 9.4. Despite this fact, it is prudent to choose efficient designs, and to strive for optimum use of the experimental units available and time for experimentation that exists. The considerations that are best for data with a continuous response variable also turn out to be best for a binary or categorical response (see Chapter 7). The consequence of this is that one can seek a design for experimentation irrespective of the nature of the response variable.

This book did not attempt to address all questions that might arise in the use of cross–over design methodology, nor did it deal with considerations that are specific to one subject area. For example, the ethical considerations that would apply in a clinical trial with humans suffering from a certain medical condition are likely to be more stringent than those in educational or psychological experiments. Our intention was not to delve into questions that are specific to a given discipline, but to provide a basis for a *general* approach to the design and analysis of cross–over trials. If we have achieved nothing else, we hope that we have succeeded in convincing the reader to select cross–over designs that allow good separation of treatment from carryover effects, and to avoid *ad hoc* analyses of the data emerging from those designs.

Exercises

9.1. Consider the milk yield data of Table 1.1, Chapter 1. Pretend that Cow 2, Square 1, did not enter the trial until the second period.

(a) Show how the coding of the data for that experiment is affected by that assumption.

(b) Analyze this altered data set, paying particular attention to the test of treatment and carryover effects.

References

Afsarinejad, K. (1985). Optimal repeated measurements designs. *Statistics* **16**: 563–8.

Agresti, A. (1990). *The Analysis of Categorical Data*. Wiley, New York.

Appelbaum, M.I. and Cramer, E.M. (1974). Some problems in the nonorthogonal analysis of variance. *Psychological Bulletin* **81**: 335–43.

Balaam, L.N. (1968). A two–period design with t^2 experimental units. *Biometrics* **24**: 61–73.

Berenblut, I.I. (1964). Change–over designs with complete balance for first residual effects. *Biometrics* **20**: 707–12.

Bishop, Y.M.M., Fienberg, S.E. and Holland, P.W. (1975). *Discrete Multivariate Analysis: Theory and Practice*. MIT Press, Cambridge, Massachusetts.

Bliss, C.I. (1967). *Statistics in Biology*. Vol. 1. McGraw Hill, New York.

Bose, R.C., Clatworthy, W.H. and Shrikhande, S.S. (1954). Tables of partially balanced designs with two associate classes. *North Carolina Agricultural Experiment Station Technical Bulletin No. 107*, 53 pages.

Box, G.E.P. (1949). A general distribution theory for a class of likelihood criteria. *Biometrika* **36**: 317–46.

Box, G.E.P. (1950). Problems in the analysis of growth and wear curves. *Biometrics* **6**: 362–89.

Brandt, A.E. (1938). *Tests of Significance in Reversal or Switchback Trials*. Agricultural Experiment Station, Iowa State College of Agricultural and Mechanical Arts, Research Bulletin 234, Ames, Iowa.

Cheng, C.S. and Wu, C.F. (1980). Balanced repeated measurements designs. *Annals of Statistics* **8**: 1272–83. Corrigendum in (1983), **11**: 349.

Chinchilli, V.M. and Elswick, R.K. (1989). Multivariate models for the analysis of crossover experiments. *SAS Users Group International, Proceedings of the 14th Annual Conference*, April 9–12, 1989, San Francisco, CA, pp. 1267–1271.

Ciminera, J.L. and Wolfe, E.K. (1953). An exampale of the use of extended cross–over designs in the comparison of NPH insulin mixtures. *Biometrics* **9**: 431–46.

Clarke, P.S. and Ratkowsky, D.A. (1990). Effect of fenoterol hydrobromide and sodium cromoglycate individually and in combination on postexercise asthma. *Annals of Allergy* **64**: 187–90.

Clatworthy, W.H. (1955). Partially balanced incomplete block designs with two associate classes and two treatments per block. *Journal of Research of the National Bureau of Standards* **54**: 177–90.

Cochran, W.G., Autrey, K.M. and Cannon, C.Y. (1941). A double change–over design for dairy cattle feeding experiments. *J. Dairy Sci.* **24**: 937–51.

Cochran, W.G. and Cox, G.M. (1957). *Experimental Designs*. Second Edition, John Wiley and Sons, New York.

Cook, R.D. and Weisberg, S. (1982). *Residuals and Influence in Regression*. Chapman and Hall, London.

Cox, M.A.A. and Plackett, R.L. (1980). Matched pairs in factorial experiments with binary data. *Biometrical Journal* **22**: 697–702.

Cramér, H. (1951). *Mathematical Methods of Statistics*. Princeton University Press, Princeton, New Jersey.

Davis, A.W. and Hall, W.B. (1969). Cyclic change–over designs. *Biometrika* **56**: 283–293.

Dey, A., Gupta, V.K. and Singh, M. (1983). *Sankhya: The Indian Journal of Statistics* **45**, Series B, Pt. 2: 233–9.

Diggle, P.J. (1988). An approach to the analysis of repeated measurements. *Biometrics* **44**: 959–971.

Draper, N.R. and Smith, H. (1981). Applied Regression Analysis. 2nd. ed., Wiley, New York.

Dunsmore, I.R. (1981). Analysis of preferences in crossover designs. *Biometrics* **37**: 575–578.

Ebbutt, A.F. (1984). Three–period crossover designs for two treatments. *Biometrics* **40**: 219–24.

Edwards, A.L. (1985). *Experimental Design in Psychological Research*. Harper and Row, New York.

Elswick, R.K., Gennings, C., Chinchilli, V.M. and Dawson, K.S. (1991). A simple approach for finding estimable functions in linear models. *American Statistician* 45: 51–3.

Federer, W.T. and Atkinson, G.F. (1964). Tied–double–change–over designs. *Biometrics* 20: 168–81.

Fleiss, J.L. (1989). A critique of recent research on the two–treatment crossover design. *Controlled Clinical Trials* 10: 237– 43.

Freeman, P.R. (1989). The performance of the two–stage analysis of two–treatment, two–period crossover trials. *Statistics in Medicine* 8: 1421–32.

Gart, J.J. (1969). An exact test for comparing matched proportions in cross–over designs. *Biometrika* 56: 75–80.

GAUSS (1988). Version 2.0, Aptech Systems Inc., Kent, Washington, U.S.A.

Gill, J.L. (1978). *Design and Analysis of Experiments in the Animal and Medical Sciences*. Volume 2, The Iowa State University Press, Ames, Iowa, U.S.A.

Gill, P.S. and Shukla, G.K. (1987). Optimal change–over designs for correlated observations. *Communications in Statistics – Theory and Methods* 16: 2243–2261.

Grieve, A.P. (1985). A Bayesian analysis of the two–period crossover design for clinical trials. *Biometrics* 41: 979–990. Corrigendum in (1986), 42: 459.

Grieve, A.P. (1987a). Applications of Bayesian software: two examples. *The Statistician* 36: 283–288.

Grieve, A.P. (1987b). A note on the analysis of the two–period crossover design when the period–treatment interaction is significant. *Biometrical Journal* 7: 771–5.

Grizzle, J.E. (1965). The two–period change–over design and its use in clinical trials. *Biometrics* 21: 467–80. Corrigendum in (1974), 30: 727.

Gunst, R.F. and Mason, R.L. (1980). *Regression Analysis and Its Application*. Marcel Dekker, New York.

Hedayat, A. and Afsarinejad, K. (1975). Repeated measurements designs I. In *A Survey of Statistical Design and Linear Models*, Srivastava, J.N. (ed.), pp. 229–42. North–Holland, Amsterdam.

Hedayat, A. and Afsarinejad, K. (1978). Repeated measurements designs II. *Annals of Statistics* 6: 619–28.

Hills, M. and Armitage, P. (1979). The two–period cross–over clinical trial. *British Journal of Clinical Pharmacology* 8: 7–20.

Hocking, R.R. (1973). A discussion of the two–way mixed model. *The American Statistician* **27**: 148–52.

Hocking, R.R. (1985). *The Analysis of Linear Models*. Brooks/Cole, Monterey, California.

Hosmer, D.W. and Lemeshow, S. (1989). *Applied Logistic Regression*. John Wiley and Sons, New York.

Huynh, H. and Feldt, L.S. (1970). Conditions under which mean square ratios in repeated measurements designs have exact F–distributions. *Journal of the American Statistical Association* **65**: 1582–9.

Jones, B. and Kenward, M.G. (1987). Modelling binary data from a three–period cross–over trial. *Statistics in Medicine* **6**: 555–564.

Jones, B. and Kenward, M.G. (1989). *Design and Analysis of Cross–over Trials*. Chapman and Hall, London, New York.

Judge, G.G., Hill, R.C., Griffiths, W.E., Lütkepohl, H. and Lee, T.–C. (1982). *Introduction to the Theory and Practice of Econometrics*. John Wiley and Sons, New York.

Kenward, M.G. and Jones, B. (1987). A log–linear model for binary cross–over data. *Applied Statistics* **36**: 192–204.

Kenward, M.G. and Jones, B. (1991). The analysis of categorical data from cross–over trials using a latent variable model. *Statistics in Medicine* **10**: 1607–19.

Kershner, R.P. (1986). Optimal three period two–treatment crossover designs with and without baseline readings. *American Statistical Association, Proceedings of the Biopharmaceutical Section*, pp. 152–6.

Kershner, R.P. and Federer, W.T. (1981). Two–treatment crossover designs for estimating a variety of effects. *Journal of the American Statistical Association* **76**: 612–18.

Kiefer, J. (1975). Construction and optimality of generalized Youden designs. In *A Survey of Statistical Design and Linear Models*, Srivastava, J.N. (ed.), pp. 333–53. North–Holland, Amsterdam.

Koch, G.G., Amara, I.A., Brown, B.W., Colton, T. and Gillings, D.B. (1989). A two–period crossover design for the comparison of two active treatments and placebo. *Statistics in Medicine* **8**: 487–504.

Laska, E.M. and Meisner, M. (1985). A variational approach to optimal two–treatment crossover designs: applications to carryover–effect models. *Journal of the American Statistical Association* **80**: 704–10.

Laska, E., Meisner, M. and Kushner, H.B. (1983). Optimal crossover designs in the presence of carryover effects. *Biometrics* **39**: 1087–91.

Le, C.T. (1984). Logistic models for crossover designs. *Biometrika* **71**: 216–7.

Le, C.T. and Cary, M.M. (1984). Analysis of crossover designs with a categorical response. *Biometrical Journal* **26**: 859–65.

Le, C.T. and Gomez–Marin, O. (1984). Estimation of parameters in binomial crossover designs. *Biometrical Journal* **26**: 167–71.

McCullagh, P. and Nelder, J.A. (1983). *Generalized Linear Models.* Chapman and Hall, London.

McDonald, D.D. (1975). Tests for the general linear hypothesis under the multiple design multivariate linear model. *Annals of Statistics* **3**: 461–6.

Mainland, D. (1963). *Elementary Medical Statistics*, 2nd ed. Saunders, Philadelphia, Pennsylvania.

Matthews, J.N.S. (1987). Optimal crossover designs for the comparison of two treatments in the presence of carryover effects and autocorrelated errors. *Biometrika* **74**: 311–20.

Matthews, J.N.S. (1990). The analysis of data from crossover designs: the efficiency of ordinary least squares. *Biometrics* **46**: 689–96.

Mauchly, J.W. (1940). Significance test for sphericity of a normal n-variate distribution. *Annals of Mathematical Statistics* **29**: 204–9.

Milliken, G.A. and Johnson, D.E. (1984). *Analysis of Messy Data.* Volume 1: Designed Experiments. Van Nostrand Reinhold, New York.

Morrison, D.F. (1976). *Multivariate Statistical Methods.* McGraw–Hill Book Co., New York.

Nelder, J.A. and Mead, R. (1965). A simplex method for function minimisation. *Computing Journal* **7**: 303–13.

Neter, J., Wasserman, W. and Kutner, M.H. (1990). *Applied Linear Statistical Models.* Third Edition. Irwin, Homewood, Illinois.

Patel, H.I. (1983). Use of baseline measurements in the two–period cross–over design. *Communications in Statistics – Theory and Methods* **12**: 2693–712.

Patel, H.I. (1986). Analysis of repeated measures designs with changing covariates in clinical trials. *Biometrika* **73**: 707–15.

Patterson, H.D. (1973). Quenouille's changeover designs. *Biometrika* **60**: 33–45.

Patterson, H.D. and Lucas, H.L. (1959). Extra–period change–over designs. *Biometrics* **15**: 116–132.

Patterson, H.D. and Lucas, H.L. (1962). *Change–over Designs.* North Carolina Agricultural Experiment Station, Tech. Bull. No. 147.

Prescott, R. J. (1981). The comparison of success rates in cross–over trials in the presence of an order effect. *Applied Statistics* **30**: 9–15.

Press, S.J. (1989). *Bayesian Statistics: Principles, Models, and Applications*. John Wiley and Sons, New York.

Quenouille, M.H. (1953). *The Design and Analysis of Experiment*. Hafner Publishing Co., New York.

Racine, A., Grieve, A.P., Flühler, H. and Smith, A.F.M. (1986). Bayesian methods in practice: experiences in the pharmaceutical industry. *Applied Statistics* **35**: 93–150.

Ratkowsky, D.A., Alldredge, J.R. and Cotton, J.W. (1990). Analyzing balanced or unbalanced Latin squares and other repeated–measures designs for carryover effects using the GLM procedure. *SAS Users Group International, Proceedings of the 15th Annual Conference*, April 1–4, 1990, Nashville TN, pp. 1353–8.

SAS Institute Inc. (1990). SAS/STAT User's Guide, Version 6, Fourth Ed., Volume 1. SAS Institute Inc., Cary, NC, USA.

Searle, S.R. (1987). *Linear Models for Unbalanced Data*. John Wiley and Sons, New York.

Shapiro, S.S. and Wilk, M.B. (1965). An analysis of variance test for normality (complete samples). *Biometrika* **52**: 591–611.

Shoben, E.J., Sailor, K.M. and Wang, M.-Y. (1989). The role of expectancy in comparative judgments. *Memory and Cognition* **17**: 18–26.

Stigler, S.M. (1986). *The History of Statistics*. The Belknap Press of Harvard University Press, Cambridge, MA.

Taka, M.T. and Armitage, P. (1983). Autoregressive models in clinical trials. *Communications in Statistics – Theory and Methods* **12**: 865–76.

Wagenaar, W. (1969). Note on the construction of digram–balanced Latin squares. *Psychological Bulletin* **72**: 384–6.

Ware, J.H. (1985). Linear models for the analysis of longitudinal studies. *The American Statistician* **39**: 95–101.

Williams, E.J. (1949). Experimental designs balanced for the estimation of residual effects of treatments. *Australian Journal of Scientific Research* A,2: 149–68.

Williams, E.J. (1950). Experimental designs balanced for pairs of residual effects. *Australian Journal of Scientific Research*. A,3: 351–63.

Williams, E.R. (1987). A note on change–over designs. *Australian Journal of Statistics* **29**: 309–16.

Zimmermann, H. and Rahlfs, V. (1978). Testing hypotheses in the two–period change–over with binary data. *Biometrical Journal* **20**: 133–41.

Solutions

<u>Exercise 1.1</u> $\chi^2 = 14.67$, $V = 0.553$, $S = 44.7$. This compares with $S = 13.4$ and $S = 71.1$ for the other 4 x 4 Latin squares in this chapter.

<u>Exercise 1.2</u>

(a)

$$X = \begin{pmatrix}
1 & 1 & 0 & 0 & 1 & 0 & 0 & 1 & 0 & 0 & 0 & 0 & 0 \\
1 & 1 & 0 & 0 & 0 & 1 & 0 & 0 & 1 & 0 & 1 & 0 & 0 \\
1 & 1 & 0 & 0 & 0 & 0 & 1 & 0 & 0 & 1 & 0 & 1 & 0 \\
1 & 1 & 0 & 0 & -1 & -1 & -1 & -1 & -1 & -1 & 0 & 0 & 1 \\
1 & 0 & 1 & 0 & 1 & 0 & 0 & 0 & 1 & 0 & 0 & 0 & 0 \\
1 & 0 & 1 & 0 & 0 & 1 & 0 & 1 & 0 & 0 & 0 & 1 & 0 \\
1 & 0 & 1 & 0 & 0 & 0 & 1 & -1 & -1 & -1 & 1 & 0 & 0 \\
1 & 0 & 1 & 0 & -1 & -1 & -1 & 0 & 0 & 1 & -1 & -1 & -1 \\
1 & 0 & 0 & 1 & 1 & 0 & 0 & 0 & 0 & 1 & 0 & 0 & 0 \\
1 & 0 & 0 & 1 & 0 & 1 & 0 & -1 & -1 & -1 & 0 & 0 & 1 \\
1 & 0 & 0 & 1 & 0 & 0 & 1 & 0 & 1 & 0 & -1 & -1 & -1 \\
1 & 0 & 0 & 1 & -1 & -1 & -1 & 1 & 0 & 0 & 0 & 1 & 0 \\
1 & -1 & -1 & -1 & 1 & 0 & 0 & -1 & -1 & -1 & 0 & 0 & 0 \\
1 & -1 & -1 & -1 & 0 & 1 & 0 & 0 & 0 & 1 & -1 & -1 & -1 \\
1 & -1 & -1 & -1 & 0 & 0 & 1 & 1 & 0 & 0 & 0 & 0 & 1 \\
1 & -1 & -1 & -1 & -1 & -1 & -1 & 0 & 1 & 0 & 1 & 0 & 0
\end{pmatrix} .$$

$(\mathbf{X}'\mathbf{X})^{-1} =$

$$
\begin{pmatrix}
0.29792 & -0.16042 & -0.06875 & 0.14167 & -0.19167 & -0.05833 \\
-0.16042 & 0.29792 & -0.06875 & -0.19167 & 0.14167 & 0.10833 \\
-0.06875 & -0.06875 & 0.29792 & 0.10833 & -0.05833 & 0.14167 \\
0.14167 & -0.19167 & 0.10833 & 0.43333 & -0.23333 & -0.1 \\
-0.19167 & 0.14167 & -0.05833 & -0.23333 & 0.43333 & -0.1 \\
-0.05833 & 0.10833 & 0.14167 & -0.1 & -0.1 & 0.43333
\end{pmatrix}.
$$

(b)

$$
(\mathbf{X}'\mathbf{X})^{-1} =
\begin{pmatrix}
0.1875 & -0.0625 & -0.0625 \\
-0.0625 & 0.1875 & -0.0625 \\
-0.0625 & -0.0625 & 0.1875
\end{pmatrix}.
$$

Note that this is the same as for the other 4 x 4 Latin squares with carryover effects omitted.

(c) VIF $= 0.29792/0.1875 = 1.589$. Efficiency $= (1.589)^{-1} = 0.629$.

Exercise 1.3
(a) $S = 13.4$.
(b) VIF $= 5.5$.
(c)

Source	df	Type I SS	Mean Square	F value	Pr > F
COW	23	1315.795	57.2085	24.51	0.0001
PERIOD	3	388.823	129.6076	55.54	0.0001
TREAT	1	2.700	2.7001	1.16	0.2859
CARRY	1	2.950	2.9502	1.26	0.2649

Source	df	Type II SS	Mean Square	F value	Pr > F
TREAT	1	5.082	5.0819	2.18	0.1447
CARRY	1	2.950	2.9502	1.26	0.2649

No evidence of a multicollinearity effect exists here, since the Type II treatment sum of squares is not less than the Type I sum of squares.

(d) One can add Period and the square of period (Period2) to the model as a continuous, rather than a "class", variable. With carryover parameters in the model, the following Type I sum of squares are obtained.

Source	df	Type I SS	Mean Square	F value	Pr > F
COW	23	1315.795	57.2085		
PERIOD	1	378.963	378.9630		
PERIOD2	1	9.818	9.8176		
TREAT	1	2.700	2.7001		
CARRY	2	2.992	1.4962	0.6411	0.5299 .

Virtually all of the Period sum of squares obtained when Period was considered to be a class variable (388.823 with 3 d.f.) is explained by a linear component (378.963 with 1 d.f.) with the quadratic component contributing little (9.818 with 1 d.f.). Since carryover is nonsignificant, one may obtain a valid statistical test of each period component by re–running the analysis without including the carryover parameters. Then, Type I and Type II sums of squares, etc., will be identical. This leads to the conclusion that the linear component of period is highly significant ($P < 0.0001$), with the quadratic effect being marginally significant ($P = 0.043$).

Chapter 2

Exercise 2.1
(a) $S = 58.2$.
(b) From Table 2.3, $S = 77.6$.

Exercise 2.2
$VIF = 0.197155/0.16 = 1.232$,
$E_d = (0.16/0.197155) \, 100 = 81.15$.

Exercise 2.4
(a) $S = 60.8$; from Table 2.3, $S = 81.7$.
(c) Neither treatment nor carryover effects are significant.

Chapter 3

Exercise 3.1
(a) $\hat{\tau}_1 = -0.12825$ in Parameterizations 2–4, and $\hat{\tau}_1 = -0.3843$ in Parameterization 1.

(b) $\hat{\lambda}_1 = -0.5121$ (Parameterization 1), $(\widehat{\tau\pi})_{11} = -0.25605$ (Parameterization 2), $\hat{\gamma}_1 = -0.25605$ (Parameterizations 3 and 4).

Exercise 3.2

(a) It is important to include the statements
        IF CARRY='0' THEN CARRY='B'
and
        IF TREAT='0' THEN TREAT='B'
in the DATA step.

(b) The portion of the analysis of variance table for treatment and first–order carryover effects is as follows:

Source	df	Sum of Squares	Mean Square	F Value	Prob > F
TREAT	1	40.00000	40.00000	25.21972	0.0000
CARRY	1	4.81667	4.81667	3.03687	0.0870

Hence, treatments effects are highly significant, but carryover effects only marginally so.

Exercise 3.3

The elements of $(\mathbf{X}'\mathbf{X})^{-1}$ for τ_1 and λ_1 are the same as those for the design of Section 3.6 when there is only first–order carryover in the model. In this circumstance, the initial baseline reading does not improve the precision of the treatment parameter. It does contribute somewhat to the precision, however, when second–order carryover is in the model. This exercise shows that the readings in the wash–out periods are more important than the initial readings.

Chapter 4

Exercise 4.1

(a) There is the possibility of a "multicollinearity" effect, where both the treatment and carryover effects are apparently not significant, when in fact one or both of them may be important. If the analysis of variance yields non-significant treatment and carryover effects, the carryover term should be dropped from the model and the analysis re–run.

(b) VIF $= 0.75/0.1875 = 4.0$.

(c)

Source	df	Type II SS	Mean Square	F value	Pr > F
TREAT	1	1.75788	1.75788	32.92	0.0001
CARRY	1	0.00365	0.00365	0.07	0.7972

Since the treatment effect is highly significant, there is no need to drop the non–significant carryover term from the model. However, if one did do that, the treatment sum of squares would be 6.48788. Thus, the poor separability of treatment and carryover parameters does result in a marked reduction of the sum of squares by a factor of about 3.7; nevertheless, the difference between diets A and B on milk protein production is so large that the treatment effect is highly significant despite the reduction.

Exercise 4.2

(b)

Source	df	Type II SS	Mean Square	F value	Pr > F
TREAT	1	3100.168	3100.168	18.16	0.0001
CARRY	1	10.412	10.412	0.06	0.8052

Treatment effects are highly significant, but there is no evidence of significant carryover effects. From the table of least–square means, Treatment A has a greater response than Treatment B by 7.9 units.

Exercise 4.3

(b) Carryover will have three degrees of freedom.

Chapter 5

Exercise 5.1

The design has the same treatments, with the letter "P" substituted for the letter "C", as Design PL1 of Appendix 5.B. The only difference is that the latter design has two groups of three subjects arranged in blocks.

Exercise 5.2

(a)

$$\begin{pmatrix}
1 & 1 & 0 & 0 & 0 & 0 & 0 & 0 & 0 & 0 & 1 & 1 & 0 & 0 & 0 \\
1 & 1 & 0 & 0 & 0 & 0 & 0 & 0 & 0 & 0 & -1 & 0 & 1 & 1 & 0 \\
1 & 0 & 1 & 0 & 0 & 0 & 0 & 0 & 0 & 0 & 1 & 1 & 0 & 0 & 0 \\
1 & 0 & 1 & 0 & 0 & 0 & 0 & 0 & 0 & 0 & -1 & 0 & 1 & 1 & 0 \\
1 & 0 & 0 & 1 & 0 & 0 & 0 & 0 & 0 & 0 & 1 & 1 & 0 & 0 & 0 \\
1 & 0 & 0 & 1 & 0 & 0 & 0 & 0 & 0 & 0 & -1 & 0 & 1 & 1 & 0 \\
1 & 0 & 0 & 0 & 1 & 0 & 0 & 0 & 0 & 0 & 1 & 0 & 1 & 0 & 0 \\
1 & 0 & 0 & 0 & 1 & 0 & 0 & 0 & 0 & 0 & -1 & 1 & 0 & 0 & 1 \\
1 & 0 & 0 & 0 & 0 & 1 & 0 & 0 & 0 & 0 & 1 & 0 & 1 & 0 & 0 \\
1 & 0 & 0 & 0 & 0 & 1 & 0 & 0 & 0 & 0 & -1 & 1 & 0 & 0 & 1 \\
1 & 0 & 0 & 0 & 0 & 0 & 1 & 0 & 0 & 0 & 1 & 0 & 1 & 0 & 0 \\
1 & 0 & 0 & 0 & 0 & 0 & 1 & 0 & 0 & 0 & -1 & 1 & 0 & 0 & 1 \\
1 & 0 & 0 & 0 & 0 & 0 & 0 & 1 & 0 & 0 & 1 & 1 & 0 & 0 & 0 \\
1 & 0 & 0 & 0 & 0 & 0 & 0 & 1 & 0 & 0 & -1 & -1 & -1 & 1 & 0 \\
1 & 0 & 0 & 0 & 0 & 0 & 0 & 0 & 1 & 0 & 1 & -1 & -1 & 0 & 0 \\
1 & 0 & 0 & 0 & 0 & 0 & 0 & 0 & 1 & 0 & -1 & 1 & 0 & -1 & -1 \\
1 & 0 & 0 & 0 & 0 & 0 & 0 & 0 & 0 & 1 & 1 & 0 & 1 & 0 & 0 \\
1 & 0 & 0 & 0 & 0 & 0 & 0 & 0 & 0 & 1 & -1 & -1 & -1 & 0 & 1 \\
1 & -1 & -1 & -1 & -1 & -1 & -1 & -1 & -1 & 1 & -1 & -1 & 0 & 0 \\
1 & -1 & -1 & -1 & -1 & -1 & -1 & -1 & -1 & -1 & 0 & 1 & -1 & -1
\end{pmatrix} .$$

(b)

	τ_1	τ_2	λ_1	λ_2
τ_1	0.7196	−0.4233	1.1852	−0.8148
τ_2	−0.4233	0.7196	−0.8148	1.1852
λ_1	1.1852	−0.8148	2.3704	−1.6296
λ_2	−0.8148	1.1852	−1.6296	2.3704

(c)

$\mathrm{Corr}(\tau_1, \lambda_1) = \mathrm{Corr}(\tau_2, \lambda_2) = 0.907$, and
$\mathrm{Corr}(\tau_1, \lambda_2) = \mathrm{Corr}(\tau_2, \lambda_1) = -0.624.$

(continued on following page)

These are rather high correlations and indicate that there may be problems in separating "direct" treatment effects from "residual" treatment effects in this design. The variance of each of the two treatment parameters and of the two carryover parameters is also rather high at 0.7196 and 2.3704, respectively.

Exercise 5.4

(a) The main difference is that Design PL55 is grouped into four blocks of three subjects each. The VIF of Design PL55 indicates that this design is likely to be a poor one for separating treatment and carryover effects. Two–period designs are always inefficient.

(b)

Analysis of Variance:

Source	DF	Sum of Squares	Mean Square	F Value	Prob > F
Model	18	826.77740	45.93208	18.97394	0.00204
Error	5	12.10399	2.42080		
Total	23	838.88138			

Source	DF	Sum of Squares	Mean Square	F Value	Prob > F
COW	11	340.59331	30.96303	12.79042	0.00566
PERIOD	1	11.40576	11.40576	4.71157	0.08208
TREAT	3	20.25015	6.75005	2.78836	0.14907
CARRY	3	4.70427	1.56809	0.64776	0.61732
Error	5	12.10399	2.42080		

Chapter 6

Exercise 6.2

(a)

$$
\begin{pmatrix}
1 & 1 & 0 & 0 & 1 & 0 & 0 & 0 & 1 & 0 & 1 & 0 & 0 & 0 & 0 & 0 \\
1 & 1 & 0 & 0 & 1 & 0 & 0 & 0 & 0 & 1 & 0 & 1 & 0 & 1 & 0 & 0 \\
1 & 1 & 0 & 0 & 1 & 0 & 0 & 0 & -1 & -1 & 0 & 1 & 0 & 0 & 1 & 0 \\
1 & 1 & 0 & 0 & -1 & 0 & 0 & 0 & 1 & 0 & 0 & 1 & 0 & 0 & 0 & 0 \\
1 & 1 & 0 & 0 & -1 & 0 & 0 & 0 & 0 & 1 & 1 & 0 & 0 & 0 & 1 & 0 \\
1 & 1 & 0 & 0 & -1 & 0 & 0 & 0 & -1 & -1 & 1 & 0 & 0 & 1 & 0 & 0 \\
1 & 0 & 1 & 0 & 0 & 1 & 0 & 0 & 1 & 0 & 0 & 1 & 0 & 0 & 0 & 0 \\
1 & 0 & 1 & 0 & 0 & 1 & 0 & 0 & 0 & 1 & 0 & 0 & 1 & 0 & 1 & 0 \\
1 & 0 & 1 & 0 & 0 & 1 & 0 & 0 & -1 & -1 & 0 & 0 & 1 & 0 & 0 & 1 \\
1 & 0 & 1 & 0 & 0 & -1 & 0 & 0 & 1 & 0 & 0 & 0 & 1 & 0 & 0 & 0 \\
1 & 0 & 1 & 0 & 0 & -1 & 0 & 0 & 0 & 1 & 0 & 1 & 0 & 0 & 0 & 1 \\
1 & 0 & 1 & 0 & 0 & -1 & 0 & 0 & -1 & -1 & 0 & 1 & 0 & 0 & 1 & 0 \\
1 & 0 & 0 & 1 & 0 & 0 & 1 & 0 & 1 & 0 & 0 & 0 & 1 & 0 & 0 & 0 \\
1 & 0 & 0 & 1 & 0 & 0 & 1 & 0 & 0 & 1 & -1 & -1 & -1 & 0 & 0 & 1 \\
1 & 0 & 0 & 1 & 0 & 0 & 1 & 0 & -1 & -1 & -1 & -1 & -1 & -1 & -1 & -1 \\
1 & 0 & 0 & 1 & 0 & 0 & -1 & 0 & 1 & 0 & -1 & -1 & -1 & 0 & 0 & 0 \\
1 & 0 & 0 & 1 & 0 & 0 & -1 & 0 & 0 & 1 & 0 & 0 & 1 & -1 & -1 & -1 \\
1 & 0 & 0 & 1 & 0 & 0 & -1 & 0 & -1 & -1 & 0 & 0 & 1 & 0 & 0 & 1 \\
1 & -1 & -1 & -1 & 0 & 0 & 0 & 1 & 1 & 0 & -1 & -1 & -1 & 0 & 0 & 0 \\
1 & -1 & -1 & -1 & 0 & 0 & 0 & 1 & 0 & 1 & 1 & 0 & 0 & -1 & -1 & -1 \\
1 & -1 & -1 & -1 & 0 & 0 & 0 & 1 & -1 & -1 & 1 & 0 & 0 & 1 & 0 & 0 \\
1 & -1 & -1 & -1 & 0 & 0 & 0 & -1 & 1 & 0 & 1 & 0 & 0 & 0 & 0 & 0 \\
1 & -1 & -1 & -1 & 0 & 0 & 0 & -1 & 0 & 1 & -1 & -1 & -1 & 1 & 0 & 0 \\
1 & -1 & -1 & -1 & 0 & 0 & 0 & -1 & -1 & -1 & -1 & -1 & -1 & -1 & -1 & -1
\end{pmatrix} .
$$

$$
\begin{pmatrix}
0.234375 & -0.046875 & -0.140625 & 0.0 & 0.0 & 0.0 \\
-0.046875 & 0.234375 & -0.046875 & 0.0 & 0.0 & 0.0 \\
-0.140625 & -0.046875 & 0.234375 & 0.0 & 0.0 & 0.0 \\
0.0 & 0.0 & 0.0 & 0.25000 & -0.06250 & -0.12500 \\
0.0 & 0.0 & 0.0 & -0.06250 & 0.25000 & -0.06250 \\
0.0 & 0.0 & 0.0 & -0.12500 & -0.06250 & 0.25000
\end{pmatrix} .
$$

Exercise 6.3

Setting up a 24 x 16 design matrix $\mathbf{X}$ for this problem leads to the following inverse of the $\mathbf{X'X}$ matrix:

$$
\begin{pmatrix}
0.464286 & -0.250000 & 0.035714 & 0.482143 & -0.375000 & 0.267857 \\
-0.250000 & 0.464286 & -0.250000 & -0.375000 & 0.482143 & -0.375000 \\
0.035714 & -0.250000 & 0.464286 & 0.267857 & -0.375000 & 0.482143 \\
0.482143 & -0.375000 & 0.267857 & 1.178571 & -0.750000 & 0.321429 \\
-0.375000 & 0.482143 & -0.375000 & -0.750000 & 1.178571 & -0.750000 \\
0.267857 & -0.375000 & 0.482143 & 0.321429 & -0.750000 & 1.178571
\end{pmatrix}
$$

Since the variance pertaining to the treatment parameter, 0.464286, is almost twice as large as that obtained using an extra period, 0.234375, the use of baselines is much less efficient than the use of an extra period.

Chapter 7

Exercise 7.1

(a) $\hat{\tau} = 0.5479$, S.E.$(\hat{\tau}) = 0.2608$. W = 4.42, df = 1, P–value = 0.0365;

$$G^2 = 4.94, \text{ df} = 1, \text{ P–value} = 0.0262.$$

(b) Gart's Test:

$$\chi_G^2 = \frac{(3 \cdot 6 - 1 \cdot 1)^2 \cdot 11}{4 \cdot 7 \cdot 4 \cdot 7} = 4.055, \text{ df} = 1, \text{ P–value} = 0.0440.$$

Prescott's Test:

$$\chi_P^2 = \frac{[(3-1) \cdot 49 - (4-7) \cdot 22]^2 \cdot 49}{22 \cdot 27 \cdot [(4+7) \cdot 49 - (4-7)^2]} = 4.186, \text{ df} = 1, \text{ P–value} = 0.0408.$$

Exercise 7.2

(a)

ANALYSIS OF MAXIMUM LIKELIHOOD ESTIMATES

Effect	Parameter	Estimate	Standard Error	Chi–Square	Prob
INTERCEPT	1	−0.8901	0.4430	4.04	0.0445
	2	−0.4734	0.3660	1.67	0.1959
SEQUENCE	3	−0.0927	0.4430	0.04	0.8343
	4	0.2795	0.3660	0.58	0.4451
PERIOD	5	−0.2753	0.4430	0.39	0.5343
	6	−0.0759	0.3660	0.04	0.8357
TREAT	7	1.2269	0.4430	7.67	0.0056
	8	0.4234	0.3660	1.34	0.2478
ASS113	9	0.4791	1.0287	0.22	0.6414
	10	0.0512	1.0590	0.00	0.9614
ASS123	11	0.04029	1.1206	0.00	0.9714
	12	0.2542	1.1276	0.05	0.8216
ASS213	13	−0.5536	1.4085	0.15	0.6943
	14	0.7437	0.9173	0.66	0.4175
ASS223	15	1.7490	1.1765	2.21	0.1371
	16	1.1004	1.0481	1.10	0.2938

(b)

Odds Ratios: $\exp(2 \cdot 1.2269) = 11.632$ for Complete Relief vs. No Relief.

$\exp(2 \cdot 0.4234) = 2.332$ for Moderate Relief vs. No Relief.

(c) Testing for a difference between the High Dose Analgesic (C) and the Placebo (A):

$W = 7.74$, df $= 2$, P–value $= 0.0208$.

$G^2 = 7.25$, df $= 2$, P–value $= 0.0266$.

These two statistics indicate at least a marginally significant difference between treatments.

Exercise 7.3

(a) The parameter estimates for the location parameters are:

ANALYSIS OF MAXIMUM LIKELIHOOD ESTIMATES

Effect	Parameter	Estimate	Standard Error	Chi-Square	Prob
INTERCEPT	1	0.3306	0.1790	3.41	0.0648
SEQ_1	2	−0.3328	0.4030	0.68	0.4089
SEQ_2	3	−0.0562	0.3981	0.02	0.8877
SEQ_3	4	0.0794	0.3816	0.04	0.8353
SEQ_4	5	0.3624	0.4205	0.74	0.3887
SEQ_5	6	−0.0962	0.3987	0.06	0.8094
PERIOD_1	7	−0.0735	0.2196	0.11	0.7378
PERIOD_2	8	0.0610	0.2387	0.07	0.7985
TREAT_B	9	0.3709	0.2837	1.71	0.1911
TREAT_C	10	0.5795	0.2829	4.20	0.0405
CARRY_B	11	−0.0693	0.4073	0.03	0.8649
CARRY_C	12	−0.4673	0.4305	1.18	0.2777

(b) Testing for a difference between the High Dose Analgesic (C), Moderate Dose Analgesic and the Placebo (A):

$$G^2 = 8.34, \text{ df} = 2, \text{ P–value} = 0.0155.$$

This indicates a significant difference between treatments. Based on the parameter estimates for the treatments, the High Dose Analgesic is more likely to produce relief than the Low Dose Analgesic, which is more likely to produce relief than the Placebo.

(c)

Odds Ratios: $\exp[2 \cdot (0.3709) + 0.5795] = 3.748$ for Treatment B.

$\exp[0.3709 + 2 \cdot (0.5795)] = 4.618$ for Treatment C.

Chapter 8

Exercise 8.1

(a)

Source of Variation	df	Sum of Squares	Mean Squares	F	Pr > F
Between–cows:					
Sequence	1	22.5234	22.5234		
Cows(sequence)	22	1293.2715	58.7851		
Period	3	388.8228	129.6076	55.54	0.0001
Treatment	1	2.7001	2.7001	1.16	0.2859
Carryover	1	2.9502	2.9502	1.26	0.2649
Within–cows:					
Treatment	1	5.0819	5.0819	2.18	0.1447
Carryover	1	2.9502	2.9502	1.26	0.2649
Residual	67	156.3594	2.3337		

Neither treatment nor carryover effects is significant. There is a significant period effect, however, the mean response decreasing with time.

(b) Testing whether the uniform covariance matrix is consistent with the raw data, one determines

$$M = 22 \log (765.8299/335.6837) = 18.1454 \ .$$

Since $A_1 = 0.07891414$, it follows that $(1 - A_1)M = 16.71$ with $f_1 = 8$ degrees of freedom. The P-value corresponding to this is $P = 0.0332$, so the observed data matrix is marginally different from a uniform covariance structure.

(c) The least squares estimates of the Type H structure parameters are obtained from the solution of the following set of linear equations:

$$
\begin{pmatrix}
1 & 2 & 0 & 0 & 0 \\
0 & 1 & 1 & 0 & 0 \\
0 & 1 & 0 & 1 & 0 \\
0 & 1 & 0 & 0 & 1 \\
1 & 0 & 2 & 0 & 0 \\
0 & 0 & 1 & 1 & 0 \\
0 & 0 & 1 & 0 & 1 \\
1 & 0 & 0 & 2 & 0 \\
0 & 0 & 0 & 1 & 1 \\
1 & 0 & 0 & 0 & 2
\end{pmatrix}
\begin{pmatrix}
\lambda \\
\gamma_1 \\
\gamma_2 \\
\gamma_3 \\
\gamma_4
\end{pmatrix}
=
\begin{pmatrix}
18.209 \\
16.005 \\
14.730 \\
11.661 \\
16.994 \\
16.008 \\
13.005 \\
17.297 \\
13.239 \\
13.343
\end{pmatrix}
$$

which results in

$$\hat{\lambda} = 2.35275, \ \hat{\gamma}_1 = 7.64875, \ \hat{\gamma}_2 = 7.68075, \ \hat{\gamma}_3 = 7.60825 \text{ and } \hat{\gamma}_4 = 5.27825,$$

leading to the following estimated covariance matrix:

$$\hat{\Sigma} = \begin{pmatrix} 17.650 & 15.330 & 15.257 & 12.927 \\ 15.330 & 17.714 & 15.289 & 12.959 \\ 15.257 & 15.289 & 17.569 & 12.887 \\ 12.927 & 12.959 & 12.887 & 12.909 \end{pmatrix}.$$

This matrix appears to agree quite well with the matrix obtained from the data set. To test formally whether the above matrix is consistent with the Type H structure, one determines

$$|C\hat{\Sigma}C'| = 6.5734,$$

$$\text{tr}(C\hat{\Sigma}C') = 7.0581,$$

from which Mauchly's W is

$$W = 6.5734/(7.0581/3)^3 = 0.5048,$$

and $d = 0.94192$.

The statistic

$$-\nu \, d \log_e W = -22 \, (0.94192) \log_e(0.5048) = 14.17$$

has $f = 5$ degrees of freedom. The P–value corresponding to this is 0.015, so that the data set only conforms marginally with the Type H structure.

Chapter 9

Exercise 9.1
Analysis of Variance:

SOURCE	DF	SUM OF SQUARES	MEAN SQUARE	F VALUE
MODEL	14	17573.63914	1255.25994	30.39512
ERROR	2	82.59615	41.29808	
TOTAL	16	17656.23529		

Source	DF	Type II SS	Mean Square	F Value	Prob > F
TREAT	2	2576.7	1288.4	31.20	0.03106
CARRY	3	729.5	243.2	5.89	0.14861

Author Index

Subject Index

Printed in the United States
by Baker & Taylor Publisher Services